DIGITAL UNDERWATER ACOUSTIC COMMUNICATIONS

DIGITAL UNDERWATER ACOUSTIC COMMUNICATIONS

TIANZENG XU

LUFEN XU

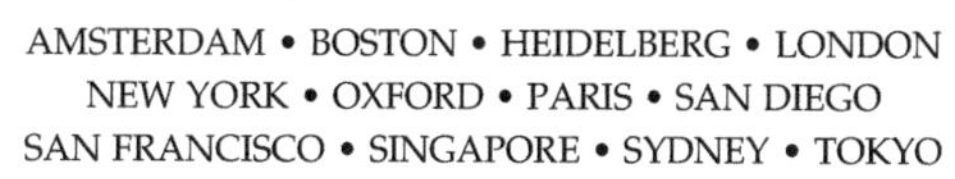
AMSTERDAM • BOSTON • HEIDELBERG • LONDON
NEW YORK • OXFORD • PARIS • SAN DIEGO
SAN FRANCISCO • SINGAPORE • SYDNEY • TOKYO
Academic Press is an imprint of Elsevier

Academic Press is an imprint of Elsevier
125 London Wall, London EC2Y 5AS, United Kingdom
525 B Street, Suite 1800, San Diego, CA 92101-4495, United States
50 Hampshire Street, 5th Floor, Cambridge, MA 02139, United States
The Boulevard, Langford Lane, Kidlington, Oxford OX5 1GB, United Kingdom

Library of Congress Cataloging-in-Publication Data
A catalog record for this book is available from the Library of Congress

British Library Cataloguing-in-Publication Data
A catalogue record for this book is available from the British Library

ISBN: 978-0-12-803009-7

For information on all Academic Press publications
visit our website at https://www.elsevier.com/

Publisher: Glyn Jones
Acquisition Editor: Simon Tian
Editorial Project Manager: Vivi Li
Production Project Manager: Jason Mitchell
Designer: Mark Rogers

Contents

About the Authors

Tianzeng Xu is a professor and the supervisor of a Ph.D. program at the College of Ocean and Earth Sciences of Xiamen University and Key Laboratory of Underwater Acoustic Communication and Marine Information Technology, Ministry of Education. He had been appointed as director of the Subtropical Marine Institute, Head of Oceanography Department, Xiamen University and vice chairman of the China Marine Physical Society.

The advanced worker of national high-new technical developing (863) program vested by Ministry of Science and Technology of the People's Republic of China.

Lufen Xu is a senior engineer and master instructor at the College of Ocean and Earth Sciences of Xiamen University and Key Laboratory of Underwater Acoustic Communication and Marine Information Technology, Ministry of Education. He is vice dean of teaching and the research section of marine physics of the college.

Foreword

Underwater acoustic communications have covered the areas of national defense and civil ocean development and exploration. Therefore, they have been highly valued by maritime nations in the past several decades.

Analog underwater acoustic communications have played an important role for many years. However, they have some drawbacks like unstable communication quality. To adapt peculiar underwater acoustic communication channels, digital underwater acoustic communications have raised vast concerns and achieved crucial developments and massive technological breakthroughs. With regard to many underwater acoustic communication fields, like underwater data transmission in an underwater acoustic network, digital communications have gradually replaced analog ones.

In principle, digital underwater acoustic communications are an extension and development from digital radio communications. However, there are many essential differences between underwater acoustic and radio communication channels, including randomly variant spatial-temporal-frequency parameters, large attenuation, severe multipath effects, a strict band-limiting property, high noise level, and low sound velocity. In particular, underwater acoustic communication channels are generally not linear nor time-invariant. Therefore, some present advanced signal processing systems, such as matched filter, or even some basic principles, like Shannon theorem, in radio communications cannot perfectly be employed, and they especially cannot be copied mechanically in underwater acoustic communications. So, exploring innovative signal processing systems to suitably employ in peculiar underwater acoustic communication channels is a valuable and challenging research topic, which is also a basic premise for establishing innovative digital underwater communication equipment that can adapt to peculiar underwater acoustic channels.

The contents discussed in this book will mainly focus on digital underwater acoustic communications for civil applications.

Military underwater acoustic communication has its own special requirements and corresponding difficulties. For example, it requires long-range and strictly confidential communication, which sometimes applies to high-speed mobile communication. However, the applied fields and operating specifications between military and civil underwater acoustic communications cross each other. In particular, the theoretical basis is generally identical. Therefore, the contents discussed in this book have a common significance.

This book has four chapters and an appendix. After reviewing the development of underwater acoustic communications, Chapter 1 is focused on describing the differences between underwater acoustic communication channels and radio ones. The peculiarities existing in the former indicate the necessity to explore some innovative signal processing systems employed effectively in digital underwater acoustic communications. After that, a communication sonar equation will be derived, and an active sonar equation against noise background will also be included. A specific example to show how to use the communication sonar equation to select the relative parameters in designing an improved FH-SS (Frequency-Hopped Spread-Spectrum) system communication sonar will be provided. Laws of transmitted sound in underwater acoustic channels will be discussed in Chapter 2, including sound transmission loss, multipath effects, sound scattering and fluctuations, and noise in the sea, especially their impacts on digital underwater acoustic communications and the possible countermeasures to adapt to them. They are the physical basis for designing innovative digital underwater acoustic communication signal processing systems. Digital underwater acoustic communication signal processing will be described in Chapter 3. After discussing some signal processing schemes employed in digital underwater acoustic communications at present, the explorations establishing innovative digital underwater acoustic communication systems are emphatically discussed. According to the peculiarities of underwater acoustic communication channels, by combining some key techniques, which urgently need to be solved in civil underwater acoustic communications, it is possible to establish an innovative, adaptive pseudo-random frequency modulation (abbreviated as APNFM) system to be employed in digital underwater acoustic communications. It may be expected to adapt to peculiar underwater acoustic communication channels and obtain an approximately optimum detection result at the criterion of maximum output SNR against a random multipath interference and noise background. Digital underwater acoustic communication equipment will be discussed in Chapter 4, including three types of civil digital underwater acoustic communication equipment developed by the authors: (1) underwater acoustic telecontrol communication equipment in which a digital time correlation accumulation decision system has been employed; (2) underwater acoustic multimedia communication equipment in which an improved FH-SS system has been used; and (3) a digital underwater acoustic communication prototype in which an innovative APNFM system has been employed. There are three innovative key parts in APNFM: (1) adaptive total time sampling processing for PN (Pseudo-random) sequences instead of the usual synchronical schemes; (2) an adaptive Rake receiver to adapt complicated and rapid varying multipath effects and utilize its energy; and (3) a rapid, adaptive channel-modifying net that can adapt to underwater acoustic communication channels having a random spatial-temporal-frequency variability.

Many notable books have been published in the field of underwater acoustics; however, there are not any monographs or systematic reports about digital underwater acoustic communication. Despite the fact that the first author is over 70 years old, he invited his daughter to write this book based on years of teaching experiences, including two doctor's degree courses: "Applied Underwater Acoustics" and "Underwater Acoustic Data Transmissions." Especially relevant was the experience of research and development, including undertaking and participating in five topics with respect to a national high, new technical developing program (since to make some contributions to this program, one has to be chosen as an advanced worker by National Science and Technology Committee and win the third prize), and it is our wish to use that to add a brick and a tile for building the edifice of *Underwater Acoustic Communications*. At the same time, the authors hope the communications develop without interruption, and become more and more mature in service to humanity to explore and use the ocean in a peaceful way.

During the process of writing this book, the authors had consulted many brilliant works and theses, and they received some inspirations from them. At the same time, the authors want to give heartfelt thanks to colleagues for their support and help during the lengthy work process.

This book is designed to serve as a reference book for postgraduate students and practicing engineers involved in the design and analysis of underwater acoustic communications, as well as relative underwater acoustic engineering. As a background, we presume that the readers have a prior knowledge of underwater acoustic physics and digital communications.

As we all know, the Monkey King in *Journey to the West* can freely travel in the ocean to visit his friend Dragon King in the Crystal Palace, but it is also a regrettable thing that he cannot carry out underwater communication with his master and apprentices.

The authors believe that with unremitting efforts we can also travel through the ocean someday and live in the man-made "Crystal Palace." Moreover, people can be guided by tortoises and dance with whales. Furthermore, like the people who live on the earth, everyone can have an underwater mobile phone to communicate with their relatives and friends in multimedia ways at anytime and anywhere. If the Monkey King knew that, he would be jealous.

The appendix includes relative ultrasonic sensing systems in the air medium. The rules with respect to ultrasonic radiation, transmission, scattering, and receiving in the air medium will first be described briefly. It is a physical basis for designing ultrasonic sensing systems employed in the air medium. Next, three new ultrasonic sensing systems developed will emphatically be discussed in this appendix, which are ultrasonic ranging and bearing sensing system employed in concrete jetting manipulators, ultrasonic terrain obstacles sensing system employed in mobile robots, and ultrasonic navigating sensing system employed in automatic guided vehicles. This appendix is suitable for practicing engineers and postgraduate students as a reference in ultrasonic sensing systems and their applications.

It should be pointed out that the authors published *Digital Underwater Acoustic Communication* in Chinese in 2010. Now, we have carried out some supplements and expect a wider range of interchanges, if it is published in English, because it has been translated by the authors.

CHAPTER 1

Introduction

Digital underwater acoustic communications for civil applications will be discussed in this book. An appendix on ultrasonic sensing systems in the air medium will also be included.

After reviewing the development of underwater acoustic communications, the emphasis in this chapter will be focused on describing the differences between underwater acoustic communication channels and radio channels. The peculiarities existing in the former mean that it is necessity to explore some innovative signal processing systems employed in digital underwater acoustic communications, which is also a basic premise to establish new digital underwater acoustic communication equipment. Also, a communication sonar equation will be derived, which provides the references for predicting the performances of present and new communication sonars being designed. An active sonar equation against noise background will also be included. A specific example to show how to use this equation to select the relative parameters in designing an FHSS system communication sonar is also introduced.

In this chapter, the general idea of the book will be summarized, and the role of each chapter in overall will be coherently defined.

1.1 OVERVIEW OF UNDERWATER ACOUSTIC COMMUNICATION DEVELOPMENT

Formation and development of a discipline come from practical needs. The urgent needs from both navy military activities and ocean resource exploitations have become a strong driving force for the development of underwater acoustic communications.

Modern undersea warfare belongs to the war of information technology. Underwater acoustic communications are the important methods of information acquisition, transmission, and control.

The main military applications for underwater acoustic communications are as follows [1,2]: the communications between submarines, submarines and surface ships, submarines and shore base stations, and submarines and underwater combating platforms, as well as military divers; in addition, there are military applications for communication among nodes in military underwater acoustic networks, submarines (now as mobile nodes) and nodes, and so forth.

To meet the urgent needs of the peaceful uses of the ocean, the applications of underwater

Digital Underwater Acoustic Communications
http://dx.doi.org/10.1016/B978-0-12-803009-7.00001-5

acoustic communications have been rapidly extended from current military to civil fields. The typical applications are as follows:

1. Underwater acoustic communications among surface command ships and divers performing underwater exploration, rescue, and salvage [3]. Generally, short-range (such as a few kilometers) voice and image communications are required. Moreover, services must be designed to be small in size and light in weight to facilitate the divers to carry and operate them. Therefore, higher operating frequencies, such as 20 kHz or more, will be selected.
2. Multimedia communications between surface ships and the various types of underwater robots, AUVs (autonomous underwater vehicle), deep submergences, etc. [4]. These communication equipments generally work in the deep sea, and where there are good conditions of vertical sound transmission, the coherent detection system can be used (such as DPSK (differential phase shift keying)). Communication distances are thus greater (eg, more than several kilometers), and transmission rates are also higher, so real-time voice or even color image communication can be achieved [5].
3. Applications of underwater acoustic communications to the exploration and utilization of marine mineral resources [4], such as underwater monitoring of offshore oil drilling platforms, underwater acoustic data telemetry of oil wells, underwater positioning, and monitoring for laying pipelines. These types of underwater acoustic communications belong to high-frequency, short-range, and point-to-point types; operating conditions are better, and thus it is easier to design corresponding communication sonars.
4. Underwater acoustic data (including remote control commands) communications for the automatic monitoring of marine environmental parameters. In this type of marine observation station, the various sensors of oceanographic parameters and a data communication sonar are arranged in an underwater platform. Sampling vessels regularly visit these stations and send commands in the vicinity of the underwater platform to acquire the relative parameters by using underwater acoustic data communications. Compared with the traditional field measurements by using surveying ships, the vessels have the outstanding advantages of saving time, effort, and cost.
5. Formation of civil underwater acoustic networks [6]. The oceanographic parameters acquired by a single buoy-based ocean observatory station mentioned earlier can only get a single point data; thus in-depth analyses of marine environmental variations and numerical predictions are very difficult. If an underwater network is formed, the multilevel parameters will be measured in every node. Then by applying underwater acoustic data communication among the nodes, the parameters will be transmitted by the antenna on the surface buoy of the gateway node into radio channels and collected by a shore station data center. Therefore, multiparameter, large area, simultaneous, continuous, and long-term valuable data can be obtained.

People are ready to capture the formation of an autonomous oceanographic sampling network [7,8]. This network will provide the exchange of data, such as control, telemetry, and video signals, among network nodes. The network nodes, stationary and mobile underwater vehicles or robots, will be equipped with various surveying instruments, such as current meters, seismometers, sonars, and video cameras. A remote user can gather various oceanographic data using direct computer access via a radio link to a central network node based on a surface buoy.

In addition, underwater acoustic communications can also be used in some other fields, such as marine disaster forecasting, the positioning of a black box from a plane crash, or the communication between underwater tour boats and the shore station. The application fields of civil underwater acoustic communications can be expected to be rapidly extended in the future.

The applied fields and performance specifications between military and civil underwater acoustic communications cross each other, whereas more are essentially different.

Military underwater acoustic communications have their own special requirements and corresponding difficulties realizing robust ones.

1. Military underwater acoustic communications generally require long-range propagation. If the information to be obtained is more distant, there is a wider range for acting opportunity.
2. Military underwater acoustic communications generally require high data rates. Once the information is received faster, actions can be obtained in an advanced amount of time.

   Compatibly satisfying the long range, r, and the high rate, R, underwater acoustic communications can be quite difficult. Based on the experimental results acquired by American scholars, an upper limit product between R (kbps) and r (km) [9] is given by

$$R_r \leq 40 \tag{1.1}$$

   Generally, it is difficult to reach this upper limit. For example, when $R = 1$ kbps is required, communication distance would be less than 40 km.
3. Military underwater acoustic communications demand high robustness to avoid accidental incidents. However, marine communication environments are complicated and varied; whereas underwater warfare and their battlefields are likely to occur anytime and anywhere. Adapting communication functions to underwater acoustic communication environments that randomly vary in spatial, temporal, and frequency aspects is very difficult.
4. Military underwater acoustic communications may be applied to high-speed mobile ones. In these cases, there exist dramatic changes in the communication environments, large Doppler frequency shifts, and the high noise levels of the ships [10]. In particular, since the usual underwater acoustic signal processing schemes, such as the matched filter and OFDM (orthogonal frequency division multiplexing), do not have frequency shift adaptability, realizing high-quality mobile underwater acoustic communications is quite difficult.
5. Military underwater acoustic communications have a requirement to be strictly confidential. If this difficult problem cannot be overcome, perhaps transmitted sound power and the corresponding communication distances have to be reduced to exchange for the improvement of confidential communication performance [1].
6. Military underwater acoustic communications generally need multimedia information to adapt to different applied fields. Installing a lot of different communication equipment (including different transducer-amplifier of power modules) in the ships with limited spaces is generally not allowed. Thus, the compatibility problems of different communication media, communication sonar, and active sonar transducers, etc., would cautiously be solved [1].

In contrast, the problems existing in military underwater acoustic communications are not as outstanding as in military ones. Civil communication equipment is usually operated at shorter distances and communication time intervals also have a great flexibility; thus the compatibility problem of either communication distances or data rates is not present. Moreover, they usually belong to fixed points or, at low-speed mobile communications, communication environments that are more stable, so there is a lighter burden on the Doppler frequency shift correction, and the noise levels are also lower. There is no request for confidentiality for such communication equipment. As long as the permissions are met for energy consumption, size, and weight, we can make the information detection under the condition of high signal-to-noise ratio (SNR). In addition, we can produce a variety of underwater acoustic communication equipment having different specifications and applied fields for different users. Generally, the complex compatibility problems mentioned earlier do not have to be considered.

Of course, the degree of difficulty between military and civil underwater acoustic communications is relative. Realizing civil underwater acoustic communications with high performances is still a difficult task. If the communication distances can be extended to be farther, this will widely adapt to practical requirements. Some communication media that have large information contents, such as in image communication, still have high required data rates. For example, we had developed a shallow water image communication prototype [11]. Although the data rates have reached 8 kbps for a simple black and white image consisting of 320 × 200 (pixels) × 16 (gray level) without using data compression, the transmission time is still 32 s. According to Eq. (1.1), the communication distances are below 5 km. In some sea areas, adverse communication environments will be encountered. A signal processing system with excellent channel adaptability is also needed for ocean developments and utilization. Some applied fields, such as the communication between a surface command ship and AUVs, are still considered mobile. Moreover, the multimedia communications and corresponding compatibility problems will also be encountered in some integral applying fields. In addition, civil underwater acoustic communication sonars have some particular requirements in size, weight, power consumption, and cost. In particular, it is difficult to acquire the prior knowledge of underwater acoustic communication channels for the civil users. Therefore, such equipment must be able to accomplish the communication without the prior knowledge of the channels. We see that there are some special difficulties for designing such civil communication sonars.

Thus, even if for civil underwater acoustic communications, achieving multimedia communications with longer distances, higher data rates, and high robustness are still of very difficult. Many core techniques critically wait to be solved, which will be discussed in this book. Some particularly difficult problems existing in military underwater acoustic communications, such as secure, high-speed mobile communications, will not be discussed in depth in this book because the contents will focus on the digital underwater acoustic communications against for civil applications.

As noted earlier, because of the complexity and variability of underwater acoustic channels, there are many practical difficulties to achieve high-quality underwater acoustic communications. However, driven by actual military and civil needs, the underwater acoustic communication discipline has advanced greatly in the process to overcome the varied difficulties.

From the layout of the communications, simple, static point-to-point communications have spread to mobile ones; now underwater acoustic networks have been formed, and we are prepared to establish land/sea/air three-dimensional mobile communication networks.

From the communication systems, the analog underwater acoustic communications have gradually been transited to digital ones. Moreover, some advanced radio communication systems, such as spread spectrum [12,13] and OFDM systems have been applied to underwater acoustic communications. We can achieve distances above 100 km in low data rate digital underwater acoustic communication [14]. In recent ten years, time reversal mirror (TRM) and phase conjugation have received much attention and extensive study [15–17].

From the communication media, single-sideband telegraph and telephone communications have been extended into digital multimedia (including voice, text, images, data, etc.) to meet the actual needs of different communication fields.

It can be expected that underwater acoustic communications will play a more important role in the exploitations and utilizations of the ocean.

1.2 PECULIARITIES OF UNDERWATER ACOUSTIC COMMUNICATION CHANNELS RELATIVE TO RADIO COMMUNICATION CHANNELS

Finding the differences between underwater acoustic and radio communication channels can be helpful to point us in the right direction to adapt to the peculiarities existing in the former.

The transmission laws of sound waves in underwater communication channels are a main research topic for underwater acoustic physics; they are also a physical base to design the digital signal processing systems suitably employed in communication sonars [18]. It would be impossible to efficiently design the systems employed in them without in-depth exploration of the peculiar transmission laws and their effects on underwater acoustic communications, in particular how to adopt the possible countermeasures to adapt to them.

Since the operating frequencies are higher (eg, above several kHz in communication sonars), the sound transmission characteristics in underwater acoustic channels may generally be described by ray acoustic theory with an intuitive physical view.

The main peculiarities of underwater acoustic communication channels relative to those of radio are as follows:

1. **Peculiarities of great transmission loss and strict band limiting due to the sound absorption and scattering in the underwater acoustic channels.**

   The sound absorption in the channels includes viscous absorption, relaxative absorption, and thermal conduction in a seawater medium. That not only causes the great transmission attenuation of sound energy but also the strict band-limited peculiarity. The latter is formed by the sound absorption coefficient β roughly proportional to the square of the operating frequencies f. According to the calculations, for example, $f = 8$, 14, and 20 kHz, which corresponds to $\beta \cong 0.6$, 1.3, and 2.4 dB/km, respectively. Provided that communication ranges r are larger, such as $r = 50$ km, the sound attenuations due to the absorption effect reach 30, 50, and 120 dB, respectively, for the three frequencies mentioned. In the case of such great transmission losses, communication ranges will be considerably decreased, and the bandwidths of the receiver will also be limited. For example, let a bandwidth be 8–14 kHz for a communication sonar: the difference of transmission losses due to sound absorption between upper and lower side frequencies reaches 35 dB at $r = 50$ km. It is difficult to compensate such a large difference by using the usual schemes of amplitude equalization. But if we select the bandwidth that is decreased to the range of 8–10 kHz, this difference will be reduced to be 9 dB. As a result, the communication ranges will be remarkably increased. Moreover, using some signal processing schemes, such as an amplitude equalizer, to make up for that is also easier.

   The inhomogeneity of the seawater medium and the roughness of the sea surface and sea bottom will generate violent sound scattering. Although sound energy is not converted to thermal energy in the process of the sound scattering, it causes the sound wave to deviate the direction toward the receiver, which is equivalent to the attenuation of sound energy. The transmission loss caused by the effect of sound scattering will also be increased with increasing f, and the more violent impact of the strict band limiting will be encountered. We see that efficient bandwidths in underwater acoustic communications are much more narrow in comparison with those in radio communications. Based on the Shannon information theorem, the maximum information transmission rate is proportional to the bandwidth when the SNR is invariable. Therefore, the data rates in underwater acoustic communications are much lower than those in radio.

It should be noted that some advanced signal processing techniques, such as SS-DS (direct sequence-spread spectrum) and OFDM systems in radio communications, require spreading spectra. There exists a fundamental contradiction to the peculiarity of the strict band limiting, and applications to underwater acoustic communications are thus confined to a certain range. Provided underwater acoustic communication ranges are nearer, such as less than 10 km, the band-limited property is not as outstanding. The bandwidths can be chosen to be larger than 5 kHz, and the OFDM system can thus be employed in underwater acoustic communications.

2. Peculiarity of the violent fluctuations of sound signals traveling in underwater acoustic channels.

The medium space of underwater acoustic channels (the sea) is much narrower than that of radio channels (the sky). Moreover, the boundaries of the former will generate sound reflection and severe scattering. The superposition between direct and reflected sound signals will cause severe signal fading (refer to Figs. 1.8 and 1.9) due to the interference effect of coherent sound waves.

The fluctuations of sound signals in the amplitude and phase generated by the sound scattering from the inhomogeneity in the body of the seawater and rough sea boundaries have remarkable impacts on the correct signal detections in communication sonars.

The fluctuation in the amplitude in underwater acoustic channels generally follows Rayleigh distribution law; sometimes fluctuating ratios approach 52%. The additional phase shifts of sound signal caused by the sound scattering from the sea surface can be described by the Rayleigh parameter:

$$K = 4\pi\frac{\xi}{\lambda}\sin\chi$$

where λ is the wavelength of a sound wave, ξ is the rms of sea-wave height, and χ is the grazing angle relative to average sea surface. When ξ/λ or χ is small enough to cause $K \ll 1$, the sea surface will be regarded as a mirror-like boundary. In such a case, the reflection coefficient $V \cong -1$ (ie, there is a 180° phase shift). With increasing K, the noncoherent component of the sound scattering from the sea surface will rise, and the phase shifts will change into a random variable. The sound scattering from sea bottom is more complicated because of its diverse and multilayered composition. We can see that the phase shifts caused by the sound reflection and scattering from the sea boundaries have a random spatial/temporal/frequency variability.

The inhomogeneity, such as thermal microstructure in the body of the seawater itself, is also a reason to generate the fluctuations of sound signals in the amplitude and phase. It depends not only on the fluctuation of the index of the sound refraction in seawater but also on the operating frequencies and communication ranges, etc.

3. Peculiarity of the sound field distributions in underwater acoustic communication channels.

The impacts of the reflection and scattering for radio communications may generally be neglected since electromagnetic waves travel in a wide space. Moreover, the gradient of the index of the refraction in air for that is small enough; its traveling path is thus along a straight line, and the attenuation of energy in distant fields obeys the spherical spreading law. On the contrary, there exist sharp gradients of sound velocity $G_c(z)$ with complicated distributions in the depth direction in the seawater that are caused by both the sun's radiation and the disturbance of sea current. In fact, there is an absence of $G_c(z) = 0$ in actual underwater acoustic channels. Strictly speaking, sound rays do not travel along a straight line: the sound field consists of the superposition of many eigen rays at the receiving point, which correspond to different radiation angles and traveling paths due to the sound reflection and refraction in the channels. As a result, complicated and variable sound fields are encountered in underwater acoustic channels due to the renewable distributions of sound energy, such as forming the sound channeling (or ducts) in mixed-layer and deep sea and the anti-channeling in the sea areas with a sharp thermocline or negative $G_c(z)$. Therefore the sound transmission laws among them have essential differences. Fig. 1.1 shows the ray diagram in deep-sea sound channeling; the sound source was located at the sound channeling axis with a minimum sound velocity where the sound reflections from the sea boundaries had been neglected [10]. We note that the distributions of sound field are quite complicated: the sound energy is concentrated in the dense areas of rays, while it is lower in sparse areas. In areas where the rays are absent, the sound energy approaches zero from the point of view of ray acoustics.

We can see that the sound attenuation laws due to the geometrical spreading in the channeling show essential differences in comparison with a free sound field. Therefore, the specifications, such as the communication ranges and data rates for a communication sonar, are thus randomly variable over time and space. The peculiarity of sound energy distribution in layered channels is therefore a chief research topic in underwater acoustic physics.

4. Peculiarity of multipath effects in underwater acoustic communication channels.

In the case of shortwave radio communications, electromagnetic waves can be reflected from an ionosphere by several times or reflected from the variable ionospheres with different heights; thus multipath propagations appear. The reflections from obstructions also exist for mobile radio communications. They will cause the time and spectra spreads of signals and thus generate ISI (inter symbol interference). When their impacts on the performances of the communications cannot be neglected, the signal processing schemes of adaptive channel equalizations, such as a linear transverse FIR filter, will be utilized to effectively suppress them.

How to adapt to the peculiarity of the multipath effects in underwater acoustic communication channels is a key technique that determines efficiently realizing expected underwater acoustic communication tasks.

The multipath effect described by ray acoustics consists of the superposition of the rays at a receiving point along different paths and corresponding traveling times. The mechanism that forms multipath propagations is shown in Fig. 1.2. A simplified sound velocity profile at a certain position in the Taiwan Strait is shown in Fig. 1.2A, which is a linear negative gradient.

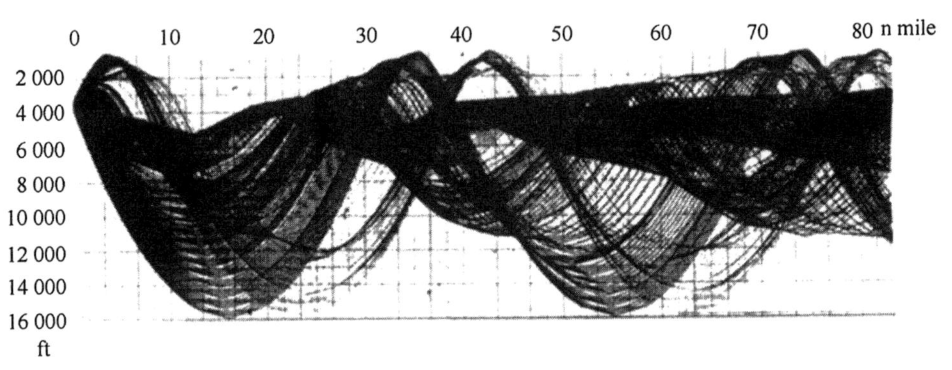

FIGURE 1.1 Ray diagrams for deep-sea sound channeling.

Fig. 1.1B shows 12 eigen rays with different radiation angles in the range of ±5 degrees, which travel along different reflected and refracted paths and then arrive at receiving point P to form the total sound field. We see that if the transmitted signal is a narrow pulse, different rays having corresponding traveling times and will thus consist of a complicated and variable pulse train in time domain: it is called a multipath structure.

We see that by examining the mechanism forming multipath propagations, the multipath structure with large delay spread (eg, up to 1 s) will be encountered in the long-range communications in deep-sea sound channeling (refer to Fig. 1.1).

The complicated and variable multipath structures in deep-sea channeling had been discovered long ago. The recorded waveforms are shown in Fig. 1.3 [19,20]. The duration of transmitted pulse was 5 ms. We see that the waveforms among transmitted and received signals have essential differences; moreover, the multipath structures are severely changed versus transmission range r. We should pay attention

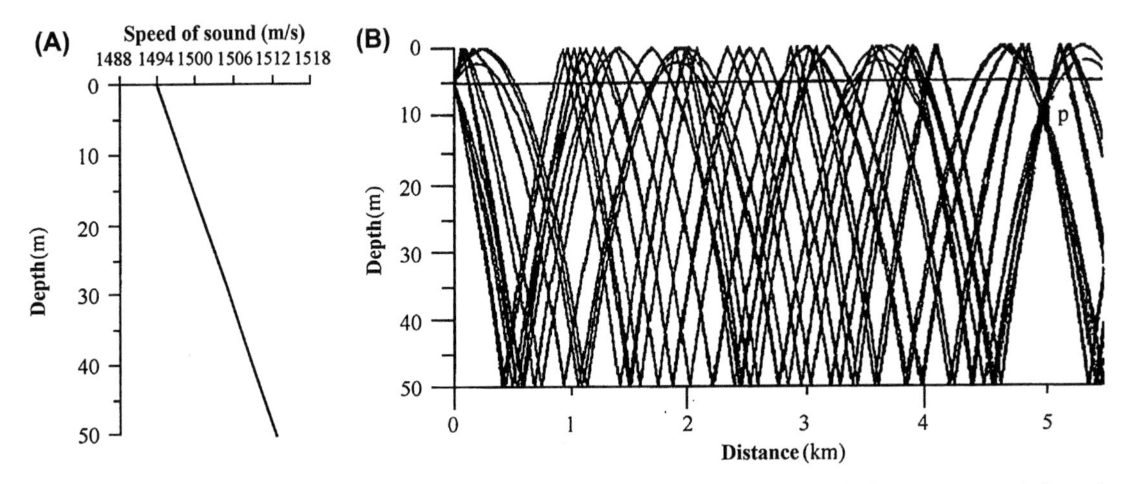

FIGURE 1.2 Superposition of eigen rays at receiving point P for a negative $G_c(z)$ in a shallow water sound channel.

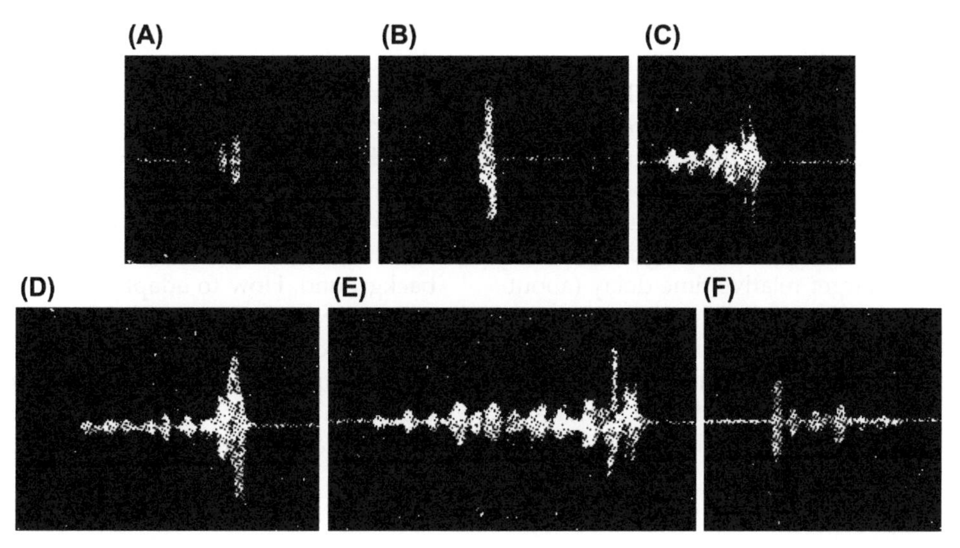

FIGURE 1.3 Waveform recorded at different traveling ranges in deep-sea channeling: (A) $r = 0.75$ km; (B) $r = 1.5$ km; (C) $r = 7.5$ km; (D) $r = 12$ km; (E) $r = 20$ km, sound source was located at the axis of channeling, and hydrophone was above that; (F) $r = 13.5$ km, hydrophone was located at the axis.

to the effect of sound interference due to the superposition of coherent multipath pulses (wave packets) with inverse phase, as shown at the final packets in the multipath structures in Fig. 1.3C and E, where the received waveforms have a double-peak structure; the concavities in the middle parts are caused by the fading of the inverse phase interference. The effect of the interference due to multipath propagations is a main characteristic relative to nonlinear underwater acoustic communication channels.

The multipath structures in shallow water acoustic channels are more complicated and variable than those in the deep-sea channeling. The waveforms of multipath structures acquired from the Yellow Sea area with a sharp thermocline are shown in Fig. 1.4 [21]. We see that the multipath structures also severely change versus operating frequencies.

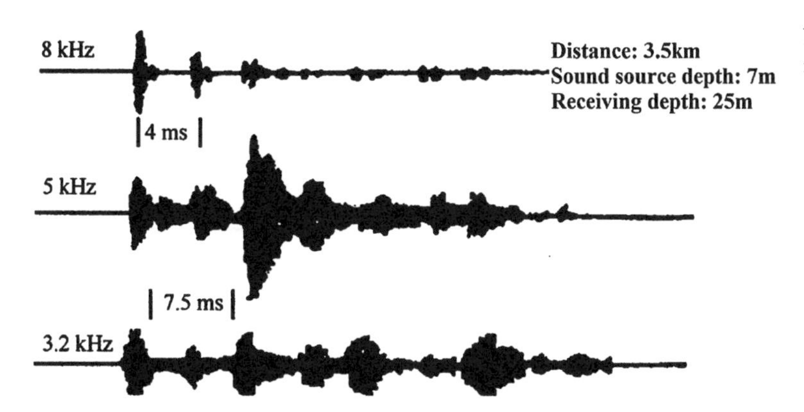

FIGURE 1.4 Multipath structures for different frequencies.

Fig. 1.5 shows the multipath structures versus the depths at which the transmitting transducer and hydrophone were locked [22]. We see that the multipath structures violently change with the depths. In particular, provided they were both placed below the thermocline, the multipath structure has a remarkably different distributing model; moreover, a larger relative time delay (about 30 ms) appears.

Multipath structures in some underwater acoustic communication channels have the characteristics of severely random spatial/temporal/frequency variability and large total delay spread, which may be much larger than the duration of transmitting pulse (eg, up to several hundred times). As well as an active sonar to detect targets against the reverberation interference background, the information signal detection for a communication sonar also has to operate against the multipath interference background. How to adapt to that is a key technique for designing a robust communication sonar.

We emphatically point out that the multipath effects process duality. We have known that the multipath effects are caused

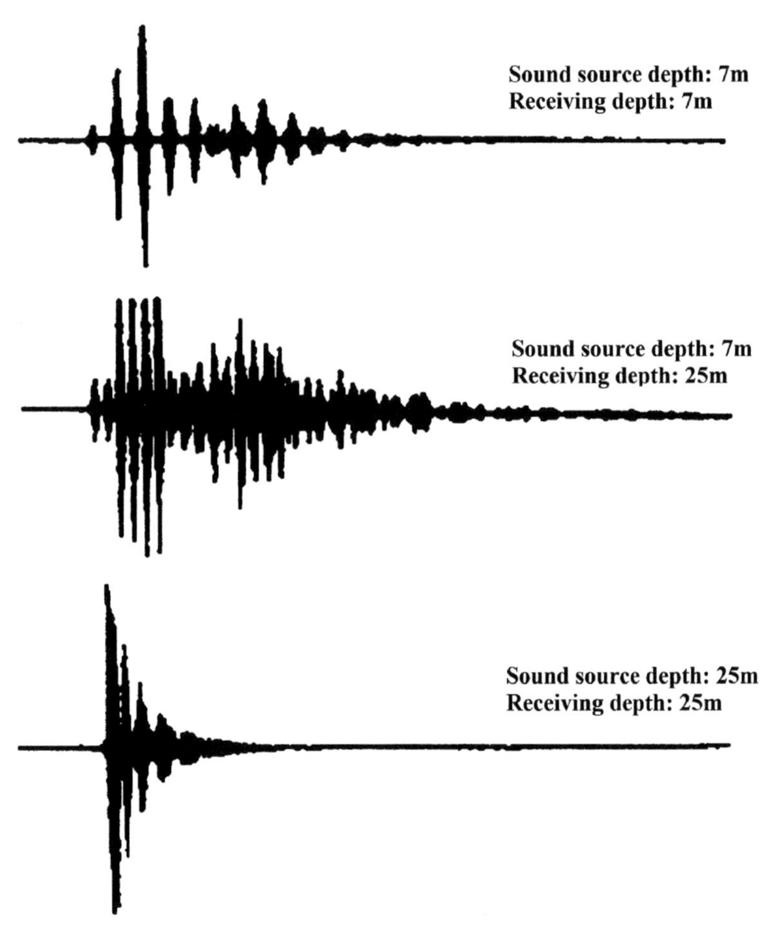

FIGURE 1.5 Multipath structures for different transmitting or receiving depths.

by the integrated influences of both the sound reflection from the sea boundaries and the refraction in seawater itself. They restrict sound transmission to obey the spherical spreading law and thus form a train of pulses in the time domain. The focusing effect of sound energy in deep-sea channeling has been analogized to the curved surface of the echo wall at Tian Tan Park in Beijing by some authors [23].

Provided that we use paths diversity (ie, Rake receiving technique) in which multipath pulses (wave packets) are first acquired and then an optimum combiner is selected, the optimum detection result would be expected at the criterion of maximum output SNR. In this case, the energy of multipath propagations would fully be used. This is the positive effect of multipath propagations.

5. Peculiarity of the background noise in the sea.

The noise (including pulse noise, narrow band noise, and fluctuating noise) in radio communications can generally be regarded as additive.

The background noise in underwater acoustic channels includes the ambient noise and the self-noise of vessels. When the communication time interval is shorter, the former may generally be thought of as stationary Gaussian noise. The signal processing schemes combating against that in underwater acoustic communications basically are the same as those employed in radio communications. Whereas, there exist some peculiarities in the background noise in the sea, such as high noise levels (*NL*) and severe spatial/temporal/frequency variability. For example, let $f = 5$ kHz, and the spectra level of ambient noise at sea state 3 is about 50 dB. Even if the bandwidth of the receiver is reduced to be 1 kHz, *NL* are still above 80 dB. It is possible that the receiver has to operate under the condition of low input SNR; thus communication ranges will be remarkably limited by the high *NL*.

In the case of mobile communications, *NL* generated by the self-noise of vessels generally are higher. For example, *NL* may achieve the same value as that at the sea state 3 even if at the lower speed of vessels, such as 5 knots. Moreover, its noise spectra are of two basically different types. One is broadband noise having a continuous spectrum. The other type of noise is tonal noise having a spectrum containing line components occurring at discrete frequencies, as schematically shown in Fig. 1.6. The components of line spectra sometimes have a spectrum level higher than average (above 20 dB). Although the line spectra generally distribute in lower frequency range (below several kHz) in which the operating frequencies for long-range underwater acoustic communications would be arranged, the modulation phenomenon among received signals and line spectra components in the self-noise of vessels appear. The noise in the sea does not always satisfy the condition of being additive. How to combat with nonadditive noise is a more difficult task.

6. Peculiarity caused by the low sound velocity in the seawater.

The sound wave velocity in seawater is much lower than the electromagnetic wave velocity in the air, so some particular

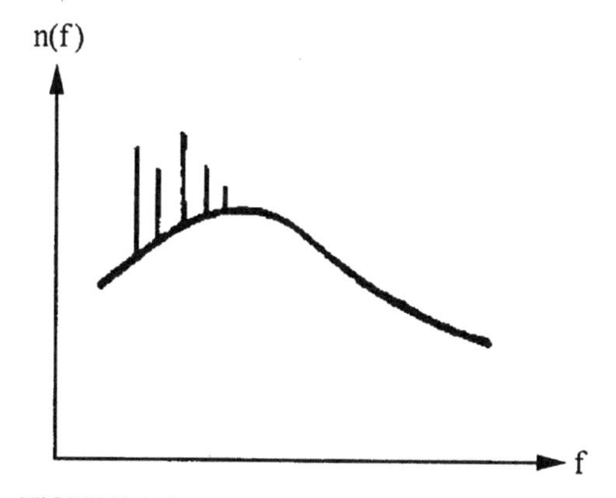

FIGURE 1.6 The spectrum for self-noise of ships.

problems in underwater acoustic signal processing thus appear.

In the case of high-speed mobile underwater acoustic communications, Doppler frequency shifts are large, which can even be analog to the bandwidth of receiver because the sound velocity is so low. As there is not the adaptability of the frequency shifts for the usual signal processing systems (such as the matched filter and OFDM) employed in underwater acoustic communications, how to adapt to sonar communication, just as in active sonar, is a difficult task.

The traveling time, t, of sound signals in underwater acoustic channels generally is large since the sound velocity is so low. For example, if a communication range is $r = 45$ km, and the corresponding $t \cong 30$ s, perhaps that would be greater than the coherent times of the channels, so the channels cannot be regarded as time invariant, and received signals would be processed as time variant. If it is necessary to exchange channel parameters between transmitting and receiving sides in real time to realize an adaptive channel modification match, it is possible that the time of exchanging information is greater than 1 min. This signal processing scheme would be inefficient. Obviously, the information feedback scheme in error detecting generally is also inefficient.

In summary, there exist many remarkable peculiarities in the underwater acoustic communication channels in comparison with radio channels, including parameters that randomly vary in spatial/temporal/frequency aspects, large attenuation, severe multipath effects, strict band-limited property, high noise level, and low sound velocity. Therefore some present advanced signal processing systems, or even some basic principles in radio communications, cannot be employed perfectly. In particular, they cannot mechanically be copied in underwater acoustic communications. Several principal aspects will be simply stated as follows:

1. Based on the Shannon theorem in information theory, maximum information transmission rates or channel tolerances in a channel are given by

$$C = B\log_2\left(1 + \frac{P}{N}\right)$$

where B is the bandwidth of signals, and P and N are the mean power of signal and that of noise, respectively. Provided we require raising C, increasing P/N or B can correspondingly be selected. The contrary is true as well.

However, there are some basic assumptions that must be satisfied for the Shannon theorem, which include the noise is white Gaussian, received signals do not distort, etc. Obviously, underwater acoustic communications do not fit these conditions: its applications are thus limited to a certain range.

We have known that there is an expression describing the relation between information rates R (kbps) and transmission ranges r (km) as shown:

$$R_r \ll 40$$

where R is inversely proportional to r.

Perhaps there are two main reasons to cause this peculiarity. One is that the sound absorption effect in underwater acoustic channels will rise remarkably with increasing operating frequencies f. Therefore, to raise r, lower f has to be selected, and R will correspondingly be lowered. The other reason is that multipath effects are severe in the channels, in particular those in which total delay spreads will be increased (refer to Fig. 1.4) with reducing f. As a simple method, decreasing R generally is used to adapt to that, such as letting the repeated periods of signal pulses that are greater than the total delay spread. In this case, the greater the r (the lower f), the larger the ISI is, and thus the lower the R (the repeated period) becomes.

2. Realizing the accurate detection of synchronizing signals is a key, though difficult, signal processing technique in digital communications. Radio communications with high performances have been realized by the support of many efficient synchronizing techniques, some of which have also been applied to digital underwater acoustic communications and held out to obtain robust performances.

   However, there exist violent and rapid fluctuations of sound signals in time domains in some adverse underwater acoustic communication channels. Experimental results have demonstrated that the fluctuation ratios of pulse repeated period T_0 are up to 0.2–1%, although T_0 is only several tens of milliseconds at the transmission range of 5.5 km in the Shou Shao shallow sea area. The fluctuation of T_0 in the Xiamen Harbor may be up to 0.2 ms for $T_0 = 10$ ms at the range of 5.5 km. Moreover, the fluctuation of T_0 has an accumulation tendency with increasing communication ranges. Therefore, the phenomenon of "out-of-synchronism" would appear in some digital underwater acoustic communications at lower input SNR, as shown in Fig. 1.7, which shows received information signals in situ for an image text communication. We see that the violent cross-interference exists between adjacent codes in the distant end of the line synchronization. It means that the usual trapping and accurate tracing for synchronical signals are inefficient in some underwater acoustic channels having severe and rapid fluctuation in time domains at lower input SNR. The authors put forward that we can adopt an innovative signal tracing scheme: adaptive total time sampling processing for pseudo-random signal sequences instead of the usual synchronical techniques, which has a prospect to accurately solve the difficult problem in detecting the synchronical signals and provides a technical support to adaptive path diversity (refer to Chapters 3 and 4).

THE UNIVERSITY

FIGURE 1.7 Record of text communication at lower SNR.

3. Because of the random scattering and reflection of sound signals from the sea boundaries, and sound scattering due to the inhomogeneity in seawater itself, the severe fluctuations of sound signals in both the amplitude and phase thus appear. In particular, the fluctuation and 180 degrees sudden change in phase have a remarkable random spatial/temporal/frequency variability. Therefore the application of a phase-coherent detection method to underwater acoustic communications will be limited, or even inefficient in some adverse communication circumstances.
4. Some channel equalization techniques employed in radio communications are unsuitable to being used in underwater acoustic situations because received information signals are multipath structures with a violent and rapid random spatial/temporal/frequency variability and large delay spreads. Provided that the multipath diversity (ie, Rake receiving technique in radio communications) is modified by an adaptive operating approach, optimum signal detection would be expected at the criterion of the maximum output SNR against the background of multipath propagations.

On the whole, by knowing in depth the essential differences between radio and underwater acoustic communication channels, exploring efficient approaches to adapt to the peculiarities existing in latter and then developing innovative signal processing systems efficiently employed in digital underwater acoustic communications would be a challenging task. It is also a basic premise to establish robust communication sonars.

1.3 EXPLORATIONS ESTABLISHING AN INNOVATIVE DIGITAL UNDERWATER ACOUSTIC COMMUNICATION SIGNAL PROCESSING SYSTEM

1.3.1 Digital Underwater Acoustic Communications

The analog communication systems, such as analog, single-sideband AM and FM had basically been employed in underwater acoustic communications at early ages [24]. Generally speaking, the robustness of the analog system is low because it cannot adapt to the complicated and variable underwater acoustic communication channels. For example, a speech communication device employing an FM system had been developed as early as 1964 [3] in which a center frequency of 120 kHz and an acoustic power of 100 mW were used. A transmitting–receiving transducer is omnidirectional in the horizontal direction, but its vertical 3 dB beamwidth is restricted to 30 degrees to prevent signals from reaching the surface of the sea; otherwise the surface-reflected signals would have traveled via a path of different length and would be shifted in a phase relative to each other and to the direct signal. As a result of the multipath propagations, the distortion of the speech signal or even losing the character of FM signal would occur. A restricted vertical beam reduces the multipath effect to a minimum and ensures reliable midwater communication. Such a device permits multiway underwater communication up to ranges of at least 1 km at depths in excess of 70 m.

Practices have demonstrated that the robustness of a digital communication system is higher than that of an analog one. Now, the digital communication system has been widely employed in underwater acoustic communications and has a tendency instead of the analog one.

There are some remarkable advantages gained by adopting a digital communication system; we would fully make use of them.

1. Data relays do not generate error accumulations for a digital communication link provided it is meticulously designed. It is quite available to be employed in underwater acoustic networks, etc., to realize long-range underwater acoustic communications and monitions.
2. Error detection and correction techniques can be used in the digital communication system, which will remarkably improve the performances combating with interference. Except the convolution code is generally used in the system, the advanced Turbo code has also been introduced in that.
3. It is more efficient to adopt privacy approaches in digital communications. It is necessary in military ones.
4. Communication media, including speech, image, text, data, etc., can be converted to digital information and transmitted in the same channel; moreover, it is convenient for storage and process. Therefore multimedia communication sonars may be more easily established.
5. Bandwidth efficiency may be remarkably improved by applying the compression coding of information sources to the system. Efficient data (eg, speech and image) compressions are key, though difficult, techniques in digital underwater acoustic

communications because there exist the peculiarities of the strict band limiting, high BER (bit error rate), etc., in underwater acoustic channels.

6. Digital communication sonars may be established by using large-scale integrated circuits. Therefore, it is expected that these sonars will be light in weight, small in size, and sparing in power, thereby being easier to establish the small mode ones for divers.
7. A digital system is more suitable to the frame of information theory; it is thus easier to develop new signal processing systems employed in communication sonars.

The advantages mentioned earlier will be reflected in detail in Chapters 3 and 4.

Note that there are costs for adopting the digital communication systems:

1. Wider bandwidths are necessary for a digital communication system. For example, a digital telephone generally requires the bandwidth to be 64 kHz, while it is only 4 kHz in an analog one. This disadvantage is much more remarkable in strict band-limited underwater acoustic channels.
2. It is necessary to realize the accurate syntonizations in a digital communication system, which is quite difficult in underwater acoustic channels that have the peculiarities of random and variable spatial/temporal/frequency parameters, violent fluctuations, etc. We realize syntonizations in an analog system is easier.
3. Digital underwater acoustic communications, as digital radio ones, have a "threshold effect." When input SNR is reduced to a certain value, the performances of the communications will severely be degraded, which is a main reason to cause the robustness to be reduced in digital underwater acoustic communications. The reduction of performances for analog communication is generally relatively smooth.
4. Some complicated signal processing techniques must be utilized in a digital communication system. Perhaps it is difficult to use them for civil underwater acoustic communications.

These disadvantages are outstanding in establishing digital communication sonars. This is why the analog communication system is still employed in some communication sonars at present.

1.3.2 Main Civil Digital Underwater Acoustic Communication Systems at Present

Because there exist many basic differences between radio and underwater acoustic communication channels, some advanced and effective signal processing techniques in the former are not always adapted to be used in the latter. Therefore, exploring innovative signal processing systems has become a key task in digital underwater acoustic communications.

The main civil digital underwater acoustic communication systems at present are as follows:

1. Incoherent system represented by MFSK (multiple frequency shift keying)

   The incoherent system represented by MFSK had generally been employed early in most underwater acoustic communications because this system has a simple structure and reliable performances. Moreover, the MFSK system has a potential to obtain higher data rates in a shorter range (as little as 10 km) underwater acoustic communications because it permits selecting higher operating frequencies. In this case, the scheme of the amplitude threshold decision can be used to combat the weak multipath interference (refer to Section 3.2.1). However, bandwidth efficiency is lower in the MFSK system, which is difficult to adapt to the long-range underwater acoustic communications with severe multipath effects at low operating frequencies.

In fact, the incoherent systems still have their place and are still employed in underwater acoustic communication. Their outstanding advantage is algorithmic simplicity, requiring modest processors and thus reducing cost and power. A number of commercial systems are available, and while their bandwidth efficiency may be low, they are satisfactory for many applications where low bit rates (and perhaps low power efficiency levels) are acceptable.

2. Coherent detection system represented by MPSK [25]

   MPSK (DPSK, QPSK (quadriphase shift keying), etc.) belongs to a highly bandwidth efficient, phase-coherent modulation. This system would adapt to short ranges, high data rates, and especially deep-water vertical link underwater acoustic communications. MPSK, which is a system based on the accurate detection of signal phases, is generally difficult to adapt to the underwater acoustic channels with severe and rapid signal fluctuation in phases and large delay spreads due to multipath effects (ie, its channel adaptivity is less). Whether or not using the coherent detection system must be analyzed in detail according to specific channel characteristics and operating conditions of communication sonars.

3. Spreading spectra (SS) system

   An SS system has successfully been employed in radio communications. Considering it has some unique behaviors, such as detecting weak signals, having high auto-jamming, and realizing CDMA communications, it is particularly suitable to being used in underwater acoustic communications. Therefore, how to introduce this system into underwater acoustic communications has received much attention, and many researches have been performed and have obtained some significant results. Field experiments have demonstrated that when combining an SS system with the path diversity, spatial diversity, etc., robust underwater acoustic communications can be realized (refer to Section 3.1.4).

   However, there exists a basic contraction between the peculiarity of strict band limiting in underwater acoustic channels and spreading frequency spectrum in an SS system. To obtain higher SS processing gains, sometimes low data rate (as little as several tens of bits per second) underwater acoustic communications have to be selected. Provided they are allowed (as text communication), we may realize long-range (eg, 100 km) underwater acoustic communications by using DS-SS because it enables one to detect the information signals at a very low input SNR (eg, below −15 dB).

4. Optimum linear (matched) filter

   The matched filter may realize the optimum linear detection based on the criterion of maximum output SNR. It is an efficient signal processing system in radio communications.

The premise conditions for using the matched filter are as follows:

1. The channel is equivalent to a linear filter, which performs the linear transformation of transmitted signals. A linear channel can be expressed by impulse response function $h(t,\mathbf{r},\tau)$ in the time domain or frequency response (transfer) function $H(t,\mathbf{r},\omega)$ in the frequency domain. Generally speaking, they are spatial−temporal functions.
2. A channel belongs to a time-invariant system. If the input signal is $x(t)$, then the output signal is $y(t)$, and the relation between them can be expressed in the time domain as

$$y(t) = \int_{-\infty}^{\infty} x(t)h(t-\tau,\mathbf{r})\mathrm{d}\tau \qquad (1.2)$$

That can also be expressed in the frequency domain as

$$y(t) = \frac{1}{2\pi}\int_{-\infty}^{\infty} X(\omega)H(\omega,\mathbf{r})e^{i\omega t}d\omega \qquad (1.3)$$

where $X(\omega)$ is the spectrum function of $x(t)$.

3. Additive white noise background
   In this case, the input of channel is given by

$$x(t) = s(t) + n(t) \qquad (1.4)$$

   where $s(t)$ is the information signal, and $n(t)$ is the additive white noise. The output for a linear time-invariant channel is given by

$$y(t) = s_y(t) + n_y(t) \qquad (1.5)$$

4. Detecting regular signals
   The output waveform of channel $s_y(t)$ is certainly known. Though, its amplitude may be attenuated, and its phase may be shifted. If we utilize a cross-correlator to realize optimum linear matching, its reference signal is thus transmitted. This kind of cross-correlator is generally called a copy correlator.

Obviously, the input of the receiver is a random and variable spatial/temporal/frequency multipath structure in an underwater acoustic communication channel, and the performances of the copy correlator will be considerably reduced. Moreover, the self-noise of ships includes the nonadditive component of discrete line spectra (refer to Section 2.4), which also does not perfectly satisfy the condition of the additive white noise.

A basic problem of whether underwater acoustic communication channels obey the linear condition will be discussed first.

A system is referred to as linear if it satisfies the linear superposition theorem:

$$T\{ax(t)\} = aT\{x(t)\} \qquad (1.6)$$

$$T\{x_1(t) + x_2(t)\} = T\{x_1(t) + x_2(t)\} \qquad (1.7)$$

where T is a kind of transform or operating factor, and a is a constant. Eq. (1.6) is called homogeneity, and Eq. (1.7) is called additivity.

We must pay attention to a in Eq. (1.6), which if large enough, the acoustic wave will tend to finite amplitude, nonlinear effects, including the distortion of waveforms, and larger transmission loss will thus occur.

We have also known that the multipath propagations are a basic characteristic for underwater acoustic communication channels. The effect of the interference due to the superposition of coherent sound waves arriving at a receiving point from different paths is therefore a main reason to generate nonlinear characteristic. Magnitudes will rise for the same phase superposition and will be lowered for reverse phase due to the effect of the sound interference. Fig. 1.8 shows an experimental record with respect to an interference fading diagram generated by the superposition between direct and surface-reflected sound waves for a digital, linear FM signal. We can see that the waveforms in the left-hand side of the diagram belong to those raised by the superposition with approximately the same phase between them, and then they will gradually be transited to reverse with less amplitude. The waveforms at the right-hand side of this diagram have a tendency to symmetrically rise with that in

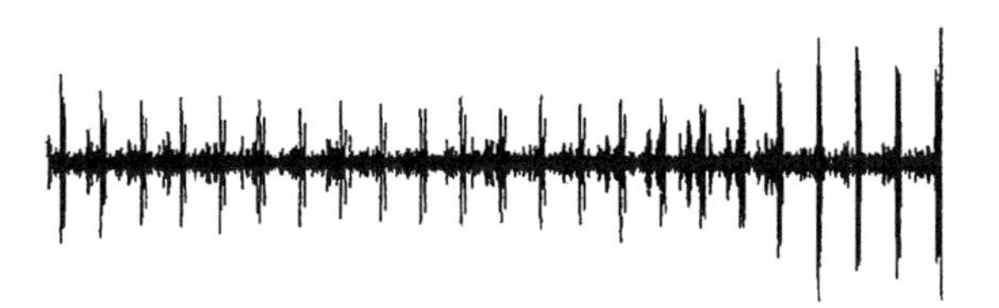

FIGURE 1.8 Record diagram for interference due to the superposition between the direct and reflected sound waves from the sea surface.

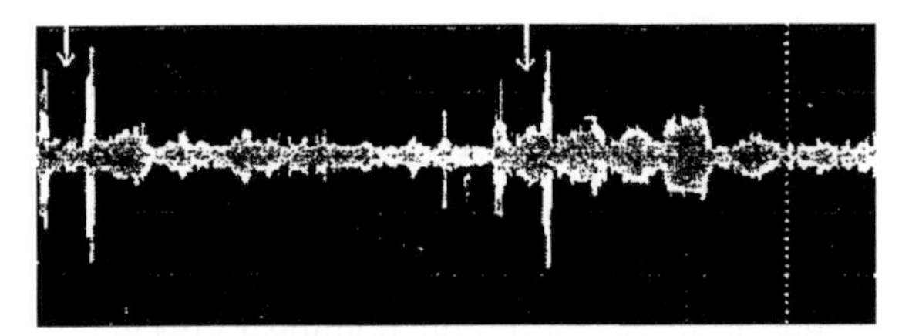

FIGURE 1.9 Spread waveform in time domain for reverse phase superposition.

the left-hand side. Fig. 1.9 shows the two spreading waveforms of reverse phase superposition in time domain for another experiment. We see that there are narrow double peaks with approximately equal durations and higher amplitude at the two sides of the concavity due to the reverse phase interference effect. This is the basic characteristic of the signal fading generated by sound interference effect between direct and surface-reflected sound waves. The durations of the peaks correspond to the time differences between direct and reflected sound waves arriving at the receiving point, and their amplitudes correspond to those of direct and reflected waves alone. In Fig. 1.8, we observe that the superposing amplitudes due to the same phase sound waves at the right-hand side of the diagram are approximately equal to two times in comparison with those of direct ones because the sea surface was smooth enough in experimental time interval: its reflection coefficient is $V \cong -1$.

The other kind of sound interference due to the superposition of coherent sound waves arriving at a receiving point from different refraction paths appears in the deep-sea sound channeling. The recorded waveforms are shown in Fig. 1.3C and E, in which similar waveform structures with double peaks also exist.

As was mentioned, underwater acoustic channels do not always satisfy the linear superposition theorem because the effect of interference caused by multipath propagations occurs widely. Obviously, provided that a pulse signal with the duration τ_s is narrow enough or a delta function with a bandwidth B that tends to infinite in theoretical conception is transmitted in the channels, the interference effect is absent. In fact, the practical τ_s or B for a specific communication sonar is chiefly determined by its data rates.

In the case of existing interference effect due to multipath propagations, whether or not an underwater acoustic channel will satisfy a linear condition depends not only on the characteristics of the channel but also on the τ_s or B of a specific communication sonar. If we spread the usual impulse response function $h(\tau,\mathbf{r})$ and frequency response function $H(\omega,\mathbf{r})$ for linear time-invariant channels to the finite bandwidth impulse response function $h_l(\tau,\mathbf{r},B)$ and frequency response function $H_l(\omega,\mathbf{r},B)$ under the condition of coherent multipath channels to describe the multipath structures excited by applied pulses with finite bandwidths, it is expected that they will further satisfy the actual conditions of underwater acoustic communication channels. Perhaps there are several pulses (wave packets) in a multipath structure in which the interference effect appears, including the fading due to the reverse phase superposition; it would adapt to that by means of adaptive Rake receiving technique as viewed from underwater acoustic communication engineering because the probability of existing multipath superposition all having the reverse phase at a specific multipath structure is small. Therefore, the output SNR for the Rake receiver still has an expected value. Using suitable schemes, such as spatial or time diversities, etc., will further reduce this probability.

Generally speaking, the underwater acoustic communication channels belong to time-variant channels. They may be processed by separating total communication time interval, in which the channel belongs to a time-variant one, to several subintervals in which they may be regarded as time-invariant channels. Then the subintervals are transited from one to another by means of an adaptive correction net (refer to Chapter 3).

For some underwater acoustic channels, as deep-water vertical links, they satisfy the linear superposition theorem: $h_l(\tau,\mathbf{r},B)$ and $H_l(\omega,\mathbf{r},B)$ will reduce to $h_l(\tau,\mathbf{r})$ and $H_l(\omega,\mathbf{r})$, respectively (refer to Chapter 4).

As aforementioned, the underwater acoustic communication channels do not always satisfy the assumptions of the linear system and the additive noise. The received signals are randomly variable spatial/temporal/frequency multipath structures. Therefore the applications of the matched filter or of the cross-correlator to underwater acoustic communications have some essential limitations.

5. Multicarrier modulation techniques, such as OFDM, have been used to overcome ISI caused by multipath effects in underwater acoustic communication channels. By dividing the available bandwidth into a number of narrower bands, OFDM systems can perform the equalization in the frequency domain and eliminate the need for complex time domain equalizers. Moreover, OFDM modulation and demodulation can easily be accomplished by using FFT (fast fourier transformation).

   We would note that, when using coded OFDM, consecutive symbols are often striped across subcarriers to reduce the error correlation due to fading. However, impulse noise present in some environments can affect multiple subcarriers simultaneously, generating correlated errors. Moreover, OFDM systems are very sensitive to Doppler shift due to the small bandwidth of each subcarrier as compared to the Doppler shift.
6. TRM and the phase conjugation processing scheme have also been developed in digital underwater acoustic communications [15–17].

Due to the symmetry of the linear wave equation, sound transmitted from one location and received at other locations, reversed, and retransmitted, focuses back at the original source location. This is the principle behind TRM or its frequency domain equivalent: active phase conjugation. The temporal compression effect of TRM reduces the delay spread of the channel, while the spatial focusing effect improves SNR and reduces fading.

In a TRM-based communication system, a probe signal has to be first transmitted from the receiver to the transmitter. The transmitter then uses a time-reversed version of this signal to convey information. As the channel changes in the time domain, the probe signal has to be retransmitted to sample the channel, but decoherence times up to several tens of minutes were observed at the frequency of 3.5 kHz during experiments.

Although TRM helps to reduce the delay spread of the channel, it does not eliminate ISI completely.

1.3.3 Explorations Establishing an Innovative Digital Underwater Acoustic Communication System

As aforementioned, there are many essential differences between underwater acoustic communication channels and radio ones. How to establish the effective signal processing systems that may adapt to the peculiar underwater acoustic communication channels is a valuable and challenging research topic.

According to the peculiarities of underwater acoustic communication channels, by combining with some key techniques that urgently need to be solved in civil underwater acoustic communications, the basic principles to establish an innovative digital underwater acoustic communication system are as follows:

1. Selections of modulation methods. Generally speaking, AM and PM are restricted from being employed in digital underwater acoustic communications with violent signal fluctuations; whereas FM is more available to be employed in them because it is not as sensitive to the fluctuations of signals in the amplitude and the phase as AM and PM.
2. Nimble and effective applications of varied adaptive signal processing schemes to adapt to the underwater acoustic communication channels with severely variable spatial/temporal/frequency parameters are naturally selected.
3. Because there exist the violent fluctuations in time–amplitude domains in underwater acoustic communication channels, the accurate detection of synchronical signals is quite difficult, in particular at a lower input SNR (refer to Fig. 1.7). The authors put forward an innovative scheme: total time sampling processing for PN sequences instead of the usual synchronical schemes would solve this difficult problem.
4. Realizing the suitable match to the real-time outputs of channels, ie, randomly variable spatial/temporal/frequency multipath structures (rather than transmitted signals), is a key technique for robust digital underwater acoustic communications. Therefore it is necessary to differentiate and analyze the sparse characters of multipath structures. Based on that, an adaptive path diversity (Rake receiver) with an optimum combiner would be an optimum selection.
5. How to realize the adaptive channel suitable match without prior knowledge of underwater acoustic channels is the basic reflection of applicability in civil underwater acoustic communications.
6. Multimedia information can be converted to bit streams by means of the processing of quantization format. This is the "generality" of digital multimedia communications. By combining with the particularities of different communication media, such as the different requirements of data rates, BER, etc., and then implementing the efficient information source coding and channel coding, it is possible to establish a compatible multimedia communication sonar.

By organically connecting with these several aspects mentioned earlier, it is possible to establish an innovative, adaptive, pseudo-random frequency modulation (abbreviated as APNFM) system to be employed in digital underwater acoustic communications. It would be expected to obtain an approximately optimum detection result at the criterion of maximum output SNR against the multipath interference and noise background.

There are the two specific approaches to implement this APNFM system. One is a perfectly adaptive processing scheme. That is to say, it is unnecessary to provide any prior knowledge of channels in the whole signal detecting process. In this case, the functions of both adaptive total time sample processing and adaptive Rake receiver will fully be utilized. Based on that, a key part of a rapid adaptive modifying net is adopted in the system. Its adaptive adjusting time depends on operating frequencies, data rates, and channel characteristics, while it would finish within 1 s, which is much less than the coherent time of the channels.

This approach is quite complicated and is more suitable to be employed in military underwater acoustic communications, in particular for the communication circumstance with severe random and variable spatial/temporal/frequency parameters, such as in high-speed mobile communications (refer to Figs. 1.3–1.5).

Since the focus in this book is on civil underwater acoustic communications, this approach will not be discussed in depth.

The other specific approach to implement the APNFM system is when the field input signal of the channels (ie, the multipath structures) is first acquired in the communication receiver, and the relative parameters of the system are then adaptively adjusted to realize a suitable field channel match. This approach will simply be described next (refer to Chapters 3 and 4).

There are four key parts in the APNFM system receiver, as shown in Fig. 1.10. They will brief be introduced as follows:

1. Before transmitting formal communication information, a suitable calling signal is first transmitted by the side requiring communication to let the receiving side respond; it is also used as a channel exciting source. By passing through preset, preamplifying, and present total time sample processing parts, followed by the detection-processing part of channel parameters, the PNFM pattern suitably matched to the channels can therefore be obtained; moreover, the field channel parameters, including multipath structures, Doppler frequency shifts, etc., will also be obtained and used as reference data for following preset adaptive parts. By passing through the preset processing, the corresponding parts will operate as adaptive ones in the system.

   Once the receiving side receives the calling signal, it transmits a reply signal that is agreeable with the calling one. When the transmitting side receives this signal, the detection and processing for the channel parameter, just like the receiving side, will correspondingly be performed. Moreover, the distance information between them may also be acquired.
2. Based on the signal processing, the corresponding PNFM bit streams suitably matched to field communication channels are transmitted. They travel over the channels and are converted to randomly variable spatial/temporal/frequency multipath structures mixed with the background noise in the sea; this is the input of the communication receiver.
3. General speaking, the noise levels in the sea are high. Therefore effective methods to suppress the noise at the preamplifying part would be adopted to avoid the excessive impact on the corrective signal detection for following parts.
4. The idea and name for signal processing by the adaptive total time sampling processing part refer to the total range of sample processing in active sonars, where the information of target ranges is unknown. The usual synchronical conception is absent for them, but the target ranges can accurately be acquired by means of total range sample processing. It would be expected that this total time sample processing can replace conventional synchronical signals to correctly detect the PNFM sequences, which has a

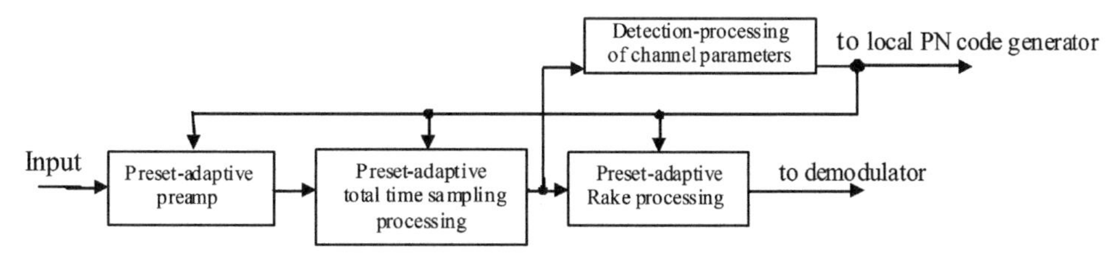

FIGURE 1.10 Four key parts in APNFM system receiver.

characteristic of multifunction, especially combating against fluctuations and noise (refer to Chapter 3).

5. The adaptive Rake processor is a key part in the APNFM system. To adapt to the peculiarity of randomly variable spatial/temporal/frequency multipath structures, which are the input signals of communication receiver, in underwater acoustic communication channels, the numbers of the lobes of the "rake," the width of the lobe, the space between adjacent lobes, and the ratio of "rake up" in the Rake processor may adaptively adjust according to channel parameters acquired in situ. Based on that, the multipath pulses (wave packets) may first be "rake up" and then combined with an optimum mode (refer to Section 3.1.3). It would be expected to obtain approximately optimum detection result at the criterion of the maximum output SNR against the multipath interference background.

Field experiments had demonstrated that the multipath structures have a high stability in some underwater acoustic channels (refer to Fig. 2.3.17). In such a case, the APNFM system underwater acoustic communication will obtain excellent results. The stability is different for varied underwater acoustic communication channels. It is thus necessary to adopt an adaptive channel modifying part for severely variable parameter channels to adapt to them (refer to Chapter 4).

1.4 COMMUNICATION SONAR EQUATION

The performances of sonars are closely related to many factors, including underwater acoustic equipment, detected targets, and, in particular, complex and variable underwater acoustic channels. To integrally consider the effects caused by them, different types of sonar equations have been introduced. They are the expressions tying together the effects of the channels, targets, and equipment as the references for the predictions of the performances, such as transmission ranges, data rates, etc., for present and new sonars being designed.

First, we will derive a communication sonar equation, which is independent of the targets, and then we will provide a specific example to show how to implement the prediction of the performances for an FHSS system communication sonar. Moreover, an active sonar equation against the noise background will also be derived.

The sound spreading model traveling in underwater acoustic channels will be simplified, ie, considering that it obeys spherical spreading law. Therefore sound intensity $I_0 \sim 1/r^2$, where r is communication distance. The in situ experiments had demonstrated that the sound geometrical attenuation law in the sea is generally agreeable with this simplified model [13]. Moreover, it can be used as a standard: when actual sound intensity I is higher than I_0, that is called the focusing effect of a sound field. On the contrary, when I is lower than I_0, it is called the divergent effect of the sound field. The effects can specifically be described by a focusing factor:

$$F = \frac{I}{I_0} \tag{1.8}$$

This factor is randomly variable according to space, time, and frequency, and it will be discussed in detail in Chapter 2.

According to the acoustic principle, the received sound intensity I_0 at a certain distance r in a sound field with spherical spreading law is given by

$$I_0 = \frac{W_a \gamma}{4\pi r^2} R^2(\theta, \varphi) \times 10^{-0.1\beta r}$$

where W_a is transmitted acoustic power, γ is the aggregate coefficient, $R(\theta,\varphi)$ is the directivity factor of the transducer, and β is the attenuation coefficient due to sound absorption and scattering in the sea. The direction of maximum sound intensity will be taken in actual operations, thus

$$R(\theta, \varphi) = 1$$

Therefore,

$$I_0 = \frac{W_a \Upsilon}{4\pi r^2} \times 10^{-0.1\beta r} \tag{1.9}$$

The received signal will be mixed with some different kinds of interferences, I_N. Generally speaking, I_N includes both background noise intensity (I_n) and multipath interference intensity (I_M).

Now the communication sonar equation against the noise background ($I_N \cong I_n$) will be derived first, which corresponds to the condition of weak multipath propagations. When a communication sonar operates at a higher operating frequency, it will fit such a condition.

Whether or not a specific communication receiver can realize the reconstruction of transmitting information at the preset specifications, such as BER, communication ranges, and data rates, is determined by the input SNR. In other words, it is determined by the output SNR for a receiver with a definite signal processing gain.

Since the sound intensity encountered in underwater acoustic channels has a wide dynamic range, the sonar equation is thus represented by logarithmic mode in dB. Thus Eq. (1.9) becomes

$$10\lg\left(\frac{I_0}{I_n}\right)_i = 10\lg\frac{W_a \Upsilon}{4\pi r^2} - 10\lg\left(10^{0.1\beta r}\right) - 10\lg\left(\frac{I_n}{I_r}\right) \tag{1.10}$$

where I_r is a reference intensity. Briefly writing Eq. (1.10) is as follows:

$$(DT)_i = SL - TL - NL \tag{1.11}$$

This is the communication sonar equation against the noise background, where $(DT)_i$, SL, TL, and NL are called sonar equation parameters, whose physical significances and their estimation will be discussed subsequently.

1. Input detection threshold, $(DT)_i$

$$(DT)_i = 10\lg\left(\frac{I_0}{I_n}\right)_i$$

   Provided the actual received input detection threshold is equal or larger than a preset $(DT)_i$ for a specific communication receiver, the transmitting information will be correctly detected at an expected BER P_e by means of efficient spatial–temporal signal processing, and the digital underwater acoustic communications will be accomplished reliably.

   There are different requirements for P_e for different applied fields of underwater acoustic communications and corresponding communication media. For example, $P_e = 5 \times 10^{-2}$ for an image communication without data compression, while $P_e = 5 \times 10^{-4}$ for voice communications is necessary; therefore the required $(DT)_i$ are different. The requirements in civil or military communication sonars are also different, especially as the latter generally need very low $(DT)_i$ to realize maintaining secrecy communications.

   It should be noted that there are different requirements of $(DT)_i$ for different signal processing systems employed in underwater acoustic communications. For example, an advanced DS-SS system (refer to Section 3.2.2) can normally detect a signal at a $(DT)_i$ below −10 dB, which would generally be mashed by the background noise. A *DS* system receiver has to offer a great processing gain by the aid of a matched filter or equivalent cross-correlator (refer to Chapter 3).

2. Source level, SL

$$SL = 10\lg\frac{W_a \Upsilon}{4\pi I_r} = 10\lg\frac{W_a}{4\pi I_r} + 10\lg\gamma \tag{1.12}$$

   where $10\lg\gamma = DI_T$, and it is called the directivity index, which is used to quantitatively describe the influence of the focusing effect of the directivity of transducers on SL. SL represents the ratio between sound intensity at 1 m from the acoustic center of a sound source in the direction of the main axis and reference intensity I_r in dB. Generally, the measurements of the sound intensity are accomplished at a far distance and then converted into 1 m.

Let the unit of W_a be watt, and the corresponding reference sound pressure be 1 μPa; SL is given by

$$SL = 171 + 10\lg W_a + DI_r \tag{1.13}$$

First, we should pay attention to the effect of DI_T on SL, which is equivalent to raise the transmitting sound power W_a with the same values in dB. The energy consumption for an underwater acoustic communication sonar is generally restricted, so carefully designing an available DI_T is quite valuable. Second, DI_T (ie, γ) for a transmitting transducer is determined by the shape and sizes of its radiant surface, operating frequencies, mounted models, etc. Therefore their value is usually found by actual measurement. Moreover, to decrease the additional energy consumption and additional interference, side lobes in the directivity must effectively be suppressed. Third, provided the positions are unknown from each other for two-way communications, the transducer with omni-directivity in the horizontal direction would generally be selected; moreover, considering the rocking of the transducer, the limitation of its sizes, etc., the vertical directivity cannot be selected to be very narrow. Therefore DI_T is generally less for communication sonars. Generally speaking, a tubular transducer is more suitable to be employed in underwater acoustic communications (refer to Chapter 4).

3. Transmission loss

$$TL = 10\lg\left(10^{0.1\beta' r} r^2\right) = 20\lg r + \beta' r, \quad r: \mathrm{m}, \beta': \mathrm{dB/m} \tag{1.14}$$

Transmission loss includes the sound geometrical spreading loss, $TL_g \equiv 20 \lg r$, and the attenuation loss, $TL_a \equiv \beta' r$. The latter includes the effects of the sound absorption and scattering. The acoustic energy absorbed by seawater is converted into heat energy; whereas the scattering effect causes the acoustic wave to change the direction pointing at the receiver, and the result is equivalent to reducing the acoustic energy. Both effects are quantitatively described by the attenuation coefficient β'. Generally, the absorption effect is much larger than scattering in underwater acoustic communication channels; moreover, the latter is difficult to accurately estimate. Therefore the absorption effect is only considered in the usual underwater acoustic communication engineering. The attenuation coefficient β' may be replaced by absorption coefficient β' that can be determined by empirical formulas or the in situ measurements.

TL_g has been simplified to follow the spherical spreading law, $I \sim 1/r^2$, as mentioned earlier. In fact, the geometrical spreading laws in underwater acoustic channels are complex and variable, which is a chief research topic in underwater acoustic physics. General mathematical models can be expressed by $I \sim 1/r^n$, where n may be 1, 3/2, 2, and so on. Therefore $TL_g = 20\lg r$ would be modified according to specific underwater sound channels. For example, when the sound wave travels through the surface sound channeling, it will obey the cylindrical spreading law, and thus $n = 1$.

Provided the focusing factor F can be found, the geometrical spreading attenuation may be modified as

$$TL_g = 20\lg r - 10\lg F \quad (1.15)$$

When $F > 1$, TL_g will lower; on the contrary, when $F < 1$, TL_g will rise. The experiments in situ had demonstrated that there is a higher sound energy, up to 25 dB, than that obeying the spherical spreading law in deep-sea sound channeling where the convergence zones are encountered (refer to Section 2.2.9). This focusing effect is usually used as an approach to raise the operating ranges for communication sonars. On the contrary, sound shadow zones would appear in sound anti-duct propagations, where F is very small and the corresponding TL_g is very large.

When propagation measurements are made at sea, it has been found that the spherical spreading, together with absorption, provides a reasonable fit to the measured data under a wide variety of conditions [13]. The in situ experiments for the underwater acoustic communications and fishing sonars [26] developed by our researching group generally had similar results.

4. Noise levels, *NL*

$$NL = 10\lg \frac{I_n}{I_r}$$

where I_r is the reference sound intensity, which is consistent with that in Eq. (1.10).

The ambient noise and the self-noise of ships are included in the noise levels. *NL*, as a parameter of sonar equation, is considered an isotropic and additive noise in a narrow bandwidth.

First, we should note that I_n is dependent on the directivity of the underwater acoustic transducer. I_n received by a directivity transducer is less than that of an omni-directivity one. Most transducers employed in the civil underwater acoustic communications are made of a single piezoelectric ceramic element, so their directivity and corresponding spatial processing gain can be determined by their size, operating frequency, etc. To quantitatively describe the effect of directivity on the *NL*, a directivity index is introduced:

$$DI = 10\lg \frac{I_{n1}}{I_{n2}} \quad (1.16)$$

where I_{n1} is the sound intensity received by an omni-directivity transducer, while I_{n2} is received by a directivity one at the same receiving sensitivity. Eq. (1.16) can be changed to

$$DI = 10\lg \frac{I_{n1}}{I_r} - 10\lg \frac{I_{n2}}{I_r} = NL_1 - NL_2$$

or

$$NL_2 = NL_1 - DI \quad (1.17)$$

We see that NL_2 by means of a directivity transducer is equivalent to decreasing *DI* in dB relative to NL_1 by using an omni-directivity one. So *DI* must carefully be designed, which can be expected to obtain a result equivalent to raising $(DT)_i$ by *DI* in dB. It may roughly be considered as $DI = DI_T$ for a transmitting–receiving compatible transducer.

It is possible to form a proper directivity pattern by means of a hydrophone array as in underwater acoustic communication networks, etc. In this case, using a spatial processor and corresponding spatial processing gain instead of the single transducer and corresponding *DI* is suitable (refer to the active sonar equation later).

Second, we must pay attention to the bandwidth effect on *NL*. Generally speaking, the background noise does not fit the white noise condition; however, the operating bandwidth *B* in underwater acoustic communications is narrower, in which the noise levels, *NL*, may be considered to be approximately uniform. Thus

$$NL = NL_{01} + 10\lg B$$

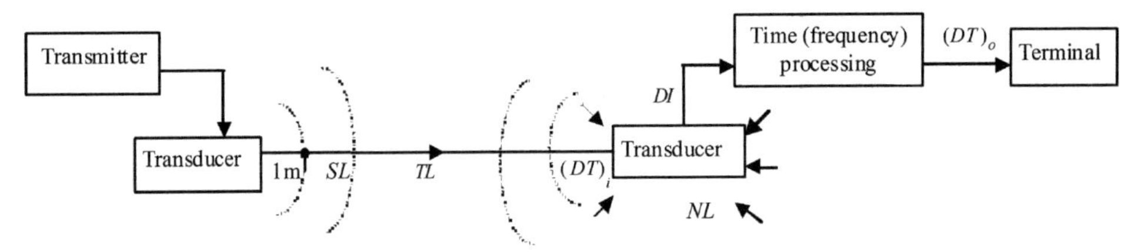

FIGURE 1.11 Physical processes of communication sonar operations.

where NL_{01} is the noise intensity spectrum level, which is received by an omni-directivity hydrophone in 1 Hz bandwidth.

To sum up, NL for a communication receiver with a directivity transducer and a bandwidth B can be found as

$$NL = NL_{01} + 10\lg B - DI \qquad (1.18)$$

The diagrammatic view of a communication sonar, illustrating the sonar parameters, is shown in Fig. 1.11. The suitable electric power acts on a transmitting transducer, which is transformed to sound power by a definite acoustic-electric efficiency, such as 70%, and then the corresponding information signals are transmitted into the underwater acoustic communication channels. Combining γ, there is an SL at unit distance on the main axis. When the signals arrive at the hydrophone at a distance r, SL will be attenuated by TL (dB). The attenuated signals are added by local isotropic background noise to form $(DT)_i$. NL is first reduced by DI (dB) due to the directivity effect of the hydrophone (ie, a spatial processing gain is obtained). Then, a large time–frequency processing gain will be obtained by means of an available signal processor as the matched filter, and so on. Finally, $(DT)_o$ will thus fit the requirement of terminal decision.

The communication sonar equation against the multipath background will briefly be discussed as follows. This corresponds to a communication situation with severe multipath interference (ie, $I_M \geq I_n$). In this case, converting the multipath detection background into a noise detection one is an effective approach.

1. We see from the mechanism forming multipath propagations, I_M is proportional to W_a; moreover, received direct signal and multipath pulses do not have considerable frequency shifts. Therefore it is inefficient to increase W_a, and adopt usual band-pass filters, or select optimum operating frequencies to suppress the multipath interference. If effective approaches have not been used to combating against this, the multipath interference with large amplitudes will destroy the correct detection of information signals (refer to the multipath structures for 3.2 and 5 kHz shown in Fig. 1.4).
2. It is an efficient scheme to reduce data rates to avoid ISI due to the multipath interference with large amplitudes. A simple method to realize that is letting the signal pulse repetitive period T_0 to be greater than the total delay spread of the multipath effect T_M (ie, $T_0 > T_M$). Provided an FHSS system is employed in digital underwater acoustic communication, the condition to separate ISI with the signals is by increasing the numbers of frequency codes and forming an FH pattern in which n frequency codes are not repetitively used (ie, let $nT_0 > T_M$; refer to Chapter 3). In these cases, the multipath detection background will therefore be changed into one of noise detection. Increasing T_0 means reducing the data rates. Similarly, increasing the numbers of frequency codes means increasing whole bandwidths of communication sonars. These

methods only fit for low data rate underwater acoustic communications.

3. The operating frequencies are low for long-range underwater acoustic communications. Therefore T_M is large (eg, up to several hundred milliseconds). On the other hand, T_0 is smaller with higher data rates. Perhaps, the condition of $T_0 > T_M$ or $nT_0 > T_M$ in an FH system to combat with ISI due to the multipath effect cannot be satisfied. In this case, an adaptive signal processing scheme combating with the multipath effect (refer to Chapter 3) may be used, with which the frequency codes will be remarkably compressed, and the signal detection is still carried out against the noise background.
4. We must pay attention to the duality of multipath effects. Certainly, the performances of underwater acoustic communications may be severely impaired by multipath interference as mentioned earlier; however, the pulse train caused by multipath propagations can be thought of as a time diversity signal that does not consume additional acoustic energy. Therefore the optimum approach to adaptively match to the background of multipath propagations would be Rake receiver, or paths diversity, in which multipath energy will be acquired and combined, so $(DT)_o$ can be remarkably improved. In fact, direct signals disappear in long-range underwater acoustic communications due to the sound refraction, and the sound signals being processed by means of a Rake receiver are a multipath pulse train. Because the multipath structures in underwater acoustic channels have the peculiarity of random spatial/temporal/frequency variability (refer to Figs. 1.4 and 1.5), a Rake receiver with adaptive operating mode would be adopted. Obviously, in this case, signal detection is still implemented against the noise background.

We see that the initiative for an underwater acoustic communication engineer is how to design the communication sonars that can match complex and variable underwater acoustic channels and realize optimum or quest-optimum detection results. It is a quite difficult task.

In some particular underwater acoustic communication fields, such as obstacle target telemetry in the shallow sea (refer to Chapter 4), active sonars have also been used. Therefore, an active sonar equation against a noise background that is relative to the targets will also be derived subsequently.

Let us introduce an effective scattering section, S_e, which is defined as a certain section of a target from which all incident sound energy will be reflected. This S_e is less than the geometrical section vertical to the sound incidence direction, and it depends upon the operating frequency: the higher the frequencies, the closer the geometrical section is. It can theoretically be estimated for some regular targets or found by actual measurements.

The incident sound intensity on a target in a spherical sound field at a distance r is given by Eq. (1.8). Therefore the acoustic power reflected from the target equals $S_e \times I_0$. In this case, the target can be thought of as a secondary sound source with an acoustic power of $S_e \times I_0$. The echo intensity at a transmitting-receiving transducer is given by

$$I_e = \frac{W_a \gamma S_e}{(4\pi)^2 r^4} \times 10^{-0.2\beta r} \tag{1.19}$$

Therefore the active sonar equation against the noise background can be given by

$$(DT)_i = SL + TS - 2TL - NL \tag{1.20}$$

where $2TL$ state the double-way TL, TS is the target intensity defined as

$$TS = 10\lg \frac{I_1}{I_i} \tag{1.21}$$

where I_i is incident intensity at the acoustic center of the target, I_1, is reflected sound intensity of the target at 1 m relative to the center. If S_e is found, TS will be given by

$$TS = 10\lg\frac{S_e}{4\pi} \tag{1.22}$$

To obtain a suitable directivity, an array established by means of beam forming techniques is generally employed in active sonars. Therefore using the spatial processing gain, GS, of the array instead of DI is suitable. Then, by means of time (frequency) processors, corresponding time processing gain, GT, will be obtained. Finally, $(DT)_o$ will satisfy the requirement of the target decision. Therefore the active sonar equation can also be given by

$$(DT)_0 = SL - 2TL + TS + GS + GT - NL \tag{1.23}$$

The diagrammatic view of echo ranging illustrating the sonar parameters is shown in Fig. 1.12.

The hydrophone array may sometimes be employed in communication sonars, as in underwater acoustic communication networks. The communication sonar equation can also be represented as

$$(DT)_0 = SL - TL + GS + GT - NL \tag{1.24}$$

where GS and GT are the same as in Eq. (1.23).

A typical example to show how to forecast the performances of a digital communication sonar will be provided as follows.

An FHSS digital underwater acoustic data communication prototype system that was developed will be described in Chapter 4. It has the specifications of moderate communication ranges (eg, up to 15 km) and moderate data rates, R (eg, up to 600 bps). It can especially adapt to some peculiarities in underwater acoustic communication channels, including randomly variable spatial/temporal/frequency multipath structures. It is expected that a compatible multimedia (including voice, text, image, data, etc.) communication sonar may be established, which would have a wide prospect of being employed in civil underwater acoustic communications.

Because the prototype must be operated in the channels with multipath interference background, it is necessary to first transform that

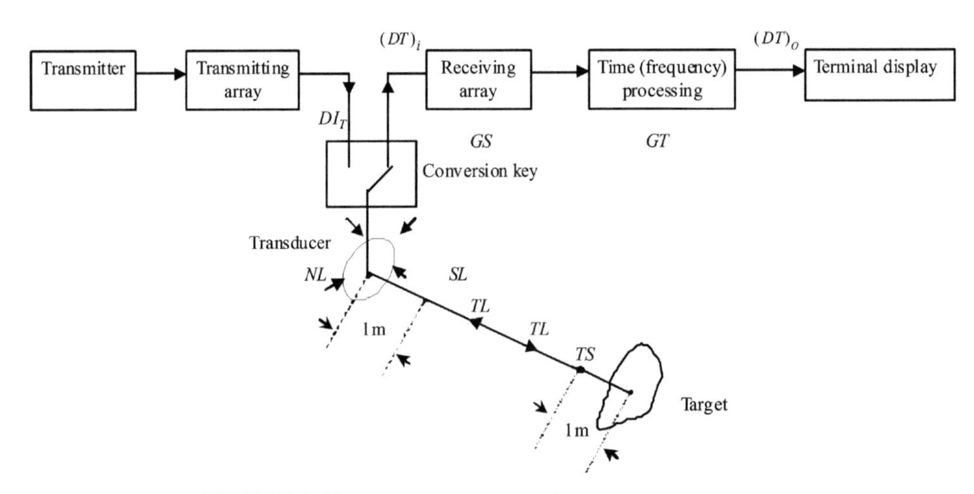

FIGURE 1.12 Physical processes of active sonar operations.

into noise background (refer to Chapter 3); then the relative parameters, such as required acoustic power, can be pre-estimated by means of the communication sonar equation.

The higher operating frequencies f (eg, 15 kHz) may be selected for moderate communication ranges. Therefore the total delay spreads T_M of multipath effects are less than 50 ms, which has been found from known data. Whenever the scheme of anti-multipath frequency code compression (refer to Chapter 3) is employed in the FHSS system underwater communication, the required numbers of frequency codes are less than 10; moreover, the prototype still operates against the noise background.

If the scheme of FFT discriminating frequencies is utilized, frequency resolutions will be 600 Hz for $R = 600$ bps; if CZT (chirp z-transform) is used, the resolutions may be raised by a fraction of 2/5 (refer to Chapter 4) obtained by our original studies. Therefore the frequency gap in FH pattern equals 360 Hz; thus the total bandwidth is only 3.2 kHz, which is more suitable for the expected moderate communication ranges. To avoid the broad bandwidth in a usual FH system raising the NL, a tracking filter has been adopted, so the efficient bandwidth B is still 360 Hz. The sound spectrum level NL_{01} for sea state 6 is about 45 dB. The self-noise of ships at the higher operating frequency (15 kHz) may generally be neglected for civil communication sonars. Considering adapting to possibly higher sea state and the mobile communication with higher speeds of ships, let $NL_{01} = 55$ dB. The transducer with omni-directivity in horizontal may be used, as transmitting and receiving positions in underwater acoustic communication are generally unknown to each other; moreover, the sharp vertical directivity would not be adopted either, and $DI_T = DI = 0$ dB is tolerantly selected. For an improved FH system with a weak signal detecting technique, taking $(DT)_i = 10$ dB would obtain information reconstruction at EBR $P_e = 3 \times 10^{-2}$, which is suitable for an image communication at effective data rates $R_e = 600$ bps by means of a data compression processing the small ratios (eg, less than 8; refer to Chapter 4). $P_e = 3 \times 10^{-4}$ is necessary for voice communication. In this case, the convolution code scheme with coding ratios of 2:1 is adaptively used, with a corresponding $R_e = 300$ bps. For data communication, $P_e = 10^{-5}$ would be necessary, and the Turbo code scheme with 3:1 coding ratios can be adopted. Therefore the corresponding $R_e = 200$ bps.

Based on the estimations of the relative parameters as mentioned earlier, the required W_a can be determined by means of communication sonar equation against the noise background:

$$SL = (DT)_i + TL + NL$$

where $TL = 20\lg r + \beta r = 106$ dB ($\beta \cong 1.5$ dB/km at $f = 15$ kHz) and $NL = NL_{01} + 10\lg B - DI = 81$ dB. Therefore $SL = 197$ dB (ie, $W_a = 350$ W). Let the acoustic-electric efficiency be 70%, and the required electric power equals 500 W, which is a more suitable value for a civil communication sonar mounted on ships.

Provided the innovative APNFM system is employed instead of the improved FH system in digital underwater acoustic communications, the operating frequencies may be decreased, and TL_a will correspondingly be reduced. Moreover, a Rake receiver can be used to acquire the multipath pulse train, and by combining with an available model, it is able to obtain a path diversity processing gain (eg, up to several decibels). By connecting both approaches, the communication distances would be remarkably increased.

Let the operating frequency be 8 kHz in an APNFM system underwater acoustic communication, and $\beta \cong 0.55$ dB/km. NL_{01} at this frequency is raised about 5 dB higher than that of 15 kHz in an FH system. Since $DI_T = DI = 0$ dB to be selected, as mentioned earlier, the directivity effect by reducing the operating frequency on DI_T and DI may be disregarded. Assume the processing gain acquired by means of path diversity is 4 dB, which is equivalent to $(DT)_i$ to be reduced from 10 dB in an FH system to 6 dB here. Therefore

$$TL = SL - (DT)_i - NL = 107 \text{ dB}$$

The corresponding predictive communication distance $r = 30$ km for $SL = 197$ dB.

It should be noted that the estimated W_a or r is based on the simplified assumption: TL_g fits the spherical spreading law. Generally, it is necessary to make some modifications according to actual communication channels, as shown by Eq. (1.15).

References

[1] A.D. Waite, Sonar for Practicing Engineer, third ed., National Deference Industry Press, Beijing, 2004.

[2] N. Falmouth, E-mail from submarine, Ocean Space 244 (June 2000).

[3] D.G. Tucker, B.K. Gazey, Applied Underwater Acoustics, Pergamon Press, 1977.

[4] M. Stojanovic, Recent advances in high-speed underwater acoustic communications, IEEE Journal of Oceanic Engineering 21 (2) (1996) 125–136.

[5] M. Suzuki, T. Sasaki, Digital acoustic image transmission system for deep sea research submersible, in: Proc. Oceans' 92, 1992, pp. 567–570.

[6] G.A. Car, A.E. Adams, ACMENet: an underwater acoustic sensor network for real-time environmental monitoring in coastal areas, IEE Proceedings Radar, Sonar and Navigation 153 (4) (2006) 365–380.

[7] D.P. Brady, J.A. Catipovic, Adaptive multi-user detection for underwater acoustical channels, IEEE Journal of Oceanic Engineering 19 (2) (1994) 158–165.

[8] A. Bessios, Compound compensation strategies for wireless data communications over the multimode acoustic ocean waveguide, IEEE Journal of Oceanic Engineering 21 (2) (1996) 167–180.

[9] L. Qihu, Introduction to Signal Processing of Sonar Written, China Ocean Press, Beijing, 2000.

[10] R.J. Urick, Principles of Underwater Sound, McGraw-Hill Book Company, 1975.

[11] X. Keping, X. Tianzeng, et al., Underwater acoustic wireless communications, Journal of Xiamen University (Natural Science) 40 (2) (2001) 311–319.

[12] M.D. Green, J.A. Rice, Channel-tolerant FH-MFSK acoustic signaling for undersea communications and networks, IEEE Journal of Oceanic Engineering 25 (1) (2000) 28–39.

[13] C.C. Tsimenidis, O.R. Hinton, B.S. Shrif, A.E. Adams, Spread-spectrum based adaptive array receiver algorithms for the shallow-water acoustic channel, in: Oceans' 2000 MTS/IEEE Conference and Exhibition, vol. 2, 2000, pp. 1233–1237.

[14] A. Yuhui, et al., A study of m sequence spread spectrum underwater acoustic communication, Journal of Harbin Engineering University 21 (2) (2000) 15–18.

[15] D. Rouseff, et al., Underwater acoustic communication by passive-phase conjugation and experimental result, IEEE Journal of Oceanic Engineering 26 (4) (2001) 821–831.

[16] J.V. Candy, A.W. Meyer, et al., Time-reversal processing for an acoustic communication experimental in a highly reverberant environment, JASA 115 (4) (2004) 1621–1631.

[17] Y. Jingwei, et al., Application of passive time reversal mirror in underwater acoustic communication, Acta Acustica 32 (4) (2007) 362–368.

[18] W. Dezhao, S. Erchang, Underwater Acoustic, Science Press, Beijing, 1981.

[19] H.C. Areeba, The Pulse Propagation in Underwater Acoustic Channel, Physical translated collection: Underwater Acoustics, Science Press, Beijing, 1960, pp. 48–57.

[20] L.M. Brekhovskikh, Ocean Acoustic, Science Press, Beijing, 1983.

[21] Zhang Renhe, et al., The multipath structures of signal waveforms in shallow water with thermocline, Acta Oceanologica Sinica 3 (1) (1981) 57–69.

[22] Y. Zhu, R. Zhang, et al., Theoretical analysis on pulsed waveforms in shallow water with an ideal thermocline, Acta Acustica 20 (4) (1995) 289–297.

[23] L.M. Brekhovskikh, Waves in Layered Media, Academic Press Inc., New York, 1960.

[24] Q. Perliang, et al., Fundamentals of Digital Communications, Electric Industry Press, 2007.

[25] L. Freitag, M. Stojanovic, D. Kilfoyle, J. Preisig, High-rate phase-coherent acoustic communication: review of a decade of research and a perspective on future challenges, in: Proc. 7th European Conf. on Underwater Acoustics, July 2004.

[26] The Shallow Sea Fishing Sonar, The shallow sea fishing sonar, Journal of Xiamen University (Natural Science) 20 (2) (1981) 201–207.

CHAPTER

2

Underwater Acoustic Communication Channels

There exist a lot of peculiarities for underwater acoustic communication channels in comparison with radio communication ones, including random spatial-temporal-frequency variability, severe multipath interference, rapid signal fluctuations, large transmission loss, strict bandwidth limitation, high noise levels, and low sound velocity. Therefore many signal processing systems, or even some basic principles in radio communications, cannot directly be employed in digital underwater acoustic communications. In such a case, developing new signal processing systems and, based on them, forming corresponding communication sonars with excellent performances to adapt to these peculiarities is a first task. Therefore sound transmission laws in underwater acoustic communication channels must be understood in detail. Some of these laws that directly relate to digital underwater acoustic communications will first be discussed:

1. transmission loss
2. multipath effects
3. signal fluctuations
4. noise in the sea

Then the impacts of these peculiar transmission laws on the digital underwater acoustic communication and the possible countermeasures to adapt to them will emphatically be discussed in this chapter.

The sound transmission laws in underwater acoustic communication channels are closely related to each other, so separating them to form different sections will be more convenient for discussion. In fact, it is quite difficult to study these laws in the channels with both layered and randomly inhomogeneous characteristics. A feasible approach is first to separate them for discussion and then analyze their relationship. For example, the sound scattering from the sea surface and sea bottom is a factor causing sound transmission loss, but it is also a main reason for the generation of sound signal fluctuations. The reflections of sound signals from the sea surface and bottom are the chief cause of multipath propagations, though they are also modulated by the sound signal refraction due to the inhomogeneous seawater medium. Therefore the multipath structures have a random spatial-temporal-frequency variability.

It should be pointed out that the sound propagation laws to be discussed in this chapter will provide a physical basis for designing signal processing systems employed in digital underwater acoustic communications. They would have the features to adapt to the peculiarities existing in underwater acoustic communication channels.

Digital Underwater Acoustic Communications
http://dx.doi.org/10.1016/B978-0-12-803009-7.00002-7

TABLE 2.1 Main Studying Points for Acoustic Fundamentals and Underwater Acoustic Physics

Elementary Acoustics	Underwater Acoustic Physics
Sound propagation in ideal medium	Sound propagations in absorption media
Sound propagation in homogeneous medium	Sound propagations in layered, inhomogeneous media
Sound propagation in infinite medium	Effects of sea surface and bottom on sound propagations
Sound propagation in non-source medium	Sound propagations in media with sound sources and marine noise
Low-amplitude sound propagation	High-amplitude sound propagations

In fact, the discussed laws are the main contents of underwater acoustic physics.

We have known that the derivation of an elementary acoustic wave equation must satisfy some specific conditions, including ideal (lossless), homogeneous, infinite, and non-source media, the sound wave with low amplitudes, etc. Of course, actual underwater acoustic channels do not satisfy these conditions. Underwater acoustic physics just studies the peculiarities that deviate from the basic conditions required in elementary theory of acoustics, as shown in Table 2.1.

The arrangement of the transmission laws being discussed in this chapter is partly consistent with the contents of study in underwater acoustic physics.

2.1 THEORETICAL METHODS OF UNDERWATER ACOUSTIC FIELDS

2.1.1 Wave Equation and Conditions for Determining Solution [1]

Because fluid media cannot support shear force, the fundamental equations that describe the wave process of sound are the Euler equation and continuity equation:

$$\frac{d\mathbf{u}}{dt}+\frac{1}{\rho}\text{grad}p = 0 \tag{2.1}$$

$$\frac{dp}{dt}+\rho\text{div}\mathbf{u} = 0 \tag{2.2}$$

where $\mathbf{u}$ is the vector of particle vibration velocity, p is sound pressure, and ρ is the density of the fluid media.

Considering low-amplitude acoustic waves, the second-order terms may be neglected: therefore Eqs. (2.1) and (2.2) become

$$\frac{\partial\mathbf{u}}{\partial t}+\frac{1}{\rho}\text{grad}p = 0 \tag{2.3}$$

$$\frac{\partial p}{\partial t}+\rho c^2\text{div}\mathbf{u} = 0 \tag{2.4}$$

where an adiabatic thermal sound velocity is introduced:

$$c^2\equiv\left(\frac{\partial p}{\partial \rho}\right) \tag{2.5}$$

According to Eqs. (2.3) and (2.4), the differential equation relating the p can be obtained:

$$\rho\text{div}\left(\frac{1}{\rho}\text{grad}p\right)-\frac{1}{c^2}\frac{\partial^2 p}{\partial t^2} = 0 \tag{2.6}$$

Let the density be homogenous, and we can derive the wave equation:

$$\Delta p-\frac{1}{c^2}\frac{\partial^2 p}{\partial t^2} = 0 \tag{2.7}$$

where Δ is the Laplace operator $\left(\frac{\partial^2}{\partial x^2}+\frac{\partial^2}{\partial y^2}+\frac{\partial^2}{\partial z^2}\right)$.

For a harmonic process (the time factor e^{-iwt} will not appear in the following derivation), Eq. (2.7) becomes

$$\Delta p+k^2 p = 0 \tag{2.8}$$

where wave number $k = w/c$.

Because ρ and c are spatial function in a macro-inhomogeneous media, the wave equation with respect to p cannot directly be obtained. A wave function Ψ will be applied as

$$\psi \equiv \frac{1}{\sqrt{\rho}} p \tag{2.9}$$

According to Eq. (2.6), the wave equation relating the Ψ can thus be obtained:

$$\Delta\psi + K^2(x,y,z)\psi = 0 \tag{2.10}$$

where

$$K^2 = k^2 + \frac{1}{2\rho}\Delta\rho - \frac{3}{4}\left(\frac{1}{\rho}\mathrm{grad}\rho\right)^2$$

The solution of the wave equation is still undefined. Thus the conditions for determining solutions should be given for a unique determined physical process. For example, for a vibration process that is described by an ordinary differential equation of second order, the initial vibration displacements and velocity should be given, and the varied laws of the vibration displacement with respect to time can therefore be unique determined. In the case of a wave propagation that is described by partial differential equation of the second order, together with initial conditions and boundary conditions, including the displacements of wave function, its derivation in a normal direction along a certain boundary $\sum$ in arbitrary times should be given.

Initial conditions are unnecessary and boundary conditions are independent of time for harmonic processes.

The boundary conditions describe the effect of the boundary on the wave motion in solving the sound fields. Some well-known boundary conditions are summarized as follows:

1. Absolutely soft boundary. It cannot support the pressure:

$$p|_{\sum} = 0 \tag{2.11}$$

which is called a homogenous boundary condition of the first kind. A smooth sea surface satisfies this condition.

The pressure distribution is given by

$$p|_{\sum} = f_1\left(\sum\right) \tag{2.12}$$

which is called the inhomogeneous boundary condition of the first kind.

2. Absolutely hard boundary. The vibration velocity in the normal direction equals zero:

$$u_n|_{\sum} = 0 \tag{2.13}$$

which is called the homogenous boundary condition of the second kind.

When the distribution of vibration velocity on the boundary is given by

$$u_n|_{\sum} = f_2\left(\sum\right) \tag{2.14}$$

which is called the inhomogeneous boundary condition of the second kind. The underwater acoustic transducers with definite shapes generally use this kind of boundary condition.

3. The boundary with linear relation between pressure and vibration velocity

$$ap|_{\sum} + bu_n|_{\sum} = f_3\left(\sum\right)$$

This is the boundary condition of the third kind. Provided $f_3(\sum) = 0$, the impedance boundary condition is given:

$$\left.\frac{p}{u_n}\right|_{\sum} = c \tag{2.15}$$

4. The boundary with finite disconnection of ρ or c. In fact, it is not a "boundary" but a jumping variant surface for ρ or c that appears in layered media. There are wave fields in both sides of such a surface; thus the continuity condition at the surface should satisfy

$$\begin{aligned} p|_{\sum -0} + p|_{\sum +0} &= 0 \\ u_n|_{\sum -0} + u_n|_{\sum +0} &= 0 \end{aligned} \tag{2.16}$$

The first equation in Eq. (2.16) shows pressure continuity; otherwise, particle acceleration $a \to \infty$, which is not feasible in physical processes. The second equation in Eq. (2.16) shows normal velocity continuity; otherwise, media will appear "interruption." Since there exist the wave fields in both sides of the surface (or infinite point), the additive determining solution conditions at each should be given to determine the unique solution.

5. Radiation condition. To insure the unique solution for the wave equation, the radiation condition at an infinite point should also be considered.

   It can be proved [2] that the mathematical expression for the radiation condition at an infinite point in three-dimensional space ($R \to \infty$) satisfies

$$\lim_{R\to\infty} R\left(\frac{\partial\psi}{\partial R} - ik\psi\right) = 0 \tag{2.17}$$

where the time factor is $e^{-i\omega t}$.

Similarly, the radiation condition at an infinite point for two-dimensional space ($r \to \infty$) is given by

$$\lim_{r\to\infty} r^{\frac{1}{2}}\left(\frac{\partial\psi}{\partial r} - ik\psi\right) = 0 \tag{2.18}$$

2.1.2 Theory of Ray Acoustics

2.1.2.1 *Fundamental Equations of Ray Acoustics*

In the scope of fundamental ray acoustics, the sound energy is traveling along rays in a sound field. The sound field is the superposition of the rays radiating from the sound source and traveling according to some certain paths. There are different arriving times and phases for different paths. The sound energy carried by each ray tube is conservative, so ray intensity can be determined by calculating the cross-section change of the ray tube. There are two fundamental equations for ray acoustics: one is the eikonal equation to determine the ray traveling path; the other is the ray intensity equation to determine the signal ray intensity.

Both equations can be derived by the approximation of the acoustic wave equation under certain conditions. The wave equation for a harmonic process is given by

$$\nabla^2\psi + k_0^2 n^2(x,y,z)\psi = 0 \tag{2.19}$$

where n is the refraction index of media, and $k_0 = w/c_0$ is a certain value of k at a certain reference point. Assume the wave function of sound is

$$\psi \equiv A(x,y,z)e^{ik_0 S(x,y,z)} \tag{2.20}$$

By substituting Eq. (2.20) into Eq. (2.19) and separating the real and imaginary parts, we can obtain

$$(\nabla S)^2 - \frac{\nabla^2 A}{A k_0^2} - n^2 = 0 \tag{2.21}$$

$$\nabla^2 S + \frac{2(\nabla S \cdot \nabla A)}{A} = 0 \tag{2.22}$$

If the second term in Eq. (2.21) is neglected, the following condition can be satisfied:

$$\frac{\nabla^2 A}{A k_0^2} \ll n^2 \tag{2.23}$$

We then will obtain the following equation to determine an equal-phase surface $S(x,y,z)$:

$$(\nabla S)^2 = n^2(x,y,z) \tag{2.24}$$

This is the first fundamental equation of ray acoustics: eikonal equation. The traveling direction of a ray can be determined as the normal direction of $S(x,y,z) = \text{constant}$. The direction vector of the ray $\boldsymbol{\gamma}(\alpha,\beta,\gamma)$ is therefore given by

$$\boldsymbol{\gamma}(\alpha,\beta,\gamma) = \frac{1}{n}\left(\frac{\partial S}{\partial x} + \frac{\partial S}{\partial y} + \frac{\partial S}{\partial z}\right)$$

These expressions can also be derived by using direction cosine with respect to the normal direction of $S(x,y,z) = \text{constant}$ in differential geometry:

$$\alpha = \frac{\partial S}{\partial x} \bigg/ \sqrt{\left(\frac{\partial S}{\partial x}\right)^2 + \left(\frac{\partial S}{\partial y}\right)^2 + \left(\frac{\partial S}{\partial z}\right)^2}$$

$$\beta = \frac{\partial S}{\partial y} \bigg/ \sqrt{\left(\frac{\partial S}{\partial x}\right)^2 + \left(\frac{\partial S}{\partial y}\right)^2 + \left(\frac{\partial S}{\partial z}\right)^2}$$

$$\gamma = \frac{\partial S}{\partial z} \bigg/ \sqrt{\left(\frac{\partial S}{\partial x}\right)^2 + \left(\frac{\partial S}{\partial y}\right)^2 + \left(\frac{\partial S}{\partial z}\right)^2}$$

According to Eq. (2.22) and considering the following expression,

$$\nabla \cdot (A^2 \nabla S) = 2A\nabla A \cdot \nabla S + A^2 \nabla^2 S$$

the second equation of ray acoustics, that is, the intensity equation, can thus be obtained as follows:

$$\nabla \cdot (A^2 \nabla S) = 0 \tag{2.25}$$

Consider ray intensity has a proportional relationship:

$$\mathbf{I} \sim A^2 \nabla S = 0 \tag{2.26}$$

According to Eq. (2.25), we can obtain

$$\nabla \cdot \mathbf{I} = 0 \tag{2.27}$$

The ray intensity vector satisfies the condition of tube ruling field: $\text{div}\mathbf{I} = 0$. We see that the magnitude of the ray intensity determined by Eq. (2.26) is proportional to the square of amplitude A, while its direction is the same with itself.

In summary, the wave field determined by the ray fundamental Eqs. (2.24) and (2.27) is given by

$$\psi = A(x, y, z)e^{ik_0 S(x,y,z)}$$

This is the approximate solution of the wave equation provided the condition (Eq. 2.23) is satisfied.

According to Eq. (2.23), ray acoustic approximation may be used provided that both the following conditions are satisfied:

1. The relative change of sound pressure amplitudes is very small provided the sound wave has traveled a distance that may be analog to one wavelength.
2. The relative change of sound velocities is very small provided the sound wave has traveled a distance that may be analog to one wavelength.

We see that the ray acoustics is the approximation of wave acoustics at high frequencies; moreover, it cannot be used in sound shadow zones and caustics where the acoustic fields must be modified by means of wave acoustic theory (refer to Section 2.2.1).

2.1.2.2 Ray Equation in Layered Media

The refraction ratio $n(z)$ is only the function of z in layered media. Now, let us examine the ray tracings in plane (x,y). Eikonal $S(x,y)$ satisfies the following equation:

$$\left(\frac{\partial S}{\partial x}\right)^2 + \left(\frac{\partial S}{\partial z}\right)^2 = n^2(z) \tag{2.28}$$

Let us use the method of separation of variables to solve this equation. Let

$$S(x, z) = S_1(x) + S_2(z)$$

We can obtain

$$\left(\frac{\partial S_1}{\partial x}\right)^2 = \xi^2 \tag{2.29}$$

$$\left(\frac{\partial S_2}{\partial z}\right)^2 = n^2(z) - \xi^2 \tag{2.30}$$

where ξ is a separation constant. By solving Eqs. (2.29) and (2.30), we obtain

$$S_1(x) = \xi x + c$$

$$S_2(z) = \int_0^z \sqrt{n^2(z) - \xi^2} dz$$

Thus

$$S(x,z) = \xi x + \int_0^z \sqrt{n^2(z) - \xi^2} \mathrm{d}z + c$$

The wave field given by ray acoustics is as follows:

$$\begin{aligned}\psi(x,z) &= A(x,z)e^{ik_0 S(x,z)} \\ &= Ae^{ik_0\xi x + ik_0\left(\int_0^z \sqrt{n^2(z)-\xi^2}\mathrm{d}z + c\right)}\end{aligned} \tag{2.31}$$

While spatial tracings of rays are given by the direction cosine:

$$\alpha = \cos\theta = \frac{1}{n(z)}\frac{\partial S_1}{\partial x} = \frac{1}{n(z)}\xi \tag{2.32}$$

$$\gamma = \sin\theta = \frac{1}{n(z)}\frac{\partial S_2}{\partial z} = \frac{1}{n(z)}\sqrt{n^2(z) - \xi^2} \tag{2.33}$$

Eq. (2.32) is the Snell law:

$$n(z)\cos\theta = \xi = \text{const.} \tag{2.34}$$

The ray equation can be obtained by using Eqs. (2.32) and (2.33):

$$\frac{\mathrm{d}z}{\mathrm{d}x} = \tan\theta = \frac{\gamma}{\alpha} = \sqrt{\frac{n^2(z)}{\xi^2} - 1}$$

Integrating this expression gives

$$x = \int_0^z \frac{\xi}{\sqrt{n^2(z) - \xi^2}}\mathrm{d}z$$

According to Snell law, $\cos\theta$ will be written as $\cos\theta_0$ when the separation constant ξ corresponds to $n = 1$; therefore the ray tracing equation becomes

$$x = \cos\theta_0 \int_0^z \frac{\mathrm{d}z}{\sqrt{n^2(z) - \cos^2\theta}} \tag{2.35}$$

Let us derive the ray intensity equation next.

According to the Euler–Gauss theorem, we have

$$\oiint_S \mathbf{I}\cdot\mathrm{d}\mathbf{A}_S = \iiint_V \mathrm{div}\mathbf{I}\mathrm{d}V \tag{2.36}$$

Substituting Eq. (2.27) into Eq. (2.36), we get (refer to Fig. 2.1)

$$\oiint_S \mathbf{I}\cdot\mathrm{d}\mathbf{A}_S = 0$$

$$-(I_1 A_{S_1}) + (I_2 A_{S_2}) = 0$$

Therefore,

$$I_1 A_{S_1} = I_2 A_{S_2} = \cdots = I_m A_{m_2} = \text{const.} \tag{2.37}$$

According to the initial condition at a point source, the constant can be determined by the sound power included in solid angle $\mathrm{d}\Omega$, which is a narrow angle between starting angle θ_0 and $\theta_0 + \mathrm{d}\theta$ in the ray tube (refer to Fig. 2.1). If the sound power of a unit solid angle is represented by W, we can get

$$I(z)\mathrm{d}A_z = W\mathrm{d}\Omega$$

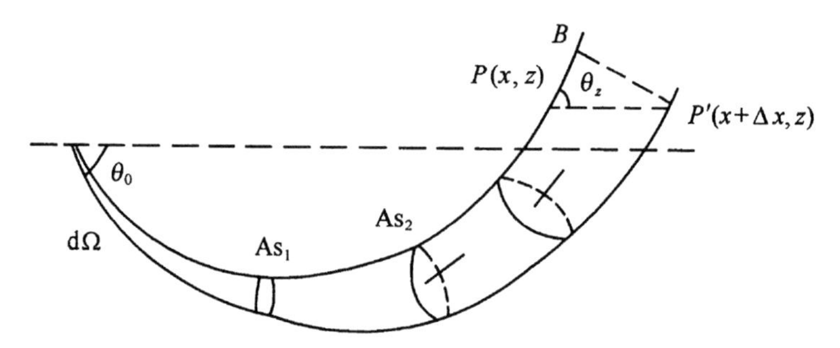

FIGURE 2.1 Schematic diagram for ray tube.

or

$$I(z) = \frac{W d\Omega}{dA_z} \tag{2.38}$$

Now, let us calculate the sound intensity at the point $P(x,y)$. First, the solid angle is given by

$$d\Omega = \frac{dS}{x^2} = 2\pi \cos\theta_0 d\theta_0$$

Then the normal cross-section of the ray tube at P is given by

$$dA_z = 2\pi x \cdot \overline{BP'}$$

moreover $\overline{BP'} = \overline{PP'} \cdot \sin\theta_z = \Delta x \sin\theta_z = \left(\frac{\partial x}{\partial \theta}\right)_{\theta_0} \Delta\theta_0 \sin\theta_z$; thus,

$$dA_z = 2\pi x \left(\frac{\partial x}{\partial \theta}\right)_{\theta_0} \Delta\theta_0 \sin\theta_z$$

Therefore,

$$I(x,z) = \frac{W 2\pi \cos\theta_0 d\theta_0}{2\pi x \left(\frac{\partial x}{\partial \theta}\right)_{\theta_0} d\theta_0 \sin\theta_z} = \frac{W \cos\theta_0}{x \left(\frac{\partial x}{\partial \theta}\right)_{\theta_0} \sin\theta_z} \tag{2.39}$$

Eq. (2.39) is the fundamental formula of the sound intensity in layered media. By combining with the ray equation Eq. (2.35), the sound field can be determined.

According to Eq. (2.31), the sound field, which is described by wave function, can be represented as follows:

$$\psi(x,z) = A(x,z) e^{\phi(x,z)} \tag{2.40}$$

where

$$\phi(x,z) = ik_0 x + ik_0 \int_0^z \sqrt{n^2(z) - \xi^2} dz$$

$$A(x,z) = (I)^{1/2} = \sqrt{\frac{W \cos\theta_0}{x \left(\frac{\partial x}{\partial \theta}\right)_{\theta_0} \sin\theta_z}}$$

Although the sound field determined by ray acoustics is the approximate solution of wave acoustics, and the former has considerable intuitive appeal and presents a picture of the propagations in the form of the ray diagram, it is still an important method to solve the sound field, particularly at higher operating frequencies. This method is usually employed in describing the sound propagations in deep-sea channeling, etc.

We introduce focusing factor F that is defined as the ratio between actual received sound intensity I and the intensity I_0 obeying spherical spreading law:

$$F(x,z) \equiv \frac{I(x,z)}{I_0} \tag{2.41}$$

Since

$$I_0 = \frac{W}{R^2}$$

where R is the slope distance between the sound source and receiving point. Therefore,

$$F(r,z) = \frac{R^2 \cos\theta_0}{x \left(\frac{\partial x}{\partial \theta}\right)_{\theta_0} \sin\theta_z} \approx \frac{x \cos\theta_0}{\left(\frac{\partial x}{\partial \theta}\right)_{\theta_0} \sin\theta_z} \tag{2.42}$$

We see from F that calculating the intensity of sound fields is not available provided $\left(\frac{\partial x}{\partial \theta}\right)_{\theta_0} \to 0$. Caustics are thus introduced where $F \to \infty$ [3], which are the envelopes of ray families. The ray diagram with a caustic AA is shown in Fig. 2.2.

We know that every ray can be described by the following equation:

$$x = x(\theta_0, z) \tag{2.43}$$

where θ_0 is initial starting angle of the sound source. The envelope of the ray family can be determined by the derivation of Eq. (2.43) with respect to θ_0:

$$\frac{\partial x(\theta_0, z)}{\partial \theta_0} = 0 \tag{2.44}$$

According to Eqs. (2.42) and (2.44), F will tend to infinity at the envelope of the ray family.

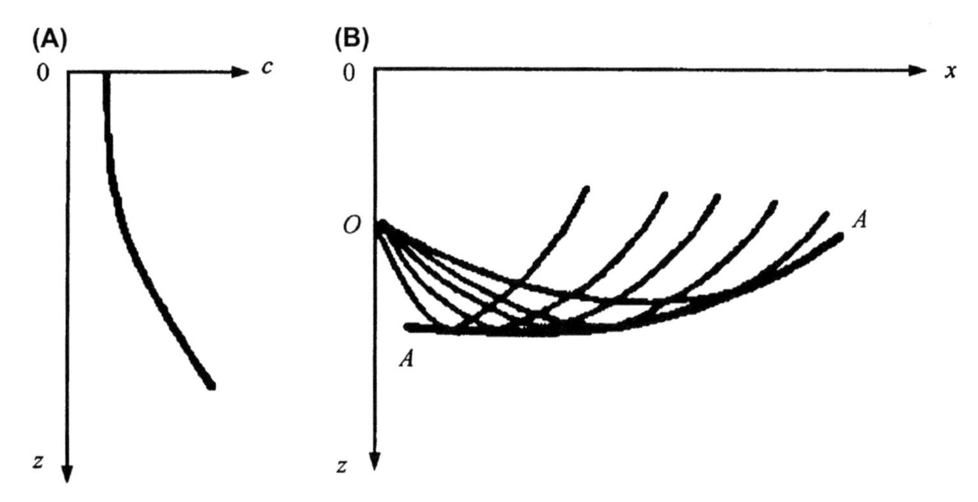

FIGURE 2.2 Example of forming a caustic: (A) sound velocity profile and (B) ray diagram with a caustic.

To calculate F at the caustic and adjacent zones, it is necessary to make some modifications to ray acoustics. A formula to calculate the focusing factor F in these areas is only given here:

$$f = 2^{5/3}\frac{\cos\theta_0(k_0\sin\theta_0)^{1/3}}{\sin\theta}x\left|\frac{\partial^2 x}{\partial\theta_0^2}\right|^{-2/3} v^2(q) \quad (2.45)$$

where $k_0 = w/c_0$, c_0 is the sound velocity at starting point, and $v(q)$ is Airy function drawn in Fig. 2.3 (refer to Section 2.3.1) where $v(0) = 0.629$.

The argument of the Airy function is shown:

$$q = \pm 2^{1/3}\left|\frac{\partial^2 x}{\partial\theta_0^2}\right|^{-1/3}(k_1\sin\theta_0)^{2/3}(x - x_0) \quad (2.46)$$

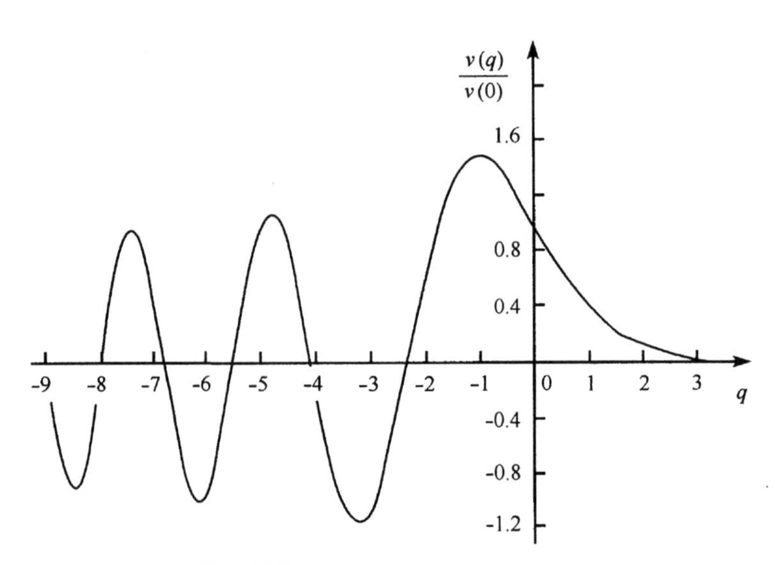

FIGURE 2.3 Diagram of Airy function.

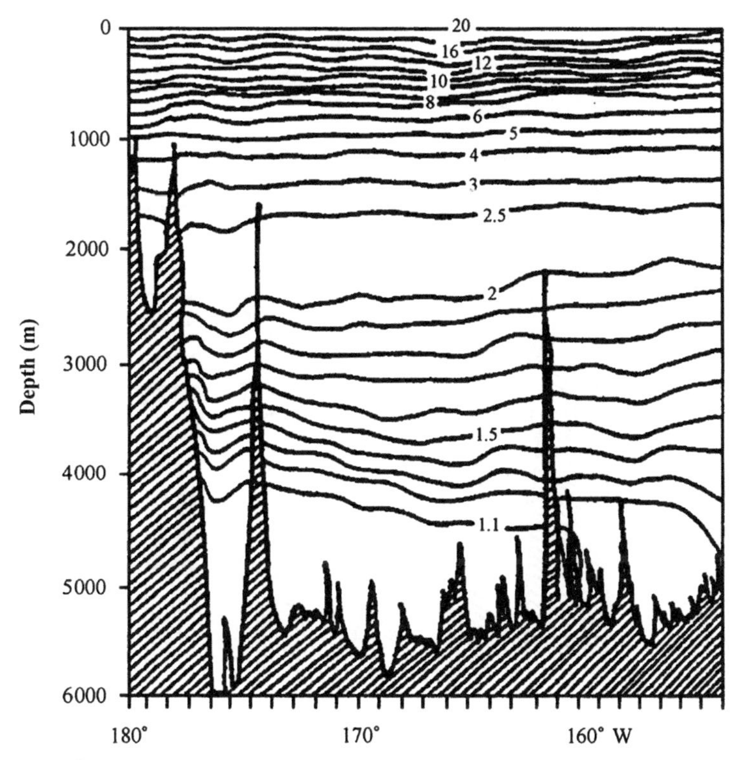

FIGURE 2.4 Profile of temperature (°C) along 28°S near the western boundary of the deep South Pacific, depths in meters.

where $(x - x_0)$ is the horizontal distance between the observing point and the caustic. When $\partial^2 x/\partial^2\theta_0 < 0$, a positive symbol in Eq. (2.46) will be selected. On the contrary, if $\partial^2 x/\partial^2\theta_0 > 0$, a negative symbol will be selected. There exist two rays to intersect each other at any of the points above the caustic; the spatial vibration of an acoustic field due to their interference effect thus appears. There are not any rays in the zone below the caustic (refer to Fig. 2.4), where an acoustic shadow zone occurs in which the sound intensity will be severely reduced with increased distance away from the caustic, as shown in Fig. 2.5.

2.1.2.3 Ray Diagrams in Layered Media

The solution of the eikonal equation in layered media follows the Snell law; moreover the ray tracing is given by Eq. (2.35):

$$x = \cos\theta_0 \int_{z_0}^{z} \frac{\mathrm{d}z}{\sqrt{n^2(z) - \cos^2\theta}} \tag{2.47}$$

For a linear distribution sound velocity,

$$c(z) = c_0(1 + az) \tag{2.48}$$

and let the ray parameter be as shown:

$$P \equiv \frac{\cos\theta(z)}{c(z)} = \frac{\cos\theta(z_0)}{c(z_0)} \tag{2.49}$$

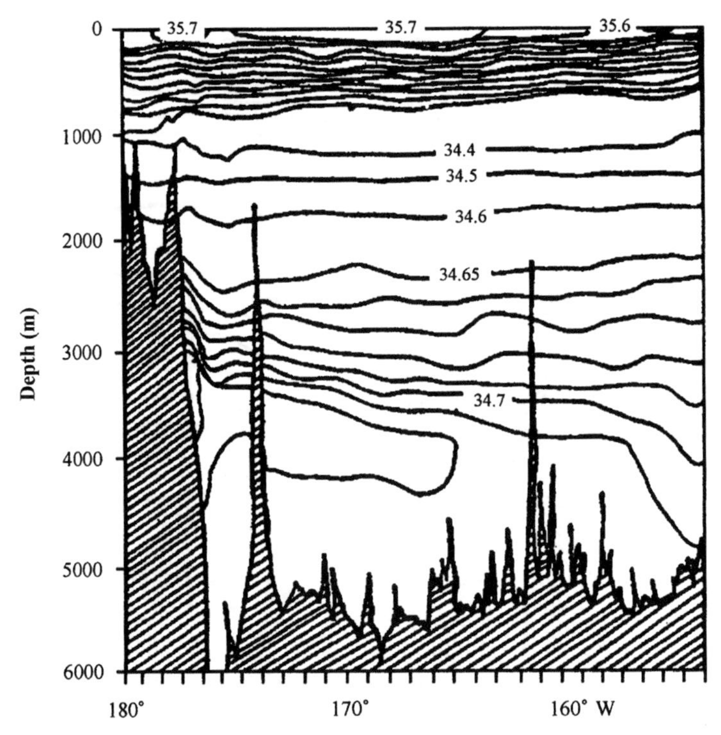

FIGURE 2.5 Profile of salinity (‰) along 28°S near the western boundary of the deep South Pacific, depths in meters.

We obtain

$$x = \int_{z_0}^{z} \frac{P_c(z)}{\sqrt{1 - P^2c^2(z)}} dz \tag{2.50}$$

by means of the following integral formula:

$$\int \frac{y}{\sqrt{1-y^2}} dy = -\sqrt{1-y^2}$$

The ray tracing can easily be determined and proven that it is circular.

According to the Snell law and the assumption of linear sound velocity, we can get

$$dz = -\frac{\sin\theta(z)}{a\cos\theta_0} d\theta$$

Eq. (2.47) becomes

$$x = \frac{-1}{a\cos\theta_0} \int_{\theta_0}^{\theta_z} \cos\theta d\theta = \frac{\sin\theta_0 - \sin\theta_z}{a\cos\theta_0} \tag{2.51}$$

where

$$\sin\theta_z = \sqrt{1 - \cos^2\theta_0 (1 + az)^2}$$

The traveling time of a ray in media with linear velocity is given by

$$t = \int_{z_0}^{z} \frac{dz}{c(z)\sqrt{1 - P^2c^2(z)}} = \int_{\theta(z_0)}^{\theta(z)} \frac{d\theta}{Pc(z) \cdot ac(z_0)}$$

$$= \left[\int_{\theta(z_0)}^{\theta(z)} \frac{d\theta}{\cos\theta(z)}\right] \frac{1}{ac(z_0)}$$

$$= \frac{1}{ac(z_0)} \frac{1}{2} \ln \frac{1+\sin\theta}{1-\cos\theta} \Bigg|_{\theta(Z_0)}^{\theta(Z)}$$

$$= \frac{1}{ac(z_0)} \ln \frac{\tan\left(\frac{\theta(z)+\pi/2}{2}\right)}{\tan\left(\frac{\theta(z_0)+\pi/2}{2}\right)}$$

According to Eq. (2.39), the intensity of a single ray may be determined by the following expression:

$$I = W\frac{\cos\theta_0}{x\left(\frac{\partial x}{\partial\theta_0}\right)\sin\theta(z)} \tag{2.52}$$

Therefore,

$$I = W\frac{\cos^2\theta_0}{x^2} \tag{2.53}$$

An important application for calculating in detail the ray traveling in a medium with linear sound velocity is when arbitrary complex velocity distributions may approximately be considered to be formed by connecting a lot of thin layers with linear distributions, ie, by using broken lines, one represents the actual $c(z)$ as generally having complex distributions. But we should pay attention to false caustics that would possibly appear when this method is used.

Letting $c_i(z)$, $\theta_i(z)$, Δx_i, and $g_i = \left(\frac{\partial c}{\partial z}\right)_i$ be the sound velocity, the refraction angle of ray, the crosswise distance of the ray, and the gradient of sound velocity in the ith layer, respectively, and letting θ_0 be the initial grazing angle at the layer of $i = 0$, according to Eq. (2.51), we get

$$\Delta x = \frac{c_i}{g_i\cos\theta_i(z)}|\sin\theta_i(z) - \sin\theta_{i+1}(z)|$$

The total horizontal distance for n layers is

$$x = \sum_{i=0}^{n}\Delta x_i = \frac{c_0}{\cos\theta_0}\sum_{i=0}^{n}\left|\frac{\sin\theta_i(z) - \sin\theta_{i+1}(z)}{g_i}\right| \tag{2.54}$$

because

$$\frac{\partial x}{\partial\theta_0} = \frac{\sin\theta_0}{\cos\theta_0}\sum_{i=0}^{n}\frac{\Delta x_i}{\sin\theta_i\sin\theta_{i+1}}$$

So, the intensity for a single ray in the nth layer can be given by

$$I = W\frac{\cos\theta_0}{x\sin\theta_n(z)\frac{\sin\theta_0}{\cos\theta_0}\sum_{i=0}^{n}\frac{\Delta x_i}{\sin\theta_i\sin\theta_{i+1}}} \tag{2.55}$$

The corresponding traveling time is shown:

$$t = \sum_{i=0}^{n}\Delta t_i = \sum_{i=0}^{n}\frac{1}{g_i}\ln\frac{\tan\left(\frac{\theta_{i+1}}{2}\right)}{\tan\left(\frac{\theta_i}{2}\right)} \tag{2.56}$$

2.2 SOUND TRANSMISSION LOSS IN THE SEA

The sound transmission loss TL in the sea includes geometrical spreading loss TL_g and the sound transmission attenuation TL_a due to sound absorption and scatting effect, ie, $TL = TL_g + TL_a$.

The geometric spreading loss TL_g in layered inhomogeneous media will first be discussed.

The distributions of energy in sound fields will be changed due to the sound refraction in layered inhomogeneous media, so the TL_g will exist with severely spatial-temporal variant peculiarity. It is necessary to analyze in depth the variant laws of TL_g and carry out corresponding predictions in underwater acoustic communication engineering.

Next, the effect of sound attenuation and the corresponding TL_a will be discussed.

Because of the severe sound absorption in underwater acoustic channels, communication distances are limited. Moreover, it is a cause of strict band-limited peculiarity in the channels.

Generally speaking, the effect of the sound absorption on TL_a is much greater than that of the sound scattering; moreover, the variant laws for the latter are difficult to determine, so the values of TL_a will generally be given by the former in applied underwater acoustic communication engineering.

2.2.1 Sound Fields in Layered Inhomogeneous Media

The distributions of sound velocity in the vertical direction are a fundamental physical parameter remarkably affecting sound propagations in the sea.

We know that the sound velocity is as follows:

$$c = \frac{1}{\sqrt{\rho K_s}} \tag{2.57}$$

where ρ is the density of seawater, and K_s is the adiabatic thermal compression coefficient. Sine ρ and K_s are functions of temperature T, salinity S, and steady pressure P, sound velocity c will thus change as they change. Note that c increases with the increase of these three factors.

Consider Fig. 2.4, which shows the temperature profiles as a function of depth at a location in the Pacific Ocean, and Fig. 2.5, which gives salinity profiles at the same place. The most prominent features of these two charts are the almost parallel, horizontal isotherms (constant temperature lines) and isohalines (constant salinity lines). Since the lines of constant depth are horizontal in stable water, all three constituents of the sound velocity are approximately horizontally stratified. Therefore the sound speed itself is approximately horizontally stratified. So the spatial variable of sound velocity can be approximately represented as $c(z)$, and the horizontal variable may be neglected. This is the layered inhomogeneous model of underwater acoustic communication channels.

By finding the first derivative of c with respect to z, the gradient of velocity will be obtained:

$$G_c = \left(\frac{dc}{dz}\right) \tag{2.58}$$

Since $c = c(T,S,P)$,

$$G_c = a_T g_T + a_S g_S + a_P g_P$$

where $g_T = \partial T/\partial z$, $g_S = \partial S/\partial z$, $g_P = \partial P/\partial z$ are the gradients of temperature, salinity, and pressure, respectively, and $a_T = \partial c/\partial T$, $a_S = \partial c/\partial S$, $a_P = \partial c/\partial P$ are the variable ratios of the velocity with respect to the temperature, salinity, and pressure, respectively.

According to the empirical formula,

$$c = 1450 + 4.21T - 0.037T^2 + 1.14(S - 35) + 0.175P$$

Thus,

$$a_T = 4.21 - 0.074T \ (\text{m/s})/^\circ\text{C}$$
$$a_S = 1.14 \ (\text{m/s})/‰$$
$$a_P = 0.175 \ (\text{m/s})/\text{atm}$$

We see that the sound velocity will approximately increase about 4 m/s by increasing the temperature by 1°C. It will increase about 0.17 m/s by increasing the pressure by 1 atm, which is equivalent to increasing the depth by 10 m. It will increase about 1.14 m/s by increasing the salinity by 1‰. However, in the case of deep-sea communication circumstances, the depths are up to several kilometers; moreover, both T and S are relatively steady, so the increase of the sound velocity is mainly contributed by the increase in depth.

The specific distribution laws of $G(z)$, as well as T and S, have a complicated spatial-temporal variability. Therefore the laws of the sound refraction are also complicated and variable, and the sound's geometric attenuation does not obey the spherical spreading law.

G_c in underwater acoustic channels may be divided into some typical modes of distributions, including ideal $G_c = 0$, positive, and negative G_c, the jumping variation in G_c due to the present of a thermocline, and in particular, the existence of a minimum value of the sound velocity in the deep sea. Correspondingly, there are some modes of sound wave ducts or sound channeling propagations, such as mixed-layer sound channeling and deep-sea sound channeling, that prevent sound waves from spreading in all directions and remain inside the boundaries of the sound channeling. On the other hand, there exist some modes of anti-channeling propagations in which TL_g will be much greater than those obeying the spherical spreading law.

Both typical sound fields, including anti-channeling in the shallow sea and the sound channeling in deep sea, will be analyzed in this section, and their TL_g will also be roughly estimated.

Since the operating frequencies in underwater acoustic communications are generally higher, such as above several kHz, it is appropriate to analyze the characteristics of sound fields by means of the ray acoustic theory. However, in some particular sound fields, such as in a sound shadow, the theory is useless. Thus it is necessary to use the wave acoustic theory to describe them.

2.2.1.1 Sound Anti-Channeling Propagations [7]

When the sound velocity linearly decreases with increasing depths, as shown in Fig. 2.6A, sound anti-channeling propagation will occur. A simplified ray diagram is shown in Fig. 2.6B. In such a case, all rays will refract downward, and the ray tangential to the sea surface is called a critical ray by which the sound field is divided into "light" and "shadow" zones. The latter is shown in Fig. 2.6B by the region denoted by hatched lines. The sound intensity in the "light" zone can be determined by Eq. (2.53). It would be zero in the shadow zone according to the ray acoustic theory.

Let z_1 and z_2 be the depths of a sound source and hydrophone, respectively; moreover, z_2 is placed at critical ray, and the horizontal distance D (refer to Fig. 2.6) between them is called a geometrical operating distance, which is the maximum operating distance as viewed from ray acoustics.

Obviously, D is given by

$$D \cong \sqrt{\frac{2c_0}{G_c}}\left(\sqrt{z_1} + \sqrt{z_2}\right) \tag{2.59}$$

where c_0 is the sound velocity at the plane placing the sound source, and G_c is the sound velocity gradient in the depth direction. D is generally several kilometers in low-latitude oceans with a sharp negative G_c.

In fact, the sound intensity in shadow zones does not equal zero due to the sound diffraction effect.

The sound field in shadow zones will be analyzed by means of the method of *WKB* (Wenzel, Kramers. Brillouin) approximation based on the wave acoustics theory.

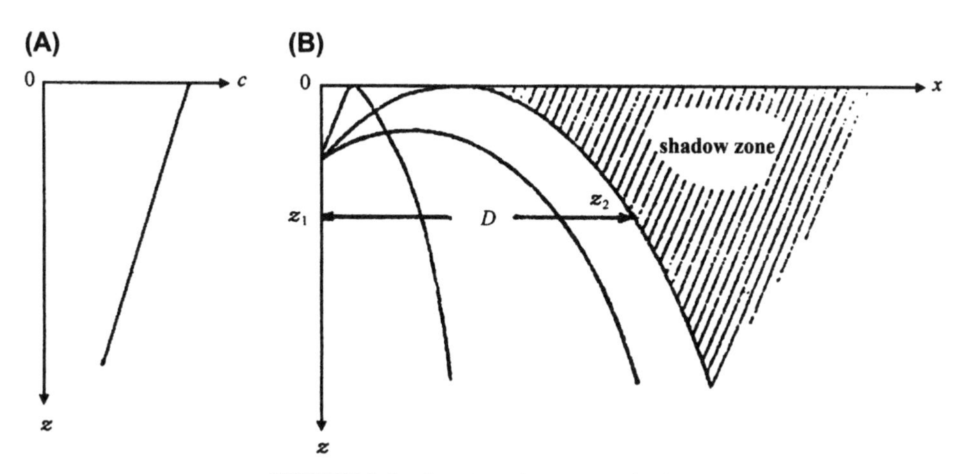

FIGURE 2.6 Forming of geometric shadow.

If the distance between a sound shadow zone and sound source is much greater than one wavelength, the wave front may be considered to be a plane, and the processing problem will thus be restricted to two dimensions. The wave equation relating to sound pressure p is given by

$$\frac{\partial^2 p}{\partial x^2}+\frac{\partial^2 p}{\partial z^2}+k^2(z)p = 0,\ k(z) = k_0 n(z),\quad k_0 = \frac{\omega}{c_0},\ n(z) = \frac{c_0}{c(z)} \tag{2.60}$$

We can use the method of separating a variable to solve this equation. Assuming

$$p = X(x)Z(z) \tag{2.61}$$

and substituting that into Eq. (2.60), we get

$$\frac{1}{Z}(\ddot{Z}+k^2Z) = -\frac{1}{X}\ddot{X}$$

where "point signs" on X and Z represent the derivative with respect to correspondingly independent variables. The term on the left-hand side of this expression only depends on z, while that on the right-hand side of the expression is only dependent on x. Obviously this condition will satisfy if both sides of the expression are independent of both x and z, and equal a constant μ^2.

Therefore we get two equations:

$$\ddot{X} = -\mu^2 X \tag{2.62}$$

$$\ddot{Z}+(k^2-\mu^2)Z = 0 \tag{2.63}$$

Eq. (2.62) has a simple solution:

$$X = A\exp(i\mu x) \tag{2.64}$$

The "separating parameter" μ (horizontal wave number) will be determined next if its real and imaginary parts satisfy following conditions:

$$\mathrm{Re}\mu > 0,\quad \mathrm{Im} > 0 \tag{2.65}$$

Thus Eq. (2.64) represents the wave starting from the sound source with attenuated amplitudes.

We may write Eq. (2.63) as follows:

$$\ddot{Z}+\vartheta^2(z)Z = 0,\quad \vartheta^2(z) = k^2(z)-\mu^2 \tag{2.66}$$

If ϑ is a constant, the solutions of Eq. (2.66) will represent the plane waves $\exp(i\vartheta z)$ and $\exp(-i\vartheta z)$ that travel in positive and negative directions on the z axis, respectively. Now, ϑ is dependent on z, assuming the variation of ϑ is very small as z changes one wavelength. Therefore the solution of Eq. (2.66) may be represented as

$$Z = F(z)\exp[i\phi(z)] \tag{2.67}$$

where $F(z)$ and $\phi(z)$ are slow variable functions.

Substituting Eq. (2.67) into Eq. (2.66), we obtain

$$\ddot{F}+2i\dot{\phi}\dot{F}+i\ddot{\phi}F+\left[\vartheta^2(z)-(\dot{\phi})^2\right]F = 0 \tag{2.68}$$

Since $F(z)$ is a slow variable function, the second derivative of that can be neglected. If both of the following equalities hold,

$$\dot{\phi} = \pm\vartheta(z),\quad 2\dot{\phi}\dot{F}+\ddot{\phi}F = 0 \tag{2.69}$$

then Eq. (2.68) will also satisfy. According to the first equality, we can get

$$\dot{\phi} = \pm\int_{z_0}^{z}\vartheta(z)dz \tag{2.70}$$

where z_0 is an arbitrary constant. By means of the relation between both equalities, the second equality becomes

$$\frac{\dot{F}}{F} = -\frac{\dot{\vartheta}}{2\vartheta}$$

Integrating this equation, we obtain

$$F = \frac{a}{\sqrt{\vartheta(z)}} \tag{2.71}$$

where a is an arbitrary constant.

Substituting the solutions for ϕ and F into Eq. (2.67) and considering ϕ with positive and negative signs, we obtain

$$Z = \frac{1}{\sqrt{\vartheta(z)}}\left[a_1 \exp\left(i\int_0^z \vartheta(z)\mathrm{d}z\right) + a_2 \exp\left(-i\int_0^z \vartheta(z)\mathrm{d}z\right)\right] \tag{2.72}$$

where let $z_0 = 0$, which only changes the values of a_1 and a_2.

Assuming $\vartheta(z)$ does not equal zero at any points, the *WKB* approximate solution of Eq. (2.66) in general conditions will be Eq. (2.72).

The following simplified conditions will be discussed:

$$\begin{aligned} c(z) &= c_0(1+az)^{1/2}, \\ n^2(z) &= [c_0/c(z)]^2 = 1+az \end{aligned} \tag{2.73}$$

When $az \ll 1$, then $(z) \approx c_0(1-az/2)$, which is the distribution of sound velocity as shown in Fig. 2.6.

In such a case, we can obtain

$$\begin{aligned} \int_0^z \vartheta \mathrm{d}z &= \int_0^z \sqrt{k_0^2 - \mu^2 + k_0^2 az}\mathrm{d}z \\ &= \frac{2}{3ak_0^2}\left[\left(g^2+k_0^2\right)^{3/2} - g^3\right] \\ &= \omega(z) - \omega_0 \end{aligned} \tag{2.74}$$

where

$$\begin{aligned} \omega(z) &= \frac{2}{3ak_0^2}\left(g^2+k_0^2\right)^{3/2}, \quad \omega_0 = \frac{2g^3}{3ak_0^2}, \\ g^2 &= k_0^2 - \mu^2 \end{aligned} \tag{2.75}$$

It is necessary to determine a_1, a_2 and μ. One of the conditions to determine these constants is that the solution for Z will represent the wave traveling in z direction when z is very large. It may be proven that the following equality will be given for this condition:

$$a_2 = e^{-i(2\omega_0+\pi/2)}a_1$$

The constant a_1 may be determined by the sound's power and the depth of the sound source. According to Eq. (2.72), for the condition to be accurate to constant factors, we get

$$Z \approx \frac{1}{\sqrt{\vartheta(z)}}\left[\exp(i\omega) + \exp\left(-i\omega - i\frac{\pi}{2}\right)\right] \tag{2.76}$$

The sound pressure would be zero at the sea surface:

$$z = 0, \quad Z = 0$$

Let $z = 0$ and the terms in square bracket in Eq. (2.76) be zero, and we obtain

$$1 + \exp\left[-i\left(2\omega_0 + \frac{\pi}{2}\right)\right] = 0 \tag{2.77}$$

Therefore, we can derive

$$-\left(2\omega_0 + \frac{\pi}{2}\right) = (2l-1)\pi, \quad l = 0, \pm1, \pm2, \cdots \tag{2.78}$$

Therefore,

$$\omega_0 = -\left(l - \frac{1}{4}\right)\pi \tag{2.79}$$

And, g can be obtained by using Eq. (2.75), as well as μ. According to Eq. (2.77), the constant μ will be in first quadrant; therefore μ^2 would be in the upper half-plane, ie, μ^2 is a positive imaginary part. Since k_0^2 is a real number, $g^2 = k_0^2 - \mu^2$ has a negative imaginary part and is thus in the lower half-plane. The argument angle of g^2 would be in the range of $(-\pi,0)$. Therefore that of g would be in $(-\pi/2,0)$:

$$\mathrm{Re}g > 0, \quad \mathrm{Im}g < 0 \tag{2.80}$$

According to Eq. (2.75), we get $g = (3ak_0^2/2)^{1/3}\omega_0^{1/3}$, where ω_0 is given by Eq. (2.79). Considering Eq. (2.80), we can obtain

$$g = \left(\frac{3}{2}ak_0^2\right)^{1/3} \exp\left(\frac{-i\pi}{3}\right)\left[\left(l - \frac{1}{4}\right)\pi\right]^{1/3},$$
$$l = 1, 2, 3, \cdots n$$

Therefore,

$$\mu_l^2 = k_0^2 - g^2$$
$$= k_0^2\left[1 - \left(\frac{3a\pi}{2k_0}\right)^{2/3} \exp\left(\frac{-i2\pi}{3}\right)\left(l - \frac{1}{4}\right)^{2/3}\right]$$

Considering $a/k_0 = a\lambda_0/2\pi$ is very small, we can obtain

$$\mu_l = k_0\left[1 - \frac{1}{2}\left(\frac{3a\pi}{2k_0}\right)^{2/3} \exp\left(\frac{-i2\pi}{3}\right)\left(l - \frac{1}{4}\right)^{2/3}\right] \tag{2.81}$$

We see that the values of μ_l are infinite that enable the solutions to satisfy all necessary conditions. The total solution is the superposition of all these solutions or normal modes:

$$p = \sum_{l=1}^{\infty} A_l p_l, \quad p_l = X_l(x)Z_l(z) \tag{2.82}$$

where A_l is the exciting coefficient of every normal mode, which is determined by the sound power and the depth of sound source; while

$$X_l(x) = \exp(i\mu_l x),$$
$$Z_l(z) = \frac{1}{\sqrt{\vartheta_l(z)}}\left[\exp(i\omega_1) + \exp\left(-i\omega_l - i\frac{\pi}{2}\right)\right] \tag{2.83}$$

where ϑ_l and ω_l are represented by μ_l by means of Eqs. (2.66) and (2.75), and the velocity and attenuation for each normal mode traveling in the x direction are determined by μ_l. Since the second term in the square bracket in Eq. (2.81) is very small, it may be considered that all normal modes have the same velocity $c_0 = \omega/k_0$. The attenuation coefficient β_l of every normal mode is given by the imaginary part of μ_l:

$$\beta_l = \mathrm{Im}\mu_l = \frac{1}{2}\left(\frac{3\pi}{2}\right)^{2/3}\left(\sin\frac{\pi}{3}\right)\left(a^2k_0\right)^{1/3}\left(l - \frac{1}{4}\right)^{2/3}$$
$$= 1.2\left(a^2k_0\right)^{1/3}\left(l - \frac{1}{4}\right)^{2/3} \tag{2.84}$$

Its amplitude is reduced by obeying the law of $\exp(-\beta_l x)$. We see that the higher the order of the normal modes, the larger the corresponding attenuation. The minimum attenuation corresponds to $l = 1$ as

$$\beta_1 = 1.0\left(a^2k_0\right)^{1/3} \tag{2.85}$$

When the distance is increased to be 1 m, the attenuation of sound intensity

$$Q = -20\lg[\exp(-\beta_1)] = 8.68\beta_1$$
$$= 8.68\left(a^2k_0\right)^{1/3} \text{ dB/m} \tag{2.86}$$

As $\beta_l/k_0 \ll 1$ by examining Eq. (2.85), it means that the attenuation will be small when sound wave travels a distance equal to one wavelength, and the sound wave can get into sound shadow zones for the ranges of many wavelengths due to the sound diffraction effect. However, traveling absolute distances are still very near as a result. For example, let $f = 1.5$ kHz and $a = 10^{-4}$/m, which is equivalent to the temperature in the sea water decreasing 5°C provided the depth increases 200 m, and we find that $Q = 0.18$ dB/m, thus entering into the sound shadow at 100 m, and the attenuation of sound intensity is up to 18 dB. The actual operating frequencies in underwater acoustic communications are higher than 1.5 kHz. If we let $f = 8$ kHz, the attenuation of the intensity is about 36 dB at the same traveling distance mentioned earlier. It means that the geometric operating distance D caused by sound shadow is a roughly limited range for underwater acoustic communications.

Since D is generally shorter (eg, several km), the received signals for long-range underwater acoustic communications are being conversed from multipath propagations (refer to Section 2.3).

2.2.1.2 *Sound Propagations in Deep-Sea Sound Channeling [1,5]*

The deep-sea sound channeling, sometimes called SOFAR (the letters denote sound fixing and ranging) channeling, is a consequence of the characteristic velocity profile of the deep sea. Recall that this profile has a minimum at a depth that varies from about 1200 m at mid-latitude to near the surface in the polar regions. This velocity minimum causes the sea to act like a kind of lens; above and below the minimum, the velocity gradient continually bends the sound rays toward the depth of minimum. A portion of the power radiated by a source in the deep sound channeling accordingly remains within the channeling and encounters no acoustic losses by reflection from the surface and bottom. Because of the low transmission loss in the channeling, very long ranges can be obtained from a source of moderate acoustic power output, especially when it is located near the depth of minimum velocity. This depth is called the axis of the sound channeling.

The deep sound channeling was utilized in the SOFAR system for the rescue of aviators downed at sea. In SOFAR a small explosive charge is dropped at sea by a downed aviator and is received at shore stations thousands of miles away. The time of arrival at two or more stations gives a "fix" for locating the point at which the detonation of the charge took place. More recently, the ability to measure accurately the arrival time of explosive signals traveling along the axis of the deep sound channeling has been used for geodetic distance determinations and for missile impact location.

2.2.1.2.1 THE SIGNAL WAVEFORMS IN DEEP-SEA SOUND CHANNELING

A typical sound ray diagram in deep sea channeling has been shown in Fig. 1.1 Another typical distribution of sound velocity forming a channel in the South China Sea and a simplified ray diagram are shown in Fig. 2.7, where the sound source is placed on the channeling axis.

Between the source and a point at a great distance away, a number of refracted propagation paths exist, each having a different travel time and crossing the channeling axis at different intervals. The path with the greatest excursion from the axis has the shortest travel time; the shortest path, straight down the axis of the

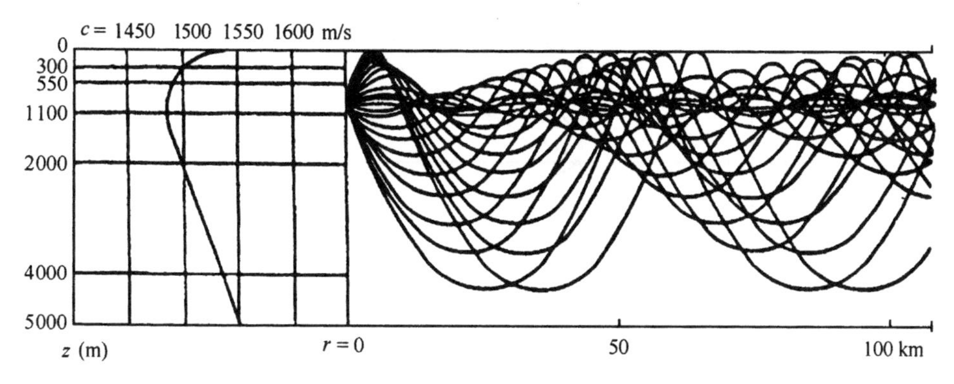

FIGURE 2.7 Distribution of sound velocity and simplified ray diagram.

channeling, has the longest travel time since it lies entirely at the depth of minimum velocity. Of all these paths, the axial path carries the greatest amount of acoustic energy and is the path of minimum transmission loss.

Because of these travel time and intensity characteristics, an explosion on the channeling axis is heard at a distance as a long, drawn-out signal. Its amplitude rises slowly to a climax (multipath spreading) and has an abrupt cutoff at the instant of arrival of sound traveling along the channeling axis. The general characteristic of signal waveform in the channeling is shown in Fig. 2.8.

A mathematical model that may explain the characteristics of waveform has been discussed [8]. Consider the following velocity profile sequence:

$$c^2 = c_a^2\left(1 - |\alpha z|^b\right)^{-1} \quad (2.87)$$

where c_a is the sound velocity at the sound channeling axis, α is a length scale factor, and b is the sequence parameter. The sound velocity profiles with different parameters b are plotted in Fig. 2.9.

The span of a half-waveguide is given:

$$L = 2\int_0^d \frac{dx}{dz} dz = 2\int_0^d \cos\theta_a \left(\frac{c_a^2}{c^2(z)} - \cos^2\theta_a\right)^{-1/2} dz \quad (2.88)$$

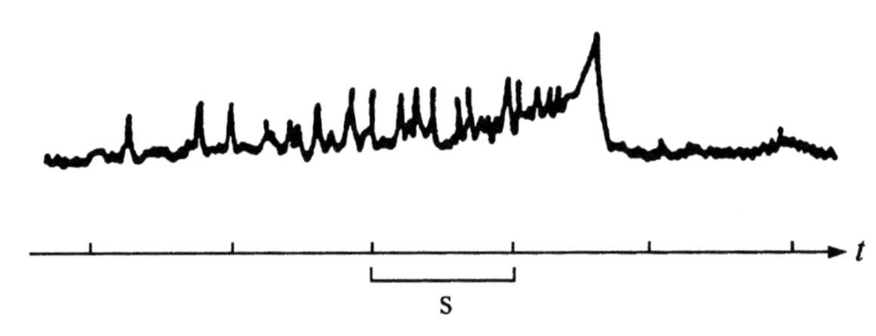

FIGURE 2.8 Multipath spreading for deep-sea sound channeling.

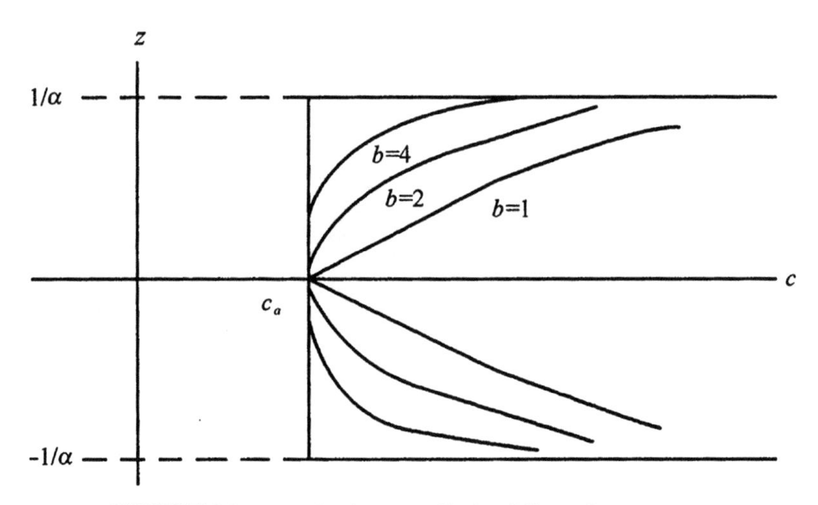

FIGURE 2.9 Sound velocity profile for different b parameters.

where θ_a is the grazing angle at the channeling axis, and d is the corresponding conversing height. By substituting Eq.(2.87) into Eq. (2.88) and then integrating, we can obtain

$$L = \frac{2\sqrt{\pi}(\sin\theta_a)^{2/b} \cdot \Gamma\left(\frac{1}{b}\right)}{ab\tan\theta_a \cdot \Gamma\left(\frac{1}{b}+\frac{1}{2}\right)} \quad (2.89)$$

Correspondingly, the time T traveling through L at θ_a is given by

$$T = 2\int_0^d \frac{1}{c}\left(\frac{dx}{dz}\right)dz$$
$$= \frac{2}{c_a}\int_0^d \frac{c_a^2}{c^2(z)}\left(\frac{c_a^2}{c^2(z)} - \cos^2\theta_a\right)^{-1/2} dz \quad (2.90)$$

Substituting Eq. (2.87) into Eq. (2.90) and taking integrative, we can get T as follows:

$$T = \frac{L}{c_a}\left\{\frac{1}{\cos\theta_a}\left[1 - \left(\frac{2}{b+2}\right)\right]\sin^2\theta_a\right\} \quad (2.91)$$

If small θ_a is only considered, we can get following approximate expression:

$$T \approx \frac{L}{c_a}\left\{1 + [(b - 2/b + 2)]\frac{\theta_a^2}{2}\right\} \quad (2.92)$$

We see that the multipath spreading will be controlled by b. When $b < 2$, T decreases with increasing Q_a; therefore the rays traveling along the axis, that is at $Q_a = 0$, will arrive at the end. When $b < 2$, on the contrary, T increases with increasing Q_a, and the rays traveling along the channeling axis will first arrive. When $b = 2$, the multipath spread is absent, all rays will arrive at the same time (refer to Fig. 2.10). We see that $1 < b < 2$ is applicable to explain roughly spreading multipath waveforms.

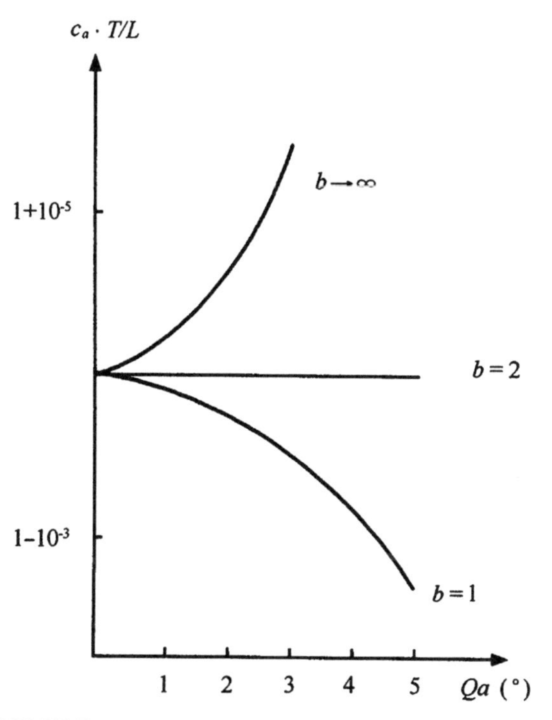

FIGURE 2.10 Multipath spread is controlled by b parameters.

2.2.1.2.2 RAY ACOUSTIC ANALYSES FOR THE SOUND FIELD IN CONVERGENCE ZONES

When the sound source and hydrophone are both played near the sea surface in a deep-sea channeling, the convergence zones will appear. As sonars are generally operated to utilize the convergence effect to realize long-range information detections, the researches on the characteristics of sound fields in convergence zones have paid attention to this.

The convergence effect can simply be explained by using ray diagrams. Consider a sound velocity profile as shown in Fig. 2.11A; if both transmitting and receiving transducers are played at depth z_1, in this case, $c_1 < c_H$, the radiated rays at angle $|\theta_1| < \arccos(c_1/c_H)$ will remain within the deep-sea channeling and thus no energy loss is caused by the reflections from either the sea surface or sea bottom. This finite ray beam will only "lighten" the corresponding finite zones of the sound field. Correspondingly, the zones that are shown in Fig. 2.11B by signs A, A', ... and B, B', ... will form sound shadow zones, which would be "lighten" only by the sound reflections from

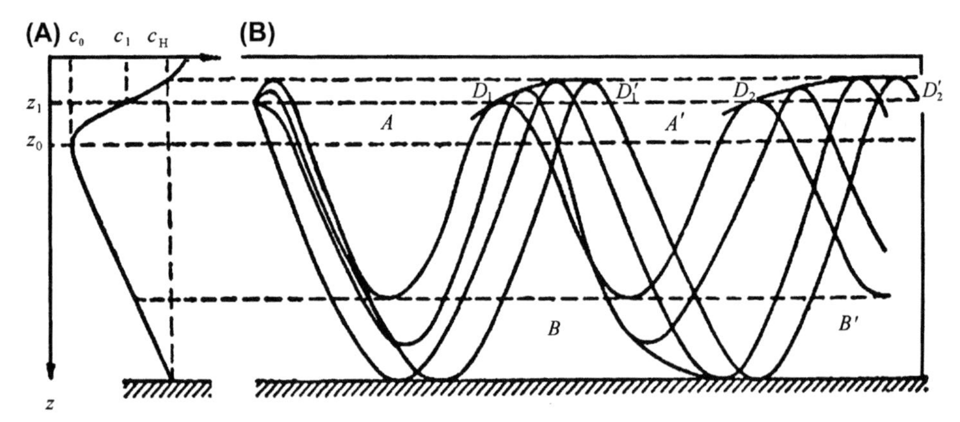

FIGURE 2.11 Depiction of convergence zones in deep-sea channeling.

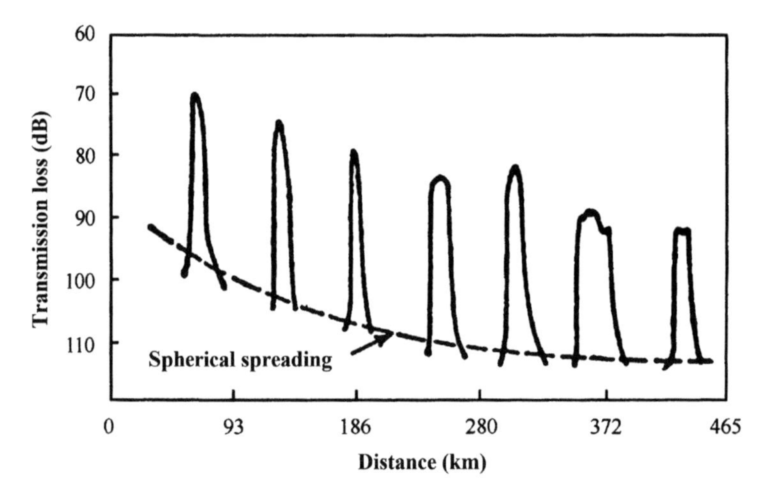

FIGURE 2.12 *TL* versus ranges in deep-sea channeling.

the sea surface and bottom, where the sound intensity is thus weak enough. Whenever a receiver departs from the sound source to far distances along $z = z_1$, the received sound intensity will pass through shadow zone A and then into "light" zone D_1D_1', following by shadow zone A' and "light" zone D_2D_2', etc. The zones D_1D_1', D_2D_2', ... are called the first, second, ... convergent zones. We can see that the widths of the convergence zones enlarge with increasing the ordinal numbers of these zones.

The characteristics of the convergence zones for a shallow source have been described by Hale [9]. An extreme example of a series of convergence zones, as reported by Hale from observed data, is given in Fig. 2.12. The peaks here extend about 25 dB above the level that would be found with spherical spreading and absorption; this increase has been called the convergence gain. Convergence gains of 10–15 dB are more common.

It is not fitted to calculate the field intensity in the caustics by using the ray acoustic theory since the focusing factor $F \to \infty$. However, it is a little phenomenon. Provided proper blurred tolerance is adopted, the available results to estimated TL can still be obtained by using ray acoustic theory [7].

2.2.2 Transmission Loss Due to Sound Absorption in the Seawater

We have known that the transmission loss $TL = TL_g + TL_a$; the latter is caused by the sound absorption and scattering in the sea.

There exist many kinds of inhomogeneity in sea water, such as fluctuations in temperature, salinity, and flow velocity, small air bubbles, small solid suspended particles, plankton, and schools of fish, from which the sound scattering appears. The sound scattering will cause the acoustic wave to deviate from the direction pointing at the receiver, which is equivalent to sound intensity attenuation.

The air bubbles formed by turbulent wave action in the air-saturated, near-surface waters will severely change their compressibility; therefore remarkable sound absorption, velocity variability, and scattering would be encountered. But the air bubbles are generally at shallow-water regions less than 10 m; moreover, the severe absorption occurs at their resonant frequencies (above 20 kHz), which are generally higher than the operating frequencies employed in underwater acoustic communications. The sizes of the solid particles and plankton are also much smaller than corresponding wavelengths. Therefore their effects on underwater acoustic communications may be neglected. Of course, once a large school of fish, deep-sea scattering layers, and wakes encounter each other, additive TL must be considered. The wakes usually were encountered when we carried out the experiments for underwater acoustic communications in Xiamen Harbor, and the experiments have to stop for several minutes.

The sound absorption in the seawater is a main reason to cause both the large TL_a and the strict band-limited peculiarity; therefore their variant laws, in particular regarding how to reduce their impacts, would carefully be analyzed.

There are three mechanisms to cause the sound absorption in fluid media:

1. Sound absorption due to the viscosity of fluid media. In this case, the sound energy will be converted into heat energy.
2. Sound absorption due to thermal conduction. The pressure variations occur during sound propagations in fluid media; consequently, thermal gradients and nonreversible thermal exchanges are produced.
3. Relaxation absorption. It is relative to nonreversible internal molecule processes consuming the sound energy.

They will briefly be introduced next.

2.2.2.1 Sound Absorption in Pure Water

Generally speaking, viscous coefficients in the fluid media contain two parts: one is the known shear viscous coefficient; the other is the volume viscous coefficient, which is generally neglected in fluid mechanics though it has an important effect on the sound propagations.

In the case of a plane sound wave with low amplitude, the viscous stress is proportional to the gradient of the vibrating velocity of fluid particles.

$$f = \eta \frac{\partial u}{\partial x}$$

where η is called the viscous coefficient; moreover,

$$\eta = \frac{4}{3}\mu_s + \mu_v$$

where μ_s is the well-known coefficient of shear viscosity, and μ_v is the coefficient of volume viscosity.

So the one-dimensional stress equation in viscous media may be written as follows:

$$p = -x_s\frac{\partial \xi}{\partial x} - \eta\frac{\partial^2 \xi}{\partial x \partial t} \tag{2.93}$$

where x_s is the volume elasticity module, which is the reciprocal of compressibility. Substituting Eq. (2.93) into motion equation gives

$$\rho_0\frac{\partial^2 \xi}{\partial t^2} = \frac{-\partial p}{\partial x}$$

We can obtain the wave equation relating to particle displacement:

$$\rho_0\frac{\partial^2 \xi}{\partial t^2} = x_s\frac{\partial^2 \xi}{\partial x^2} + \eta\frac{\partial^3 \xi}{\partial x^2 \partial t} \tag{2.94}$$

When the viscous effect is disregarded ($\eta = 0$), Eq. (2.94) will reduce to the wave question in ideal media.

The μ_v is usually disregarded in fluid mechanics. Based on that, Stokes first studied the effect of viscosity on the sound propagations. In this case, the wave equation is

$$\frac{\partial^2 \xi}{\partial t^2} = c_0^2\frac{\partial^2 \xi}{\partial x^2} + \frac{4}{3}\upsilon\frac{\partial^3 \xi}{\partial x^2 \partial t} \tag{2.95}$$

where $c_0 = \sqrt{\frac{x_s}{\rho_0}}$ is the sound velocity in ideal medium, and $\upsilon = \frac{\mu_s}{\rho_0}$ is the kinematic viscous coefficient.

In the case of a harmonic sound wave, the particle displacement is

$$\xi(x,t) = \xi(x)e^{-i\omega t} \tag{2.96}$$

Vibrating velocity is shown next:

$$\frac{\partial \xi}{\partial t} = -i\omega\xi(x,t) \tag{2.97}$$

By substituting Eq. (2.97) into Eq. (2.95), the latter becomes

$$\frac{\partial^2 \xi}{\partial t^2} = \left(c_0^2 - i\frac{4}{3}\upsilon\omega\right)\frac{\partial^2 \xi}{\partial x^2} = \tilde{c}^2\frac{\partial^2 \xi}{\partial x^2} \tag{2.98}$$

where

$$\tilde{c}^2 = c_0^2\left(1 - i\frac{4\omega\upsilon}{3c_0^2}\right) \tag{2.99}$$

The solution of Eq. (2.98) is as follows:

$$\xi(x,t) = \left(Ae^{i\tilde{k}x} + Be^{-i\tilde{k}x}\right)e^{-i\omega t} \tag{2.100}$$

where $\tilde{k} = \frac{\omega}{\tilde{c}} = \frac{\omega}{c_0}\frac{1}{\sqrt{1 - i\frac{4\omega\upsilon}{3c_0^2}}}$ is the complex wave number, and $\tilde{c}$ is the complex sound velocity. Since $\frac{4\omega\upsilon}{3c_0^2} \ll 1$ for general sound frequencies,

$$\tilde{k} \cong \frac{\omega}{c_0}\left(1 + i\frac{2\omega\upsilon}{3c_0^2}\right) = \frac{\omega}{c_0} + i\frac{2\omega^2\upsilon}{3c_0^3} \equiv k' + ik'' \tag{2.101}$$

By substituting Eq. (2.101) into Eq. (2.100) and only considering the traveling wave in a positive direction, we get

$$\xi(x,t) = Ae^{-k''x}e^{i(k'x-\omega t)} = Ae^{-\frac{2\upsilon\omega^2}{3c_0^3}x}e^{-i\frac{\omega}{c_0}(-x+c_0t)} \tag{2.102}$$

Let the displacement at $x = 0$ be $\xi(0,t) = \xi_0 e^{-i\omega t}$, thus $A = \xi_0$ in Eq. (2.102), which is the amplitude of the particle displacement. Therefore,

$$\xi(x,t) = \xi_0 e^{-\frac{2\upsilon\omega^2}{3c_0^3}x}e^{i(k'x-\omega t)} \equiv \xi_0 e^{-\beta_\mu x}e^{i(k'x-\omega t)} \tag{2.103}$$

where

$$\beta_{\mu_s} = k'' = \frac{2\upsilon\omega^2}{3c_0^3} = \frac{8}{3}\frac{\pi^2\mu_s}{\rho_0 c_0^3}f^2,\ k' = \frac{\omega}{c_0} \tag{2.104}$$

We see that the sound velocities in viscous and ideal media for a plane traveling wave can be regarded as to be the same, while the amplitudes of the displacement will be attenuated with increasing traveling distance x according to the exponential law in viscous media. β_{μ_s} is called the viscous absorption coefficient. According to Eq. (2.104), β_{μ_s} is proportional to μ_s and the square of the frequency, ie, the sound absorption due to viscosity at high frequencies is much larger than that at low ones. Because μ_s remarkably depends on the temperature, β_{μ_s} also changes along with it.

For example, $\mu_s = 1.0 \times 10^{-2}$ g/cm s, $\beta_{\mu_s}/f^2 \cong 8.2 \times 10^{-17}$ s^{-2}/cm, for pure water at 20°C.

Note that for the wave number $\tilde{k} = k' + ik''$, its real part k' reflects the change of the phase of sound wave, which is the same with the usual wave number, but the imaginary part k'' expresses an attenuation factor, ie, an absorption coefficient. Therefore we can find the absorption coefficient and phase velocity in absorption media by using following expressions:

The absorption coefficient is

$$\beta = \mathrm{Im}\left(\tilde{k}\right) = \mathrm{Im}\left(\frac{\omega}{\tilde{c}}\right) \tag{2.105}$$

The phase velocity is

$$c = \frac{\omega}{\mathrm{Re}\left(\tilde{k}\right)} = \frac{1}{\mathrm{Re}\left(\frac{1}{\tilde{c}}\right)} \tag{2.106}$$

where $\tilde{c}$ is the complex sound velocity. Provided $\tilde{k}$ or $\tilde{c}$ can be obtained, the corresponding β and c can also be determined.

We may understand the sound velocity or wave number to be complex in the viscous media: The volume change of media cannot completely follow the change of sound pressure because of the effect of viscosity, in particular, if sound frequencies are higher. The phase differences between them appear. Therefore both the compression coefficient of media and the sound velocity, so that and the wave number are complex.

2.2.2.1.1 SOUND ABSORPTION DUE TO THERMAL CONDUCTION

The pressure variations in media occur during sound propagations, so a thermal gradient appears, and heat flows from the regions of compression to the regions of rarefaction that tend to equalize the pressure differences in the medium, and the wave amplitude will be decreased accordingly as it propagates. The simplified attenuated plane wave equation in a thermally conducting medium is given by

$$\frac{\partial^2 \xi}{\partial t^2} = c_0^2 \frac{\partial^2 \xi}{\partial x^2} + \frac{q}{\rho_0 c_v} \frac{\gamma - 1}{\gamma} \frac{\partial^3 \xi}{\partial x^2 \partial t} \tag{2.107}$$

where $\gamma = c_p/c_v$, q is the thermal conductive coefficient of media.

By comparing Eq. (2.107) with Eq. (2.95), the absorption coefficient due to thermal conduction can be obtained as follows:

$$\beta_h = \frac{\omega^2}{2\rho_0 c_0^3} q \left(\frac{1}{c_v} - \frac{1}{c_p}\right) = \frac{2\pi^2 q(\gamma - 1)}{\rho_0 \gamma c_v c_0^3} \cdot f^2 \tag{2.108}$$

It is proportional to the square of the frequency, which was first derived by Kirchhoff.

Since $\gamma \cong 1$ for water medium, β_h is very small and can be neglected by comparing with β_{μ_s}. While $\gamma \cong 1.4$ in air medium,

$$\beta_h / \beta_{\mu_s} \cong \frac{3}{8}$$

By combining the sound absorptions due to both the viscosity and thermal conduction, the sound absorption coefficient is obtained:

$$\beta = \frac{\omega^2}{2\rho_0 c_0^3}\left[\frac{4}{3}\mu_s + q\left(\frac{1}{c_v} - \frac{1}{c_p}\right)\right] \equiv Af^2 \tag{2.109}$$

This is called the Stokes–Kirchhoff or classical formula of the sound absorption coefficient.

However, the measured values of the absorption coefficients in most air and fluid media are much larger than those calculated according to Eq. (2.109). Moreover, the relation of $\beta \sim f^2$ does not satisfy, ie, A in Eq. (2.109) is not a constant. The phenomenon of frequency dispersion also appears.

The volume viscous absorption, which had been neglected in the classical absorption formula, would be a reason to cause the aforementioned differences. Further researches have proven that the sound absorption in fluid media depends on internal molecule processes, which are macroscopically reflected by μ_v.

2.2.2.1.2 EXCESS ABSORPTION DUE TO RELAXATION PROCESS

One such explanation was offered by Hall [9] using a theory of structural relaxation. This theory assumes that water is capable of existing in two energy states: the state of lower energy is the normal one, and the high energy state is the one in which the molecules are more closely packed. Under static conditions most of the molecules are in the normal state, but the presence of a compressional acoustic pressure wave causes some of the molecules to pass into the more closely packed state, energy being extracted from the acoustic wave to do so. When the molecules revert to the normal state the energy is returned to the acoustic wave.

The finite time known as the relaxation time τ is required for molecules to pass from the lower state into the upper state and back again. At low acoustic frequencies when the period of the wave is long compared with the relaxation time, the reconverted energy is in phase with the energy extracted from the wave and no attenuation is experienced. At high frequencies no time is allowed for the molecules to change their states so once again no attenuation occurs.

Therefore we expect that when the period of the wave is comparable with the relaxation time, energy stored in the higher state as a result of a compressional cycle will reappear during the subsequent rarefaction cycle thereby tending to equalize the acoustic pressure difference and causing high sound attenuation.

Based on the Hall theory, the sound velocity and absorption coefficient due to relaxation process in pure water are as follows:

$$c = \frac{1}{\sqrt{\rho_0 K_0}} \tag{2.110}$$

$$\beta_R = \frac{\rho_0 c}{2}(K_0 - K_\infty)\tau_p \omega^2 \tag{2.111}$$

where k_∞ and k_0 are compression coefficients at $f \to \infty$ and $f \to 0$, respectively, and τ_p is the relaxation time.

Curve A shown in Fig. 2.13 is drawn according to Hall theory, which is consistent with

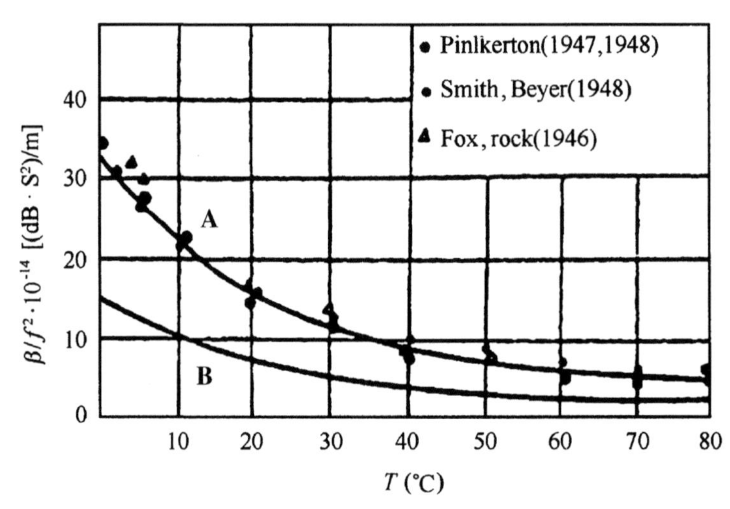

FIGURE 2.13 Sound absorption versus temperatures in pure water: (A) Hall theory absorption curve; (B) classical absorption curve.

the measured values by different authors. Curve B is calculated by the classical formula. The differences between them in vertical direction represent the excess absorption of pure water.

In summary, the total absorption coefficient in pure water is

$$\beta = \beta_{\mu_s} + \beta_h + \beta_R = \left[\frac{8\pi^2\mu_s}{3\rho_0 c_0^3} + \frac{2\pi^2 q(\gamma - 1)}{\rho_0 \gamma c_v c_0^3} + 2\pi^2 \rho_0 c(K_0 - K_\infty)\tau_p\right] f^2 \equiv Bf^2 \tag{2.112}$$

It is proportional to the square of frequency. $\beta = 2.76 \times 10^{-16}$ Nb/cm $= 2.40 \times 10^{-10}$ dB/km at 20°C.

2.2.2.2 *Sound Absorption in the Seawater*

It was soon clear that the absorption of sound in the seawater was unexpectedly high compared with that in pure water, and it could not be attributed to scattering, refraction, or other anomalies attributable to propagation in the natural environment. For example, the absorption in seawater at frequencies between 5 and 50 kHz was found to be some 30 times that in pure water.

This excess absorption can be attributed to the presence of dissolved salts, in particular magnesium sulfate, the process being known as a chemical relaxation. In water, magnesium sulfate dissociates into positive and negative ions, dissociation and recombination being in a state of equilibrium. However, the passage of a compression wave causes excess recombination of these ions. The time delays of this process and the subsequent re-dissociation lead to relaxation dissipation of acoustic energy.

The experimental results for the absorption coefficient under the conditions of 0.02 molarity of $MgSO_4$ at 20°C in low frequency ranges may be indicated as follows:

$$2\beta = L\frac{t_r\omega^2}{1 + \omega^2 t_r^2} + B\omega^2 \text{ (1/cm)} \tag{2.113}$$

The first term on the right-hand side of Eq. (2.113) is relaxation absorption; the second term in that is the absorption in pure water that is proportion to f^2, where $L = 4.6 \times 10^{-10}$ s/cm, $t_r = 1.3 \times 10^{-6}$ s.

Curve II in Fig. 2.14 represents the measured values of attenuation coefficient in seawater as a function of frequencies. Generally speaking, the actual measured values include the effects of the sound absorption and scattering; but the former has a main one on the sound attenuation

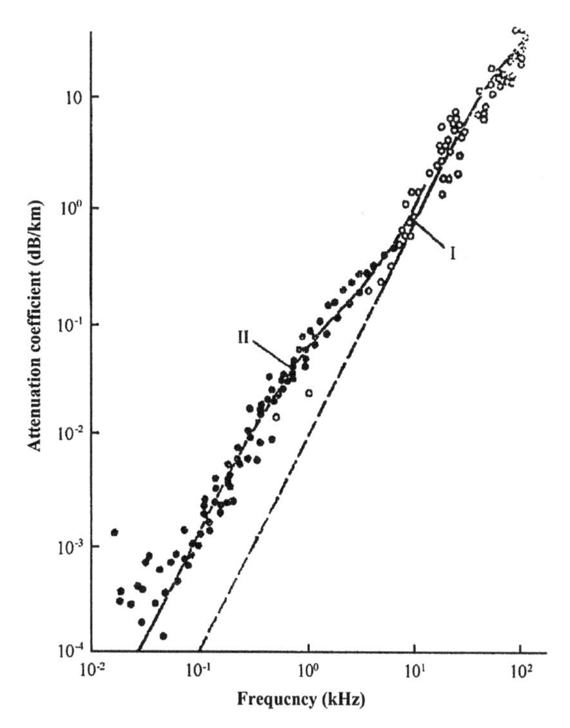

FIGURE 2.14 Measured attenuation coefficient as a function of frequencies; (◦) measured values using sine wave; (•) measured values using explosion source.

as mentioned earlier. Broken line I represents the sound absorption in pure water. The differences between curves II and I in the vertical direction may be regarded as the differences of attenuation coefficients between them.

Based on a number of experimental data in the frequency range from 2 to 20 kHz at distances less than 22 km [5], a half-empirical formula for attenuation coefficient has been acquired:

$$\beta = A_1 \frac{S f_r f^2}{{f_r}^2 + f^2} + A_2 \frac{f^2}{f_r} \text{ dB/m}$$

where $A_1 = 1.89 \times 10^{-5}$, $A_2 = 2.72 \times 10^{-5}$, S is the salinity, f is the operating frequency, and f_r is the relaxation frequency, which is the reciprocal of the relaxation time and dependent on the temperature as

$$f_r = 21.9 \times 10^{6 - \frac{1520}{T_k}}$$

where T_K is the absolute temperature (K).

At frequencies below about 5 kHz, the measured attenuation coefficients in seawater are much larger than those to be expected from the half-empirical formula. At these frequencies, an additional source of loss other than that caused by ionic relaxation of $MgSO_4$ must therefore be predominant. Further researches have stated that it is caused by chemical relaxations, including a boron–borate relaxation process.

An empirical formula for absorption coefficient caused by chemical relaxations at frequencies below 5 kHz is given by Thorp:

$$\beta_c = \frac{0.102 f^2}{1 + f^2} + \frac{40.7 f^2}{4100 + f^2} \text{ dB/km}$$

The first term on the right-hand side represents the relaxation absorption by boron–borate, and the second term represents the absorption by $MgSO_4$.

Note that the absorption coefficient β will decrease by increasing the steady pressure. The effect of the pressure on that has been investigated both theoretically and experimentally, and the absorption coefficient at a certain depth H(m) may be determined as follows:

$$\beta_H = \beta_0 \left(1 - 6.67 \times 10^{-5} H\right) \tag{2.114}$$

where β_0 is β at $H = 0$. For example, the absorption coefficient will be reduced by 6.7% provided the depth increases by 1000 m.

2.2.3 Impacts of Sound Transmission Loss on Digital Underwater Acoustic Communications and Possible Countermeasures

2.2.3.1 Impacts of TL *on Underwater Acoustic Communications*

1. Because of the large sound transmission loss in seawater, including geometric spreading and sound absorption, the sound intensity is rapidly attenuated with increasing propagating distances r. The input SNR (signal-noise ratio) for a specific receiver will correspondingly be lowered, or transmitting information cannot normally be detected. Generally speaking, that is one of the main reasons to limit long-range underwater acoustic communications.
2. *TL* in underwater acoustic channels has a spatial-temporal variability. There are different sound velocity profiles in different seasons, months, or even in a day in which "afternoon effect" possibly appears, ie, r will remarkably decrease since the negative velocity gradient will sharpen in afternoon. Moreover, weather conditions will also affect the distribution of the sound velocity, such as the depths of the mixed-layer sound channeling being relative to the conditions.

   Obviously, there are different sound velocity profiles in different sea areas, in particular at different latitudes. It is easy to form the surface sound channeling in high latitude sea areas. On the contrary, the sound anti-channeling often appears in low latitude areas.

It is valuable to note that TL_a caused by sound absorption in seawater also has a spatial-temporal variability because the absorption coefficient β is the function of salinity S and pressure P. In particular, it will remarkably reduce with increasing temperature T as the relaxation process of internal molecules is speeded up.

The variable tendencies of β versus T are shown in Table 2.2.

We see that the effect of T on TL_a is remarkable in long-range underwater acoustic communications. For example, by taking $f = 15$ kHz, the difference of β between $T = 10°C$ and $T = 15°C$ is about 0.6 dB/km. During the communication range $r = 50$ km, the corresponding difference of TL_a between both temperatures is up to 30 dB.

Generally speaking, β in shallow sea areas is larger than that in deep-sea areas. β as a function of depth has been expressed by Eq. (2.114).

3. The strict band-limited peculiarity, which is caused mostly by the sound absorption effect, restricts high data rate digital underwater acoustic communications. Moreover, the applications of some advanced communication techniques that require wider bandwidths, such as an SS (spread spectrum) system, will substantially be limited.

TL_a will rapidly rise with increasing f and r, as shown in Table 2.3, where the values of β correspond to $T = 15°C$, $S = 30\%$, and pH = 8.3 [1].

TABLE 2.2 The Variable Tendencies of β Versus T, Unit Is dB/km

f(kHz) \ T(°C)	5	10	15	20	25	30
15	1.94	1.61	1.35	1.14	0.99	0.87
20	3.29	2.72	2.27	1.90	1.62	1.39

TABLE 2.3 TL_a Versus f and r

f(kHz) \ βr(dB) \ r(km)	1	5	10	20	50	100
2	0.19	0.98	1.96	3.92	9.80	19.6
4	0.31	1.55	3.09	6.18	15.5	30.9
8	0.59	2.94	5.88	11.8	29.4	58.8
14	1.31	6.53	13.1	26.1	65.3	130.1
20	2.38	11.9	23.8	47.6	119	238

We see that lower frequencies would be selected for longer range underwater acoustic communications. For example, when $f = 4$ kHz, TL_a is 16 and 31 dB at $r = 50$ and 100 km, respectively. When $f = 14$ kHz, it is 65 and 130 dB, respectively. It is quite difficult to make up for such a large increase in TL_a by raising the sound power.

The corresponding conclusion is that the effective bandwidth B employed in communication sonars is also restricted. For example, let the efficient bandwidth be 2–4 kHz: the differences in TL_a between the upper and lower side frequencies are 5 and 10 dB at the distances of 50 and 100 km, respectively. It is possible to make up for them by raising the sound power at the upper side frequency and using a suitable amplitude equalizer. However, when B is increased to 2–8 kHz, these differences are up to 20 and 39 dB at 50 and 100 km, respectively. It is quite difficult to adapt to them. Moreover, the differences will sharpen with increasing the frequencies; such as when B is 8–14 kHz, the bandwidth is still 6 kHz, but the differences are up to 36 and 70 dB at $r = 50$ km and $r = 100$ km, respectively.

We see that it is necessary to selected low f and narrow B for long-range underwater acoustic communications. For example, when B is 4–6 kHz, it is more suitable for the communication at $r = 50$ km. In this case, the difference of TL_a between the upper and lower side frequencies is only about 5 dB, and it is easier to adapt to that by means of amplitude equalizer or appropriately raising the sound power at the upper side frequency. This is the restrict band-limited peculiarity due to sound absorption, which must be considered when designing a digital communication sonar.

It will be pointed out subsequently that the effect of sound scattering will rise with increased operating frequencies; therefore the band-limited effect on the underwater acoustic communications will sharpen.

According to the Shannonian theorem, increasing B, as well as raising the SNR (refer to Chapter 3), will improve information rates. Based on that, the advanced SS system in radio communications has been established. This system has also been employed in digital underwater acoustic communications. However, the requirement to spread frequency spectra is contrary to the strict band-limited peculiarity, so its applications to underwater acoustic communications are restricted.

The OFDM (orthogonal frequency division multiplexing) system has a number of advantages, including good abilities combating with multipath, narrow band, and frequency selecting interferences and realizing higher data rate communications; while B must be spread for this system, that is also contrary to the band-limited peculiarity in underwater acoustic channels. Therefore it is more difficult to employ the OFDM system in long-range underwater acoustic communications.

Similarly, the frequency diversity has a better ability for combating with severe frequency selecting fading. Excellent results have been obtained by employing that in underwater acoustic email transmissions (refer to Chapter 3). Of course, B has to increase several times, and information rates will thus be correspondingly reduced.

When the communication range r is nearer, B may be selected to be wider, such as above 10 kHz at $r = 5$ km. In this case, the band-limited effect on underwater acoustic communications will lower. So, we can decrease r to increase the R. The contrary is true, too. It is agreeable to the law described by Eq. (1.1).

2.2.3.2 Countermeasures to Adapt to Great TL Occurring in Underwater Acoustic Communication Channels

2.2.3.2.1 RAISING SOUND POWER W_A

It is obvious that raising W_a can compensate for TL occurring in underwater acoustic channels to remain an expected input SNR (Signal-Noise Ratio) for a specific receiver and realize information reconstructions at preset BER (Bit Error Rate).

But there are some limitations to raising W_a:

1. Limitation of energy consumed: particularly when a communication sonar is operated under the water for a long time, its energy is generally supplied by batter sets as in underwater acoustic networks. Therefore it is necessary to solve the difficult problem to accurately detect transmission information at low input SNR.
2. Cavitation limitation: when the sound power applied to a sonar transducer is increased, cavitation bubbles begin to form on the face and just in front of the transducer. These bubbles are a manifestation of the rupture of the water caused by the negative pressures of the generated sound field. These negative pressures tear the liquid apart, so to speak, when they exceed a certain value called the cavitation threshold. The cavitation threshold may be expressed as a peak pressure in atmospheres or as plane-wave intensity in watts per square centimeter.

   A cavitation threshold, if 1 atm, is therefore equivalent to a plane-wave intensity of $0.3\ \mathrm{W/cm^2}$. When multiplied by the face area of the transducer, the cavitation threshold represents the maximum working value for the power output of the transducer. When this limit is exceeded by driving the projector harder, a number of deleterious effects begin to occur, such as erosion of the transducer surface, a loss of acoustic power in absorption and scattering by the cavitation bubble cloud, a deterioration in the beam pattern of the transducer, and a reduction in the acoustic impedance into which the transducer must operate. In all these ways, the cavitation threshold represents the onset of a gradual deterioration of transducer performance.

We see that by letting the face area of transducer be 1 m^2, W_a is only equal to 3.3 kW in relation to 1 atm. To obtain a larger W_a, increasing the area is necessary, so the volume and weight of transducer will correspondingly increase. Of course, if the transducer is operated in deep-water zones or in castor oil, the cavitation threshold will rise.

Otherwise, the nonlinear effect will appear for finite amplitude sound waves that the corresponding distortion of sound fields and excess absorption will generate if the large W_a is used.

3. Limitation with maintaining secret communication: this is very important for military underwater acoustic communications. Thus W_a will sometimes be lowered to obtain this ability. Of course, weak signal detection techniques have to be employed in them.

2.2.3.2.2 LOWERING OPERATING FREQUENCIES

When operating frequencies f is lowered, TL_a will lower too. It is equivalent to raising W_a; moreover, there are not the limitations by using large W_a as mentioned earlier. Therefore it is the first selected approach to adapt to large TL_a.

The field records in Xiamen Harbor shallow-water areas for a transmitting pulse sequence with different carrier frequencies at $r = 9$ km are shown in Fig. 2.15. They are 7.5 and 9.5 kHz in (A) and (B), respectively.

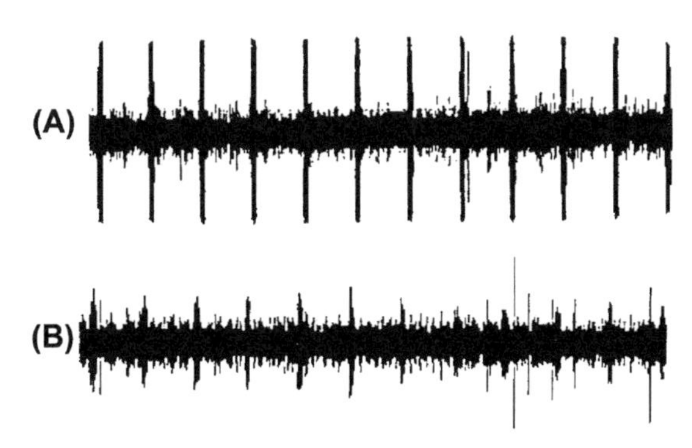

FIGURE 2.15 Recorders of waveforms for different carrier frequencies. (A) 7.5 kHz; (B) 9.5 kHz.

We see that the SNR for $f = 7.5$ kHz is much larger than that for $f = 9$ kHz because the former has a smaller sound absorption and scattering effects than that of the latter although r only equals 9 km.

Of course, there also exist some limitations to lowering f:

1. When f is lowered, the multipath effects will be more violent; in particular, the ISI (intersymbol interference) caused by them will be spread remarkably (refer to Fig. 1.4). In this case, how to combat with multipath interferences is a key technique in digital underwater acoustic communications, and it is thus more difficult.
2. The directivity of transducers will be reduced. Thus both DI_T (or SL) and DI will correspondingly be decreased, and the input SNR will also be decreased; otherwise, the size of transducer must be increased, and its volume and weight are correspondingly increased too.
3. NL will rise with lowering f; for example, the wind noise in deep-sea areas will rise up to 10 dB when f is lowered from 8 to 2 kHz.

We see that there is an optimum operating frequency to be selected for a specific communication sonar. Generally speaking, it is suitable to use the lower frequencies for longer range communications. On the contrary, a higher f

would be selected for nearer range ones. The optimum operating frequencies can roughly be estimated by the compromise calculations by means of the communication sonar equation as mentioned earlier.

2.2.3.2.3 SELECTING AVAILABLE BANDWIDTH B

We have known that there is a strict band-limited peculiarity in underwater acoustic communication channels. To get a higher R, B would generally be enlarged. The contrary is true, too. Therefore, the data compression, frequency codes compression, and the spectrum analyses having higher resolution than may be employed in digital underwater acoustic communications (refer to Chapter 3) for B to remain in a suitable width. Moreover, the adaptive amplitude equalizations that make up for the inhomogeneous frequency responses in relation to both the transducer and channel may be adopted to improve the robustness of underwater acoustic communications.

In some signal processing systems, such as the SS-FH (spread spectrum-frequency hop) system communication, a tracking filter can be used that is equivalent to compression of the B (refer to Chapter 3).

2.2.3.2.4 POSSIBLE APPROACHES TO ADAPT TO THE SPATIAL-TEMPORAL VARIABILITY IN UNDERWATER SOUND FIELDS

1. In the case of civil underwater acoustic communications (including some military communications), the excellent communication circumstances, such as deep-sea sound channeling, would be chosen. On the other hand, severe communication conditions such as sound shadow zones would possibly be avoided in carrying out the communications.
2. Predictions of sound fields.

Based on the estimations of relative marine circumstance parameters, such as sound velocity profiles, water depths, and the properties of the sea bottom and sea surface, combined with the operation conditions for a specific communication sonar, TL may generally be predicted as the references to estimate its operating specifications, such as the communication ranges r and data rates, etc.

Since the operating frequencies are higher in digital underwater acoustic communications, it is generally proper to employ ray acoustic theory in the simulations of the channels. However, the applications of ordinary ray modeling have some limitations, in particular in the sound fields with sound shadow zones and caustics.

The method of Gaussian beam tracing has received a great deal of attention in the seismological community. In comparison to standard ray tracing, the method has the advantage of being free of certain ray tracing artifacts such as perfect shadows and infinitely high energy at caustics. Based on that, M.B. Porter had successfully solved the difficult problems in calculating the sound fields at shadow zones and caustics by using the method.

The BELLHOP model based on the Gaussian beam method associates with each ray a beam with a Gaussian intensity profile normal to the ray [10]. The beam width and curvature are governed by an additional pair of differential equations, which are integrated along with the usual ray equations to compute the beam field in the vicinity of the central ray of the beam; the results are therefore more accurate.

The ray trace for a deep-sea Munk sound velocity profile, which is shown in Fig. 2.16, is plotted in Fig. 2.17. The sound reflections from both sea surface and bottom are not included.

The simulation by using the method of Gaussian beam tracing for TL is plotted in Fig. 2.18, in which a sound source and hydrophone are located at the depths of 1000 and 800 m, respectively, with operating frequency $f = 50$ Hz. We see that the variable laws for TL versus r are basically consistent with expected results, and the convergence zones and shadow zones at corresponding distances occur. We see that there are very large TL in the latter, although $TL \to \infty$ is absent in the shadow zones.

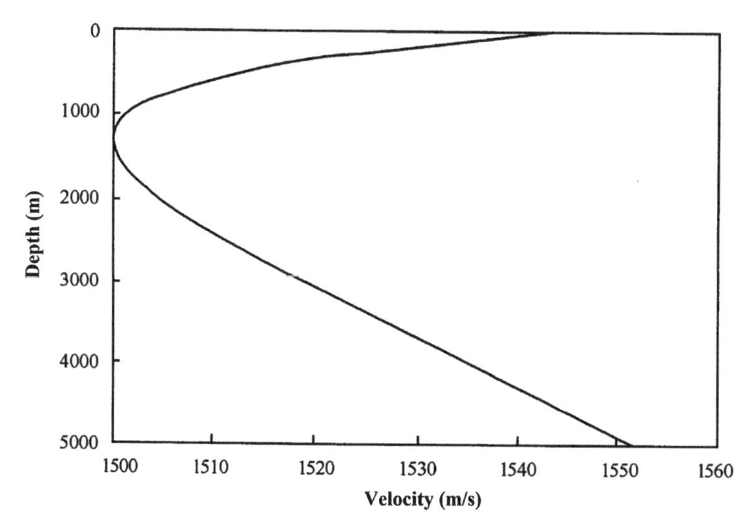

FIGURE 2.16 Munk sound velocity profile.

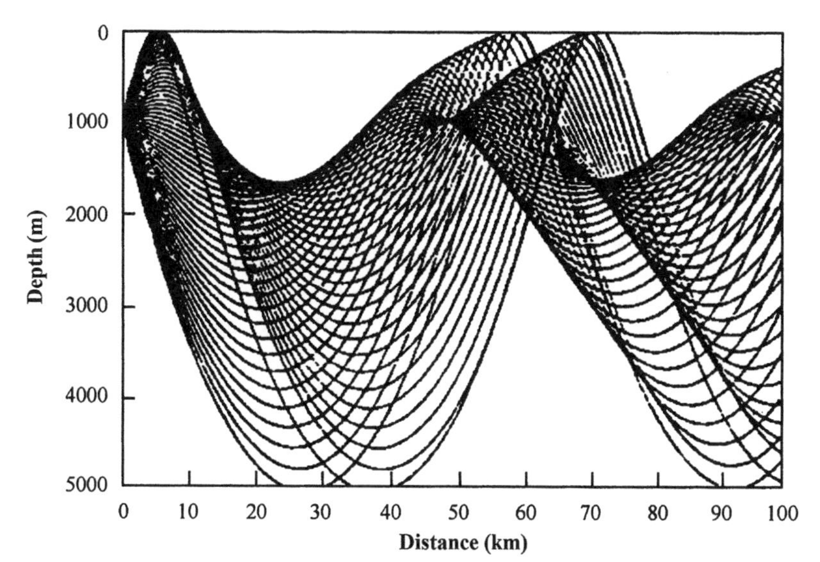

FIGURE 2.17 Ray trace for Munk sound velocity profile.

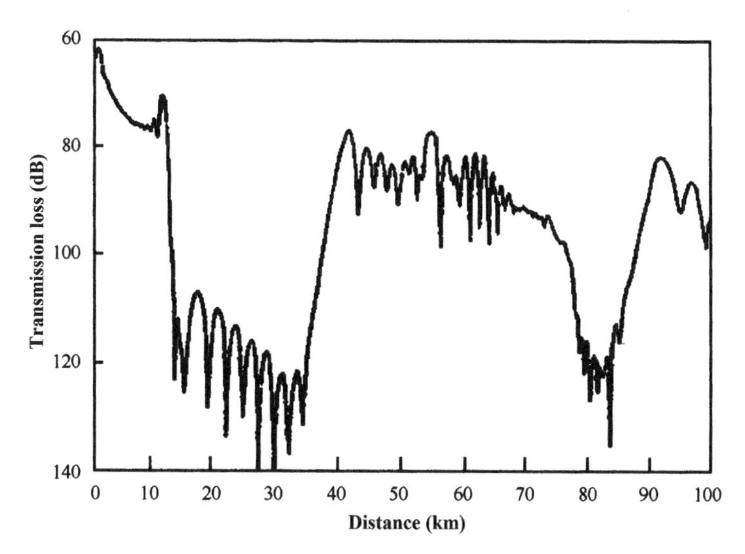

FIGURE 2.18 Simulation of *TL* by using the method of Gaussian beam tracing.

To predict shallow-water sound fields, the sound reflections from both sea surface and sea bottom must be considered. A simplified model has been adopted, in the sea surface is considered an adiabatic soft and smooth plane, and the sea bottom is equivalent to an acoustic elastic half-space where the acoustic parameters, such as sound velocities, densities, and sound absorption ratio, will be given.

A typical sound velocity profile at a certain position in the East China Sea is shown in Fig. 2.19. The simulations of *TL* for both equal-depth sea bottom and for having a changing tendency, as shown in Fig. 2.20, were accomplished; moreover, the three rays with respect to the radiation angles of 0°C and ±3°C are also plotted. The transmitting and receiving transducers are both located at a depth of 50 m. The results of simulation for *TL* are shown in Fig. 2.21. We see that *TL* for the equal-depth channel is basically agreeable to variable depth one in 0–3 km. *TL* for latter are larger than that of the former in the range from 3 to 30 km since more dense sound reflections appear for the variable-depth sea bottom. Correspondingly, sound intensity is attenuated rapidly in the ranges above 30 km.

Predicting *TL* is very important for designing and operating the communication sonars. For example, the communication circumstances for a specific sea area, in which an underwater acoustic network will be formed, must be understand in detail, and the spaces among the nodes in the network, the depths locating the transducers, the signal processing system to adapt to weak input SNR, etc., may thus be designed suitably to adapt to *TL* with severe spatial-temporal variability. Based on that, the loss of valuable data would possibly be avoided.

2.3 MULTIPATH EFFECTS IN UNDERWATER ACOUSTIC COMMUNICATION CHANNELS

Serious transmission loss is a main cause of restricting long-range underwater acoustic communications as mentioned earlier. However, we can adopt some approaches, such as raising sound power and lowering operating frequency, to roughly adapt to that.

The laws of multipath propagations in layered media and their effects on digital underwater acoustic communications will be discussed

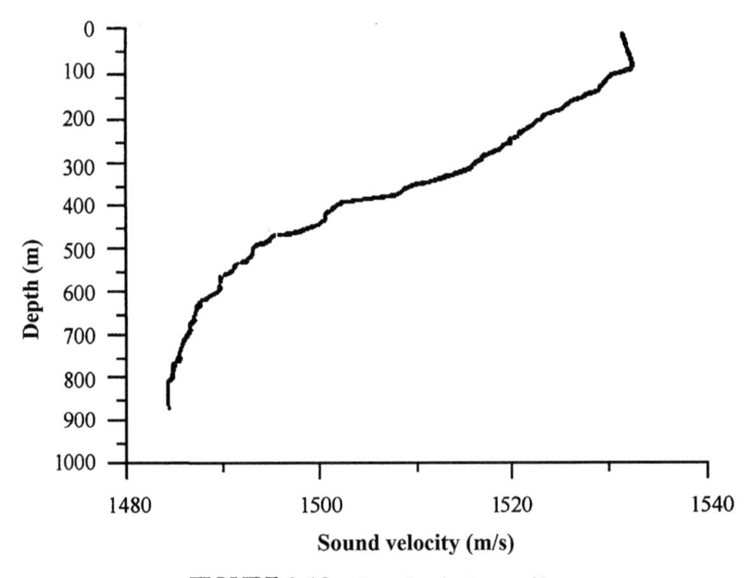

FIGURE 2.19 Sound velocity profile.

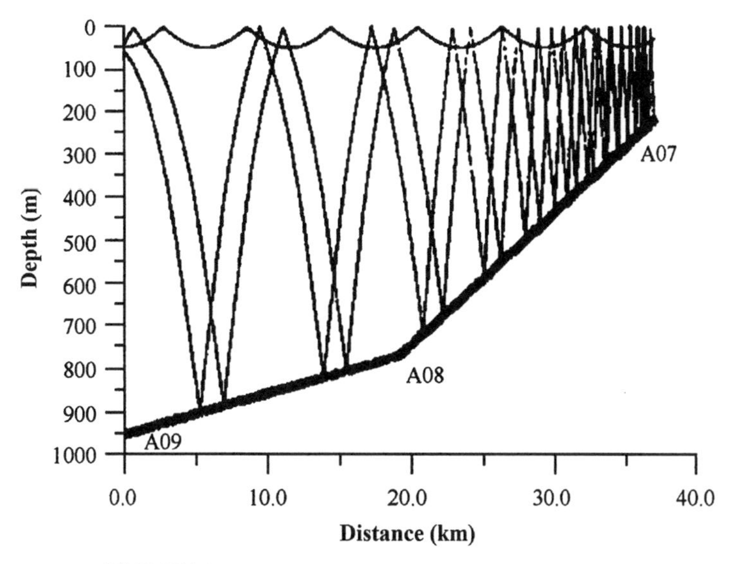

FIGURE 2.20 Three rays in a variable-depth channel.

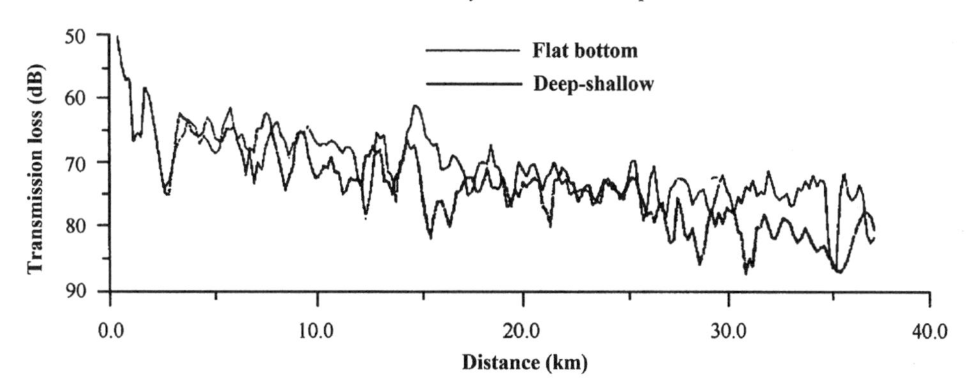

FIGURE 2.21 Simulation of *TL* for both the equal and variable depth bottoms.

subsequently. We will see that information signals have to be detected against the background of interference due to the multipath effects, just as an active sonar must detect the targets against that of reverberation. Therefore, the approaches by increasing sound power, lowering operating frequencies, using ordinary band-pass filters, and selecting the optimum operating frequency to combat against multipath interference are all ineffective.

The multipath effects in the layered media may be described by the normal mode theory, ie, sound fields may be expressed by the superposition of all normal modes with different group velocities. They may also be described by the ray acoustic theory: there are a number of rays to arrive at a receiving point along different paths with different propagating velocities and thus different arrival times. If a pulse signal is transmitted in layered inhomogeneous media, the received signal will thus be spread to a multipath structure with a spatial-temporal- frequency variability. How to overcome the multipath interference is a key technique in digital underwater acoustic communications, in particular in high data rate ones.

It should emphatically be noted that the multipath effects process duality.

The multipath effects in underwater acoustic channels are subject to the restriction of obeying the spherical spreading law due to the sound reflections from sea boundaries and the sound refraction generated by the layered property appearing in the seawater medium. Received multipath structures may be equivalent to spacial diversity signals unnecessary an additional energy consumption. Provided multipath pulses (wave packets) may be acquired and then combined with an optimum model, realization of an optimum signal detection against the background of multipath propagations would be expected. That is the positive influence of the multipath effects in which multipath energy is used.

The variable laws of the multipath effects, and the relations of the laws to marine communication circumstances, as well as the operating conditions of communication sonars will be discussed next by means of both the normal mode and ray acoustic theories. The impacts of the multipath effects on digital underwater acoustic communications and the countermeasures for how to adapt to them or even to use in full their energy will emphatically be analyzed. It is a necessary physical base to develop new signal processing systems employed in digital underwater acoustic communication equipment.

2.3.1 Solutions of Normal Modes with Respect to Sound Propagations in Two-Layer Media

2.3.1.1 The Solutions of the Normal Mode for Continuous Sound Wave Propagations

A simplified sound channel that is composed of a two-layer fluid media (as shown in Fig. 2.22) will first be discussed. A cylindrical coordinate is selected where $z = 0$ at the sea surface.

Assume the density, sound velocity, and layer thickness in the upper layer medium are ρ_1, c_1, and H, respectively. The density and velocity in the lower layer are ρ_2 and c_2, respectively, and the thickness tends to infinity. The simplified channel is approximately equivalent to a shallow sea one with a muddy sea bottom.

A unit point sound source is located at the depth z_0 on the z axis. The potential function ψ of the sound field satisfies the following wave equations:

$$\frac{1}{r}\frac{\partial}{\partial r}\left(r\frac{\partial \psi}{\partial r}\right)+\frac{\partial^2 \psi}{\partial z^2}+k_1^2\psi = -4\pi\delta(r, z-z_0) \quad 0 \le z < H$$

$$\frac{1}{r}\frac{\partial}{\partial r}\left(r\frac{\partial \psi}{\partial r}\right)+\frac{\partial^2 \psi}{\partial z^2}+k_2^2\psi = 0 \quad H < z$$

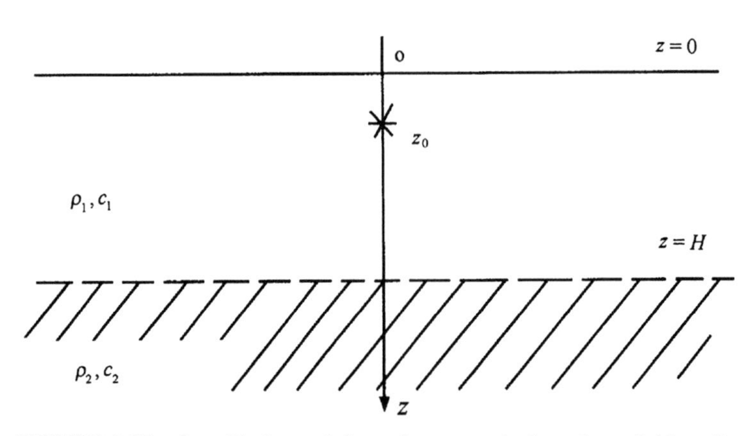

FIGURE 2.22 Simplified sound channel composed of two-layer fluid media.

We write ψ by means of the Fourier–Bessel integral as shown:

$$\psi(r,z) = \int_0^{\infty} Z(z,\xi)J_0(\xi r)\xi d\xi$$

According to the wave equation, the unknown function $Z\ (z,\xi)$ in the integral expression would satisfy the following differential equation:

$$\frac{d^2Z}{dz^2} + \left(k_1^2 - \xi^2\right)Z = -2\delta(z - z_0) \quad 0 \le z < H \tag{2.115}$$

$$\frac{d^2Z}{dz^2} + \left(k_2^2 - \xi^2\right)Z = 0 \quad H < z \tag{2.116}$$

where k is only the function of z, and $\delta\ (r,z)$ is the delta function of two dimensions. But the potential function of the sound field is continuous at both the upper and lower boundaries of the sound source:

$$Z|_{z=z_0^+} = Z|_{z=z_0^-} \tag{2.117}$$

Integrating both sides of Eq. (2.115) in the z direction between limits z_0^- and z^+ and by using the continuity condition of the potential function, we get

$$\left.\frac{dZ}{dz}\right|_{z=z_0^+} - \left.\frac{dZ}{dz}\right|_{z=z_0^-} = -2 \tag{2.118}$$

Eqs. (2.117) and (2.118) are called the point source conditions.

It is convenient to write this:

$$\beta_1 = \sqrt{k_1^2 - \xi^2}, \quad \beta_2 = \sqrt{k_2^2 - \xi^2}$$

If the sea surface is an absolutely soft boundary, then $\varphi = 0$ at this boundary. Moreover, ψ satisfies the radiation condition at an infinite point when $z \to \infty$. Because of an existing point sound source, $Z\ (z,\xi)$ has different modes at its upper and lower boundaries. $Z\ (z,\xi)$ will be selected as following modes according to the differential equation that $Z\ (z,\xi)$ must satisfy:

$$Z(z,\xi) = \begin{cases} A \sin \beta_1 z & 0 \le z < z_0 \\ B \sin \beta_1 z + C \cos \beta_1 z & z_0 < z < H \\ De^{i\beta_2 z} & H < z \end{cases}$$

where A, B, C, and D are the constants to be determined. By using the point source and boundary conditions for which ψ would satisfy both sides of the sea bottom, we can get a linear equation series that these constants would satisfy:

$$\begin{cases} A \sin \beta_1 z_0 - B \sin \beta_1 z_0 - C \cos \beta_1 z_0 = 0 \\ A \cos \beta_1 z_0 - B \cos \beta_1 z_0 + C \sin \beta_1 z_0 = \dfrac{2}{\beta_1} \\ B \sin \beta_1 H + C \cos \beta_1 H - \dfrac{1}{b} De^{i\beta_2 H} = 0 \\ B \cos \beta_1 H - C \sin \beta_1 H - i\dfrac{\beta_2}{\beta_1} De^{i\beta_2 H} = 0 \end{cases}$$

where $b = \frac{\rho_1}{\rho_2}$.

By solving these simultaneous equation series, we obtain the following:

$$A = \frac{2}{\beta_1}\left[\frac{\beta_1 \cos \beta_1(H - z_0) - ib\beta_2 \sin \beta_1(H - z_0)}{\beta_1 \cos \beta_1 H - ib\beta_2 \sin \beta_1 H}\right]$$

$$B = \frac{2 \sin \beta_1 z_0}{\beta_1}\left[\frac{\beta_1 \sin \beta_1 H + ib\beta_2 \cos \beta_1 H}{\beta_1 \cos \beta_1 H - ib\beta_2 \sin \beta_1 H}\right]$$

$$C = \frac{2 \sin \beta_1 z_0}{\beta_1}$$

$$D = \frac{2b \sin \beta_1 z_0}{\beta_1 \cos \beta_1 H - ib\beta_2 \sin \beta_1 H} e^{i\beta_2 H}$$

Let the $Z\ (z,\xi)$ and the specific expressions of the coefficients by substituting into the integration expression of ψ, the specific expressions with respect to a sound field composed of a point sound source in a two-layer media can be obtained as follows:

$$\psi(r,z) = \begin{cases} \displaystyle\int_0^{\infty} \frac{2\sin\beta_1 z}{\beta_1}\left[\frac{\beta_1\cos\beta_1(H-z_0) - ib\beta_2\sin\beta_1(H-z_0)}{\beta_1\cos\beta_1 H - ib\beta_2\sin\beta_1 H}\right] J_0(\xi r)\xi d\xi & 0 \le z < z_0 \\ \displaystyle\frac{2\sin\beta_1 z_0}{\beta_1}\left[\frac{\beta_1\cos\beta_1(H-z) - ib\beta_2\sin\beta_1(H-z)}{\beta_1\cos\beta_1 H - ib\beta_2\sin\beta_1 H}\right]_0 J_0(\xi r)\xi d\xi & z_0 < z < H \\ \displaystyle\frac{2b\sin\beta_1 z_0}{\beta_1\cos\beta_1 H - ib\beta_2\sin\beta_1 H} e^{i\beta_2(z-H)} J_0(\xi r)\xi d\xi & H < z \end{cases} \tag{2.119}$$

To further calculate the integration values of Eq. (2.119), ξ will be considered a complex variable, and we will use the method of a contour integral; moreover, we will pay attention to the following:

$$J_0(\xi r) = \frac{1}{2}\left\{H_0^{(1)}(\xi r) + H_0^{(2)}(\xi r)\right\}$$

$$H_0^{(2)}(\xi r) = -H_0^{(1)}\left(\xi r e^{i\pi}\right)$$

Taking the first expression in Eq. (2.119) as an example, and by using the transforming relation between the Bessel function and the Hankel function mentioned earlier, we get

$$\psi(r,z) = \int_{-\infty}^{\infty} \frac{\sin\beta_1 z}{\beta_1}\left[\frac{\beta_1\cos\beta_1(H-z_0) - ib\beta_2\sin\beta_1(H-z_0)}{\beta_1\cos\beta_1 H - ib\beta_2\sin\beta_1 H}\right] H_0^{(1)}(\xi r)\xi d\xi \quad 0 \le z < z_0$$

The integrand in this expression is still written simply as $Z(z,\xi)H_0^{(1)}(\xi r)\xi$.

There will be five ambiguous points in the integrand when the integration line on the complex plane ξ to be formed is

$$\xi = 0, \pm k_1, \pm k_2$$

where an ordinary point is the ambiguous point of the Hankel function. Considering that the Hankel function of the first kind will tend to zero at an infinite point in the upper half-plane of the complex plane ξ, the ambiguous lines may go downward from the ordinary point; moreover, the integration line will be above the ordinary point. The ambiguous points $\pm k_1$ and $\pm k_2$ respectively correspond to the many-valued nature of the integrand as the presence of β_1 and β_2. Considering the actual media exist sound absorption, both k_1 and k_2 have positive imaginary parts, the integration line along the real axis will around above ambiguous points $-k_1$ and $-k_2$, and then around below k_1 and k_2. To avoid entering another leaf of the Riemann plane by passing through ambiguous lines in forming contour integrals, we must increase the integrals around ambiguous points along two banks of the ambiguous lines, as shown in Fig. 2.23.

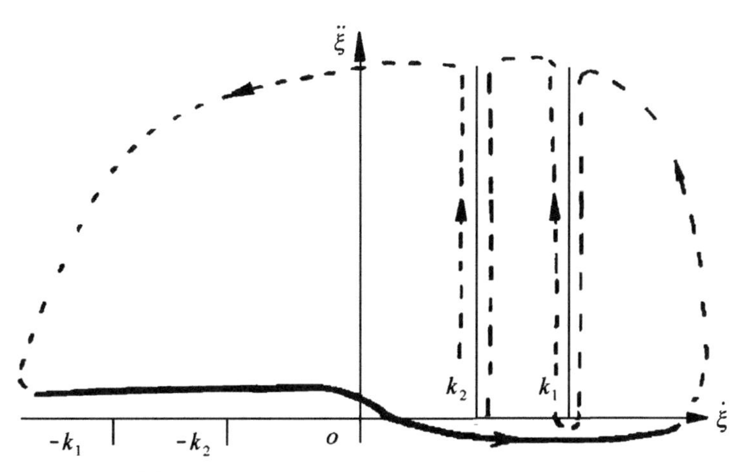

FIGURE 2.23 Integral circuit on ξ complex plane.

According to the relative theorem of the contour integral with respect to the complex variable function, we obtain

$$\begin{aligned}&\int_{-\infty}^{\infty} Z(z,\xi)H_0^{(1)}(\xi r)\xi d\xi + \int_{(k_1)} Z(z,\xi)H_0^{(1)}(\xi r)\xi d\xi\\&+\int_{(k_2)} Z(z,\xi)H_0^{(1)}(\xi r)\xi d\xi\\&+\int_{(R_\infty)} Z(z,\xi)H_0^{(1)}(\xi r)\xi d\xi\\&=\text{Res}\left\{Z(z,\xi)H_0^{(1)}(\xi r)\xi\right\}\end{aligned}$$

Based on the property of the Hankel function of the first kind, the integral would be zero at the infinity circuit in the upper half-plane. Moreover, according to the property of the even function relative to β_1 for the integrand, the integral passing through the two banks of the ambiguous line of point k_1 would also be zero. We terminally obtain

$$\begin{aligned}\psi(r,z) &= \text{Res}\left\{Z(z,\xi)H_0^{(1)}(\xi r)\xi\right\}\\&\quad-\int_{(k_2)} Z(z,\xi)H_0^{(1)}(\xi r)\xi d\xi \qquad (2.120)\end{aligned}$$

The two terms on the right-hand side of Eq. (2.120) will be discussed, respectively, as follows.

2.3.1.1.1 PART OF NORMAL MODES

The part indicated by residues of pole points in the contour in Eq. (2.120) is called a normal mode, which is written by

$$\psi_N(r,z) = \text{Res}\left\{Z(z,\xi)H_0^{(1)}(\xi r)\xi\right\}$$

According to Eq. (2.119), the equation to determine the positions of the pole points in the contour is given by

$$\beta_1 \cos \beta_1 H - ib\beta_2 \sin \beta_1 H = 0 \qquad (2.121)$$

Eq. (2.121) is generally called the frequency dispersion equation. Letting the nth roof of the equation be ξ_n and writing

$$\beta_{in} = \sqrt{k_i^2 - \xi_n^2} \quad i = 1,2$$

The integrand can be simplified by using the frequency dispersion equation; moreover, according to the residue theorem, we get

$$\begin{aligned}\psi_N(r,z) &= 2\pi i \sum_n \frac{\sin \beta_{1n}}{\beta_{1n}} \cdot \frac{\beta_{1n} \cos \beta_{1n}(H-z_0) - ib\beta_{2n} \sin \beta_{1n}(H-z_0)}{\dfrac{\partial}{\partial \xi_n}[\beta_{1n} \cos \beta_{1n}H - ib\beta_{2n} \sin \beta_{1n}H]} H_0^{(1)}(\xi_n r)\xi_n \\ &= \sum_n \frac{2\pi i \beta_{1n} \sin \beta_{1n}z \cdot \sin \beta_{1n}z_0}{\beta_{1n}H - \sin \beta_{1n}H \cdot \cos \beta_{1n}H - b^2 \tan \beta_{1n}H \cdot \sin^2 \beta_{1n}H} H_0^{(1)}(\xi_n r) \\ &= \sum_n U(z,\xi_n)U(z_0,\xi_n)H_0^{(1)}(\xi_n r)\end{aligned} \tag{2.122}$$

By examining Eq. (2.122), we can see that the obtained result with respect to z and z_0 is reciprocal. This is the requirement of the reciprocal theorem. $U\ (z,\xi_n)$ in Eq. (2.122) are referred to as normal modes, which will be determined by the characteristics of sound channels and have an unvaried model when they propagate in certain sound channels. Moreover, the model is independent of the position of the sound source. Different positions only affect the normal modes of different orders that have different excited sound intensities.

We can first analyze qualitatively that the roots of the frequency dispersion in Eq. (2.121) would be in the first quadrant of the complex plane ξ.

We see that by examining Eq. (2.122), normal modes corresponding to the complex roots of the frequency dispersion equation have an additional exponential attenuation factor with respect to increasing propagating ranges. On the contrary, this factor is absent for the normal modes corresponding to real roots. Therefore the latter will play a major role for longer range underwater sound propagations.

If $k_1 > k_2$ (ie, the sound velocity in the sea bottom is higher than that in the seawater), provided that sound waves are reflected from the sea bottom, the total reflection will appear when grazing angles are small enough, and thus the sound energy will totally return to the seawater. In this case, the real roots would possibly exist in the frequency dispersion equation. When we let

$$x = \beta_1 H, \quad \sigma^2 = \left(k_1^2 - k_2^2\right)H^2 \tag{2.123}$$

then the frequency dispersion in Eq. (2.121) becomes

$$x \cos x - ib\sqrt{x^2 - \sigma^2}\sin x = 0 \tag{2.124}$$

If real roots are only considered, we will see that there exist real solutions in Eq. (2.124) as x is a real number and smaller than σ. Therefore, the real roots of ξ must be in the interval of (k_1,k_2). We rewrite Eq. (2.124) as

$$\frac{x}{\tan x} = -b\sqrt{\sigma^2 - x^2}$$

The functions at both sides of this expression versus x are respectively plotted in Fig. 2.24. The values corresponding to the intersection points between two series of curves represent the roots of this equation.

By observing Fig. 2.24, we see that if following conditions are satisfied

$$\left(n - \frac{1}{2}\right)\pi \le \sigma < \left(n + \frac{1}{2}\right)\pi$$

then the frequency dispersion equation will exist with n real roots. Since σ is proportional to sound wave frequencies, the numbers of normal modes with real eigenvalues for lower frequencies are

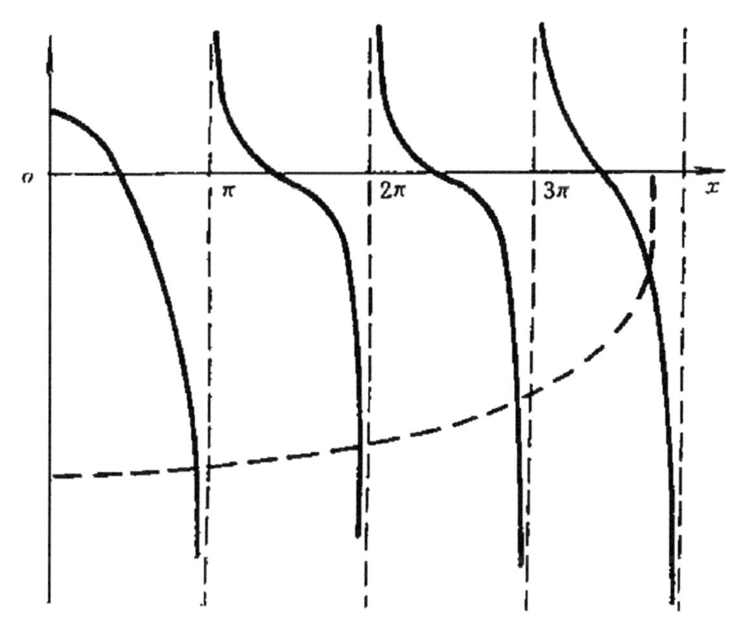

FIGURE 2.24 Graphic method for the frequency dispersion equation.

thus less than those for higher ones. In the case of the nth-order normal mode, once the frequencies are lower than

$$\omega_n H\sqrt{\frac{1}{c_1^2}-\frac{1}{c_2^2}}=\left(n-\frac{1}{2}\right)\pi$$

or

$$\omega_n=\frac{\left(n-\frac{1}{2}\right)\pi c_1 c_2}{H\sqrt{c_2^2-c_1^2}} \tag{2.125}$$

then the eigenvalue of the normal mode of this order will change into a complex number. The lowest frequency that has an eigenvalue that is still a real number for every normal mode is called the critical frequency. When sound frequencies are lower than the critical frequency of the first-order normal mode, there will be an additional attenuation that obeys the exponential law with increasing transmission ranges to all normal modes.

There are a series of normal modes with real numbers in actual underwater sound channels. For example, let the sound velocities in the seawater and sea bottom be 1480 m/s and 1600 m/s, respectively, and the depth be 50 m, then there will exist 26 order real number normal modes at the frequency of 1 kHz.

2.3.1.1.2 ORTHOGONAL PROPERTIES OF NORMAL MODES

The normal modes of different orders are orthogonal to each other in the z (depth) direction. The two-layer media are taken as a typical example to describe this property. Let the density and sound velocity be ρ_1 and c_1 in the upper layer, respectively, and ρ_2 and c_2 correspond to those in the lower layer, where c_1 and c_2 may be the arbitrary functions of depths. Assume the thickness of the upper layer is H, according to the definition of normal modes, and the normal modes are U_m and U_n, which correspond to the two roots ξ_m and ξ_n of the frequency dispersion equation; will obey the following equations:

$$\begin{cases} \dfrac{d^2U_m}{dz^2} + (k^2 - \xi_m^2)U_m = 0 \\ \dfrac{d^2U_n}{dz^2} + (k^2 - \xi_n^2)U_n = 0 \end{cases} \tag{2.126}$$

where

$$k^2 = \begin{cases} k_1^2 & 0 \le z < H \\ k_2^2 & H < z \end{cases}$$

Let

$$\rho = \begin{cases} \rho_1 & 0 \le z < H \\ \rho_2 & H < z \end{cases}$$

Two expressions in Eq. (2.126) are multiplied by ρU_m and ρU_n, respectively; by taking the difference between them and then integrating with respect to z, we obtain

$$\begin{aligned} &(\xi_n^2 - \xi_m^2)\int_0^\infty \rho U_m U_n dz \\ &= \int_0^\infty \left[\rho U_n \frac{d^2U_m}{dz^2} - \rho U_m \frac{d^2U_n}{dz^2}\right] dz \\ &= \rho_1 \left[U_n \frac{dU_m}{dz} - U_m \frac{dU_n}{dz}\right]_0^H \\ &\quad + \rho_2 \left[U_n \frac{dU_m}{dz} - U_m \frac{dU_n}{dz}\right]_H^\infty \\ &= 0 \end{aligned}$$

That is to say, the normal modes with different orders will be orthogonal to each other if $n \neq m$. The result can be extended to an arbitrary layered media. Based on the orthogonal property of the normal modes, we would obtain their values by using a linear hydrophone array hung vertically in underwater acoustic channels.

The normal modes possess the model of standing waves in the z (depth) direction. The amplitudes of the first-order normal mode versus depths for different frequencies f are shown in Fig. 2.25.

Each standing wave in the layer may be resolved into an up-traveling wave and a down-traveling one. As an example given by Eq. (2.122), the nth-order normal mode can be written as follows:

$$A(z_0, \xi_n) \sin \beta_{1n} z H_0^{(1)}(\xi_n r)$$

Let the sine function be expressed by the exponential function, and the Hankel function is replaced by its asymptotic expansion expression, so the nth-order normal mode can be written as

$$\sqrt{\frac{1}{2\pi\xi_n r}} A(z_0, \xi_n)\left\{e^{i\left(\xi_n r + \beta_{1n} z - \frac{3\pi}{4}\right)} + e^{i\left(\xi_n r - \beta_{1n} z + \frac{\pi}{4}\right)}\right\}$$

There is a 180 degree phase difference between the up- and the down-traveling waves at the sea surface. Their grazing angles are $\tan^{-1}(\beta_{1n}/\xi_n)$. We see that the higher the order for a normal mode is, the larger the grazing angle becomes. That will slough off to an inhomogeneous wave when ξ_n is a complex number.

2.3.1.1.3 FLANK WAVE PART

By calculating the approximate value of the line integral on the right-hand side of Eq. (2.120), we can obtain

$$\psi_L(r, z) \approx \frac{2bk_2 \sin\left(\sqrt{k_1^2 - k_2^2}\, z\right) \sin\left(\sqrt{k_1^2 - k_2^2}\, z_0\right)}{\left(k_1^2 - k_2^2\right)\cos^2\left(\sqrt{k_1^2 - k_2^2}\, H\right)} \cdot \frac{1}{r^2} e^{i\left(k_2 r - \frac{\pi}{2}\right)} \tag{2.127}$$

We can see that the phases of sound waves are variable with respect to the sound velocities in the sea bottom, ie, their traveling paths are mostly in the bottom in propagation processes and therefore called flank waves.

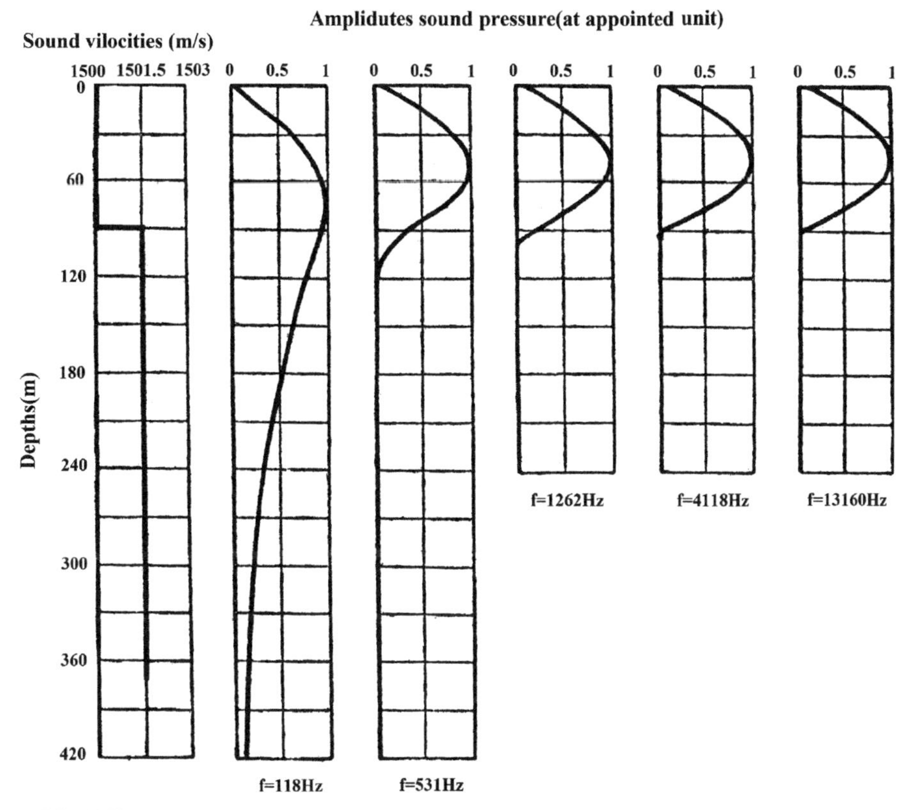

FIGURE 2.25 Amplitudes versus the depths for the first-order normal mode at different f.

According to Eq. (2.127), the amplitudes of the flank wave is attenuated more severe than that of the spherical spreading law, so the effects of the flank wave may generally be neglected for long-range underwater acoustic communications.

2.3.1.2 *Propagation of a Sound Pulse in the Two-Layer Media*

According to linear acoustics, an underwater acoustic channel can be thought of as a linear system as viewed from signal processing. The transmission property may be described by the impulse response function or transfer function, the former is the response of the channel to δ function. However, in fact, the δ function is only an ideal concept; the most approximate waveform would be the exponential pulse exited by an explosive sound source:

$$f(t) = \begin{cases} e^{-\alpha t} & t > 0 \\ 0 & t < 0 \end{cases}$$

where α is large enough. Now, let us consider the transmission characteristics in the two-layer media. It is expected that sound fields would be formed by a series of pulses (packets).

The spectrum function $Q(\omega)$ of $f(t)$ can be represented by a Fourier transform:

$$Q(\omega) = \frac{1}{2\pi}\int_{-\infty}^{\infty} f(t)e^{i\omega t}dt = \frac{1}{2\pi(\alpha - i\omega)}$$

The part of the normal modes is only considered for long-range communications. According to Eq. (2.122) and using Fourier integral, the sound field exciting by the exponential pulse can be represented as

$$F_N(r,z,t) = \sum_n \int_{-\infty}^{\infty} \frac{i\beta_{1n}\sin\beta_{1n}z\cdot\sin\beta_{1n}z_0\cdot e^{-i\omega t}}{\beta_{1n}H-\frac{1}{2}\left(1+b^2\tan^2\beta_{1n}H\right)\sin 2\beta_{1n}H} \times \frac{H_0^{(1)}(\xi_n r)}{\alpha-i\omega}\mathrm{d}\omega \approx \sqrt{\frac{2}{\pi r}}\sum_n \int_{-\infty}^{\infty} S_n(\omega)\frac{e^{i\left(\xi_n r-\omega t+\frac{\pi}{4}\right)}}{\alpha-i\omega}\mathrm{d}\omega \tag{2.128}$$

where

$$S_n(\omega) = \frac{\beta_{1n}\beta_{2n}^2\sin\beta_{1n}z\cdot\sin\beta_{1n}z_0}{\sqrt{\xi_n}\left[\beta_{1n}\beta_{2n}^2H+\left(k_1^2-k_2^2\right)\sin\beta_{1n}H\cdot\cos\beta_{1n}H\right]}$$

$S_n(\omega)$ can be thought of as a slowly varying function of the angular frequency ω. We see that the exponential part of the integrand will be a rapidly oscillating function when r and t are greater. Thus, except the zones in which the exponential part is slowly variable versus ω, the integral with respect to other zones can be neglected. We can then write

$$\Omega(\omega) = \xi_n r - \omega t + \frac{\pi}{4}$$

Then, we can use the method of implicit phases to approximately calculate the integral in Eq. (2.128). According to

$$\dot{\Omega}(\omega) = \frac{\mathrm{d}\xi_n}{\mathrm{d}\omega}r - t = 0 \tag{2.129}$$

the point of implicit phase ω_0 can be found. Let $\omega=\omega_0+s$; moreover expanding $\Omega(\omega)$ into the Taylor series at ω_0, and keeping the values of the first and second orders, we can obtain

$$\Omega(\omega)\approx\Omega(\omega_0)+\frac{1}{2}r\ddot{\xi}_{n0}s^2$$

By substituting that into Eq. (2.128), we get

$$F_N(r,z,t)\approx\sqrt{\frac{2}{\pi r}}\sum_n\frac{S_n(\omega_{0n})e^{i\Omega(\omega_{0n})}}{\alpha-i\omega_{0n}}\int_{-\infty}^{\infty}e^{i\frac{1}{2}r\ddot{\xi}_{n0}s^2}\mathrm{d}s = \frac{2}{r}\sum_n\frac{S_n(\omega_{0n})}{\left\{\left(\alpha^2+\omega_{0n}^2\right)\left|\ddot{\xi}_{n0}\right|\right\}^{1/2}}e^{i\left(\xi_{n0}r-\omega_{0n}t+\tan^{-1}\frac{\omega_{0n}}{\alpha}\right)} \times\begin{cases} e^{i\frac{\pi}{2}}, & \ddot{\xi}_{n0}>0 \\ 1, & \ddot{\xi}_{n0}<0\end{cases} \tag{2.130}$$

In the case of a single frequency sound wave, the variable law of phases versus time obeys $e^{i(\xi_n r-\omega t)}$ by examining Eq. (2.122), ie, the phase velocity of the sound wave $c^{(n)}=\omega/\xi_n$. However, according to (2.129), the sound pulse signal will propagate with group velocity $u^{(n)}=\mathrm{d}\omega/\mathrm{d}\xi_n$. But either the phase velocity or group velocity will change with frequencies. In the case of $c_2>c_1$, the group and phase velocities versus frequency for the first three orders are shown in Fig. 2.26.

The group velocities equal c_2 at the critical frequency for each normal mode, and they will tend to c_1 at high frequency and have a minimum value at a certain moderate frequency. However,

$$\ddot{\xi}_n = \frac{\mathrm{d}}{\mathrm{d}\omega}\left(\frac{1}{u_n}\right) = -\frac{1}{u_n^2}\dot{u}_n$$

In other words, the second derivative of ξ_n equals zero at the minimum group velocity, and thus Eq. (2.130) does not hold once more.

Writing the frequency corresponding to the minimum group velocity is ω_{na}, which is also

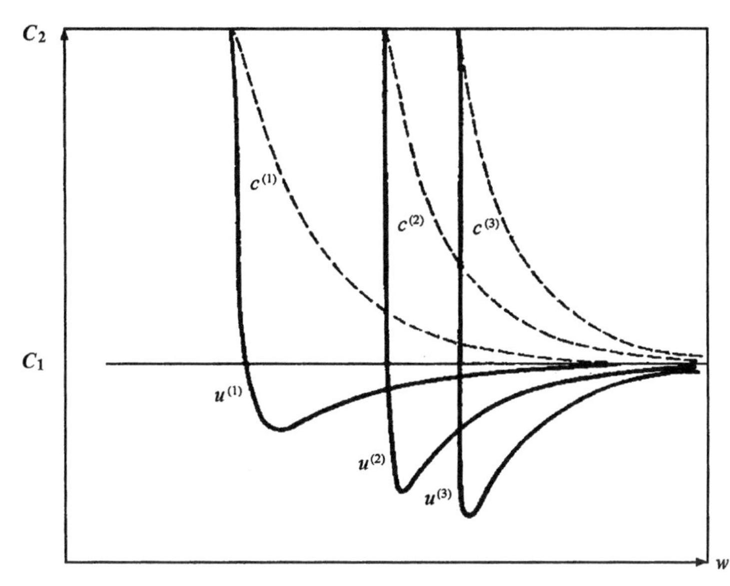

FIGURE 2.26 Group velocity $u^{(i)}$ and phase velocity $c^{(i)}$ versus frequencies for the normal modes of the first three orders.

called the frequency of Airy wave. Let $\omega = \omega_{na} + s$, and the series of $\Omega(\omega)$ at ω_{na} can approximately be represented by

$$\Omega(\omega) = \Omega(\omega_a) + (\dot{\xi}_a r - t)s + \frac{1}{6}\dddot{\xi}_a r s^3$$

Thus, Eq. (2.128) can approximately be written as

$$\begin{aligned} F_N(r,z,t) &= \sqrt{\frac{2}{\pi r}}\sum_n \frac{S_n(\omega_{na})e^{i\Omega(\omega_{na})}}{\alpha - i\omega_{na}} \int_{-\infty}^{\infty} e^{i\left[(\dot{\xi}_a r - t)s + \frac{1}{6}\dddot{\xi}_a r s^3\right]} ds \\ &= \frac{2^{3/2}}{r}\sum_n \frac{S_n(\omega_{na})e^{i\Omega(\omega_{na})}}{\alpha - i\omega_{na}} \left(\frac{\dddot{\xi}_{na} r}{2}\right)^{-1/3} \gamma\left[\sqrt[3]{\frac{2}{\dddot{\xi}_{na} r}}(\dot{\xi}_{na} r - t)\right] \end{aligned} \tag{2.131}$$

where

$$\begin{aligned} \gamma &= \frac{1}{\sqrt{\pi}}\int_0^\infty \cos\left(sx + \frac{1}{3}s^3\right) ds \\ &= \begin{cases} \frac{1}{3}(\pi x)^{1/2}\left[I_{-1/3}\left(\frac{2}{3}x^{3/2}\right) - I_{1/3}\left(\frac{2}{3}x^{3/2}\right)\right] & x > 0 \\ \frac{1}{3}(-\pi x)^{1/2}\left[J_{-1/3}\left(\frac{2}{3}(-x)^{3/2}\right) + J_{1/3}\left(\frac{2}{3}(-x)^{3/2}\right)\right] & x < 0 \end{cases} \\ &\approx \begin{cases} \exp\left(-\frac{2}{3}x^{3/2}\right)\Big/\left\{\sqrt{2}\left[2.152 + 1.619x + 16x^2\right]^{1/8}\right\} & x > 0 \\ \sin\left[\frac{2}{3}(-x)^{3/2} + \frac{\pi}{4}\right]\Big/\left[2.152 + 1.619x + x^2\right]^{1/8} & x < 0 \end{cases} \end{aligned} \tag{2.132}$$

This Airy function has been plotted in Fig. 2.3.

In summary, in the case of narrow pulse sound waves, since there are different group velocities for them with different frequencies propagating in the two-layer media, the received signal wave forms for each order normal mode will be spread, as roughly shown in Fig. 2.27. The first received signal travels from such a wave with the critical frequency

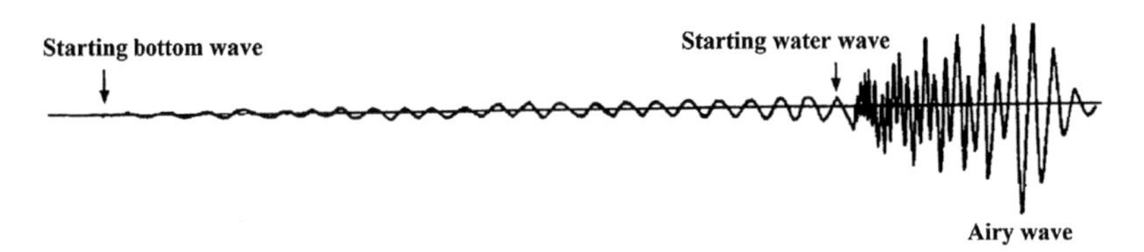

FIGURE 2.27 Received waveform for an exponential pulse traveling in two-layer media.

and has a group velocity approximately equal to c_2, which is called the bottom wave. The frequencies will gradually be increased in following received signals, and the group velocities will correspondingly be decreased and finally tend to c_1. At the same time, the signal with higher frequencies appears, which is called a water wave. We can see that the complicated interference phenomenon due to the superposition between the bottom and water waves appears. Then the frequencies of the water wave will gradually lower, and those of the bottom wave will gradually increase; finally, they all equal the frequency of the Airy wave. When the Airy wave with the minimum group velocity arrives, the pulse waveform will thus end. Since the group velocities are different for the sound waves having different frequencies, if a very narrow exponential pulse is traveling in shallow-water acoustic channels at ranges such as 10 km, the waveform for each order normal mode will possibly be spread to several hundred milliseconds, even if only calculating the time interval from the water wave with higher amplitudes to the end of the Airy wave. In the case of the normal modes of different orders, in spite of having the same frequency, whose group velocities are also different, a pulse train will also appear. That is to say, shallow-water acoustic channels essentially belong to such ones with frequency dispersions and multipath propagations. They will be limited in a certain range only if frequencies are higher and bandwidths are narrower, and thus all normal modes have the group velocity approximately equal to the sound velocity in the seawater.

The solutions of normal modes for sound propagations (including CW (continuous wave) and exponential pulse wave) in two-layer fluid media have been discussed. It is valuable to qualitatively analyze the sound propagations in shallow-water acoustic channels. Experiments have proven that the propagation laws calculated by the normal mode theory are basically identical with the measured results in shallow sea zones where the variations of both depths and sound velocities in the horizontal direction are less.

2.3.2 Expression of Virtual Source Images for Sound Propagations in Layered Media [4]

The solutions of the normal modes for continuous sound waves, as well as the multipath effects with respect to the sound pulse traveling in the two-layer fluid media, have been introduced in the preceding section. Now, we provide an alternative approach to describe the multipath propagations: they can be thought of as the superposition of the direct and a series of "virtual" sources, which is called virtual source chain, caused by successive sound reflections from both upper and lower boundaries.

The expression of virtual source images for sound propagations in layered media has some remarkable advantages, such as clearly physical significance, intuitive appeal pictures, etc. It is valuable for designing the signal processing systems employed in digital underwater acoustic communications against the background of multipath propagations.

Assume that there is a homogeneous layer whose lower boundary is at $z = 0$, upper boundary is at $z = h$, sound velocity is c, and a pulsating sound source O_{01} is placed at z_0, as shown in Fig. 2.28.

The sound potential function ψ in the layer will satisfy the wave equation:

$$\Delta\psi + k^2\psi = 0, \quad k = \frac{\omega}{c} \tag{2.133}$$

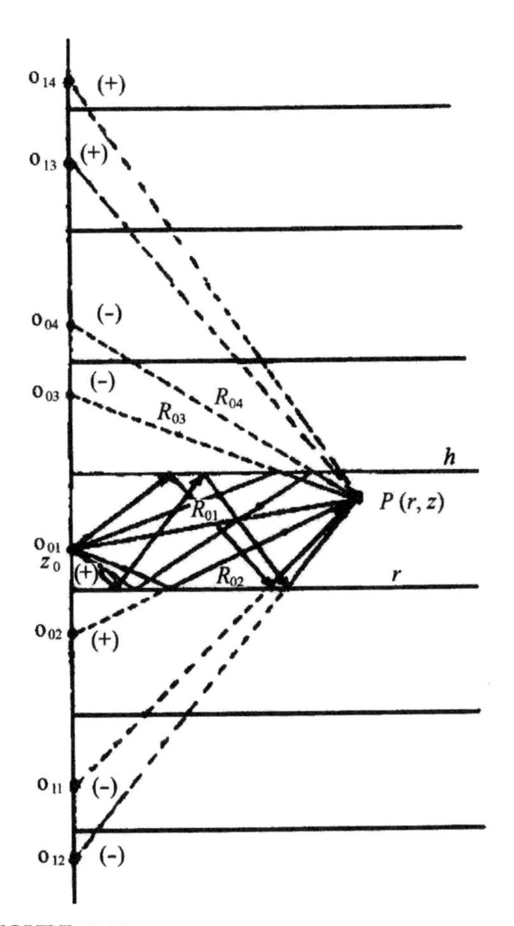

FIGURE 2.28 Expression of sound field by a virtual source chain due to the sound successive reflections from the boundaries.

Let the two boundaries be absolutely hard planes; therefore

$$\frac{\partial\psi}{\partial z} = 0\begin{cases} z = 0 \\ z = h \end{cases} \tag{2.134}$$

The sound field at an arbitrary received point $P(r,z)$ can be represented by the sum of the direct wave and radiant waves from a virtual source chain, which is formed by successive specula reflections of O_{01} from both upper and lower boundaries.

O_{02} represents a virtual sound source that is caused by the sound reflection of O_{01} from the lower boundary (see Fig. 2.28). The composing sound field is

$$\psi = \psi_{01} + \psi_{02} = \frac{e^{ikR_{01}}}{R_{01}} + \frac{e^{ikR_{02}}}{R_{02}} \tag{2.135}$$

where

$$R_{01} = \sqrt{r^2 + (z - z_0)^2}, \quad R_{02} = \sqrt{r^2 + (z + z_0)^2} \tag{2.136}$$

are the distances from O_{01} and O_{02} to the received point $P(r,z)$, respectively. ψ, expressed by Eq. (2.135), satisfies the wave equation Eq. (2.133) and the boundary condition at $z = 0$ because O_{01} and O_{02} are symmetric with respect to $z = 0$. Therefore, $\frac{\partial\psi}{\partial z} = 0$ at this boundary.

However, Eq. (2.135) does not satisfy the upper boundary condition at $z = h$. Thus, a pair of virtual sources O_{03} and O_{04} are added, which are formed by the reflections of O_{01} and O_{02} from the upper boundary, respectively. The obtained resolution will thus satisfy the wave equation and the upper boundary condition. But now the boundary condition at the lower boundary is no longer satisfied. To make up for both O_{03} and O_{04} existing, their virtual sources O_{11} and O_{12} in relation to the lower boundary must be added. But now the upper boundary is not satisfied. Therefore, it is necessary to add another pair of virtual sources with respect to the upper boundary. By this back-and-forth process, an infinite virtual source chain is built up, with high-order virtual sources tending to become insignificant because of weakening by repeated reflections and by great distance. We see that the two boundary conditions can thus be satisfied.

Therefore, the total sound field is given by

$$\psi = \sum_{l=0}^{\infty}\left(\frac{e^{ikR_{l_1}}}{R_{l_1}} + \frac{e^{ikR_{l_2}}}{R_{l_2}} + \frac{e^{ikR_{l_3}}}{R_{l_3}} + \frac{e^{ikR_{l_4}}}{R_{l_4}}\right) \quad (2.137)$$

where

$$\begin{aligned} R_{l_1} &= \sqrt{r^2 + (2lh + z - z_0)^2} \\ R_{l_2} &= \sqrt{r^2 + (2lh + z + z_0)^2} \\ R_{l_3} &= \sqrt{r^2 + [2(l+1)h - z - z_0]^2} \\ R_{l_4} &= \sqrt{r^2 + [2(l+1)h - z + z_0]^2} \end{aligned} \quad (2.138)$$

and $l = 0,1,2\ldots\infty$.

Eq. (2.137) satisfies the wave equation because it consists of a series of spherical waves; moreover, it also satisfies the source condition since $\frac{e^{ikR_{01}}}{R_{01}}$ in Eq. (2.137) will be highest term when P tends to the basic wave source O_{01}, and that requires a singular point.

It is necessary to note that each virtual source corresponds to a ray that starts from the basic radiant source O_{01}, then it travels through a certain number of reflections from both the upper and lower boundaries, and it finally arrives at the received point P, as shown in Fig. 2.28.

The actual sea surface may be considered as an absolutely soft boundary; thus $\psi = 0$ and the reflection coefficient $V = -1$ at this boundary. Therefore, the phase of the sound wave has a jumping variation of 180 degrees provided that the numbers of reflections from the sea surface are singular, and corresponding rays will be signed as a negative sign on the right sides of corresponding virtual sources, as shown in Fig. 2.28.

In this case, the expression of the sound field is given by

$$\psi = \sum_{l=0}^{\infty}(-1)^l\left(\frac{e^{ikR_{l_1}}}{R_{l_1}} + \frac{e^{ikR_{l_2}}}{R_{l_2}} - \frac{e^{ikR_{l_3}}}{R_{l_3}} - \frac{e^{ikR_{l_4}}}{R_{l_4}}\right) \quad (2.139)$$

Let V_1 and V_2 be the reflection coefficients at the upper and lower boundaries, respectively, and they may be selected as ± 1. The extended expression of the sound field is given by

$$\psi = \sum_{l=0}^{\infty}(V_1V_2)^l\left(\frac{e^{ikR_{l_1}}}{R_{l_1}} + V_1\frac{e^{ikR_{l_2}}}{R_{l_2}} + V_2\frac{e^{ikR_{l_3}}}{R_{l_3}} + V_1V_2\frac{e^{ikR_{l_4}}}{R_{l_4}}\right) \quad (2.140)$$

The expression of the sound field by means of the virtual source images strictly holds for absolutely soft and hard boundaries. In the case of V versus grazing angles, the obtained sound field is generally not a correct solution. However, it is possible to use the concept of virtual sources if the layer thickness h is great enough in comparison with the sound wavelength. This condition is generally satisfied for digital underwater acoustic communications.

In the cases of actual underwater acoustic communications, according to the specific requirements for data rates, the durations of transmitted sound pulses are correspondingly selected. Because the distances from individual virtual sources to the received point P are different, and thus there are corresponding different arrival times, complicated and variable multipath structures including the interference effect due to the superposition of adjacent sound waves appear.

The multipath structures described by the virtual source images will help us to understand clearly and intuitively their variable laws and some significant effects on digital underwater acoustic communications. Some examples follow.

When a sound wave travels in an infinite space, only a direct wave radiating from O_{01} (refer to Fig. 2.28) arrives at P that obeys the spherical spreading law. Once the sound source O_{01} is placed in a layer, the sound field can be thought of as composed by the sum of sound waves radiated by all O_{ln} ($n = 1,2,3,4$; $l = 0,1,2,3\ldots\infty$). There are the contributions of the reflected sound waves from the sea boundaries to the sound field, and we would use in full this additional sound energy in signal detections.

When sound wave frequencies are increased, TL_a will rise remarkably; moreover, the reflection coefficients for the actual boundaries of the sea will also rise (refer to Section 2.4). Therefore the amplitudes of multipath pulses (wave packets) will be reduced, and the total delay spread of multipath effects will also be decreased. So, we can increase the operating frequency to suppress ISI caused by the multipath effects. The contrary is true, too. Thus, long-range (by using low frequency) underwater acoustic communications are quite difficult.

If the directivity of transmitting or receiving transducers in the vertical direction is sharpened, the virtual sources of high orders will be absent, and the total delay spread of the multipath effect will also be reduced, thus combating against ISI will be easier.

Provided the transmitting or receiving positions and communication ranges are changed, the virtual source chains are also different. Therefore, the multipath structures have different distributions. Therefore, the multipath effect exists as spatial variability.

In particular, provided that the duration of the transmitting pulse is greater than the time difference between adjacent sound pulses arriving at $P(r,z)$, the interference effect appears due to the multipath propagations. So, the underwater acoustic communication channels no longer satisfy the additive theorem and thus belong to a nonlinear system as viewed from signal processing. How to adapt this peculiarity in digital underwater acoustic signal processing will be discussed in Chapters 3 and 4.

2.3.3 Experimental Researches on Multipath Structures in Underwater Acoustic Communication Channels

Sound propagations in simplified layered media can be described by means of the normal mode theory or expressed by the virtual source images as discussed earlier. In particular, for a narrow pulse sound source, complicated and variable multipath structures appear.

In the case of actual underwater acoustic communication channels, in particular for shallow-water ones, where the complicated and variable $G_c(z)$ and the randomly fluctuating sea surface and bottom are encountered, the theoretical analyses for the multipath structures are extremely difficult. Therefore, experimental studies have an important significance.

The multipath structures measured in situ would be used as a reference to realize the adaptable channel match in the underwater acoustic communication engineering.

2.3.3.1 Multipath Structures in Deep-Sea Sound Channeling

To analyze the signal waveforms traveling in the deep-sea sound channeling at the distances being below 20 km, a single carrier frequency sound pulse has been employed in experiments [5]. Experimental results have demonstrated that received waveforms severely change with transmission distances.

The recoded sound pulse signals for the sound pulse traveling in the channeling at different distances are shown in Fig. 1.3A-E where the sound source and hydrophone were located at the depths of 75 and 80 m, respectively. The spreading time of signals will be enlarged when the distances are increased from

0.75 to 20 km. If the duration of pulse signal is greater than the time difference arriving at a hydrophone between adjacent pulses (wave packets), the phenomenon of wave superposition appears provided that the transmission distances are larger than 7.7 km. The final wave packets in the multipath structures are split from one into two, as shown in Fig. 1.3A and E, which are caused by the wave interference effect. The signal waveforms recorded at the sound channeling axis are remarkably different in comparison with those mentioned earlier. In this case, the amplitudes of the wave packets are determined by corresponding focusing factors, as shown in Fig. 1.3F, where the distance is 13.5 km.

The received waveforms in sound shadow zones existing at the area above the plane where the sound source is located have fully different, variable laws, as shown in Fig. 2.29, where the hydrophone was located at the depth of 25 m, and the transmission distance is 14 km. We can see that the signal with maximum amplitude does not arrive finally; moreover, split structures also appear due to the interference effect of multipath propagations.

To analyze the waveforms of a sound pulse traveling in the deep-sea sound channeling, a ray diagram has been plotted under the corresponding experimental conditions; moreover, arrival times and corresponding amplitudes, including the effects of focusing factor and the directivity index of transmitting transducer on amplitudes, have also been calculated. The calculated results agree well with experimental ones [5].

The multipath structures received from the three long-range (500–700 km) transmission paths in the northeast Pacific have been examined [11]. The hydrophones were all located on a sloping bottom at depths between 1200 and 1400 m, and the sound source depth was at 450 m in deep-water areas. Runge-Kutta methods were used to solve the ray equation of motion. It allows an arbitrary ray density in launch angle, and it identifies eigenrays by searching for rays whose depths bracket the hydrophone and which were adjacent at the source. Using this procedure, it was possible to identify most of the major pulses arriving at observed points.

The sound source was a vertical air-gun array suspended at a depth of 450 m from a drifting ship. The acoustic data were originally recorded in analog form on 1/2-in. magnetic tape. The source with a mid-frequency of 120 Hz was pulsed every 15 min. Note that the sound velocity had a minimum at the depth of about 500 m,

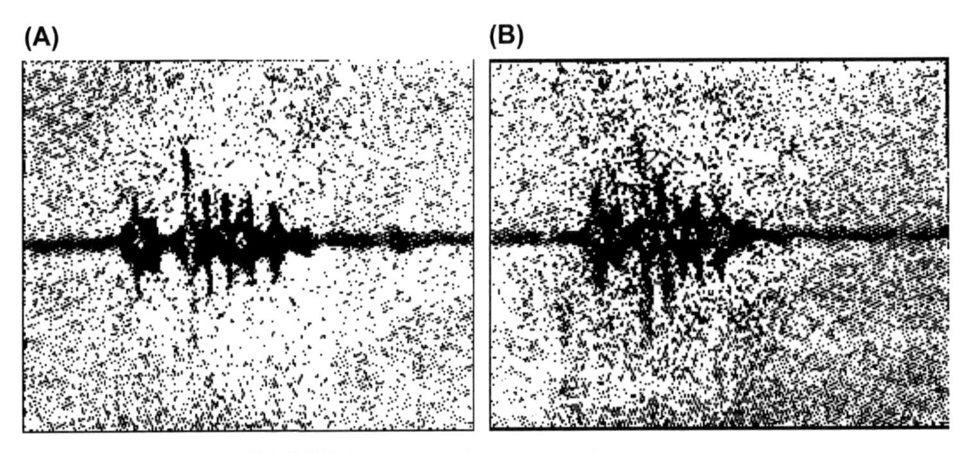

FIGURE 2.29 Received waveforms in shadow zone.

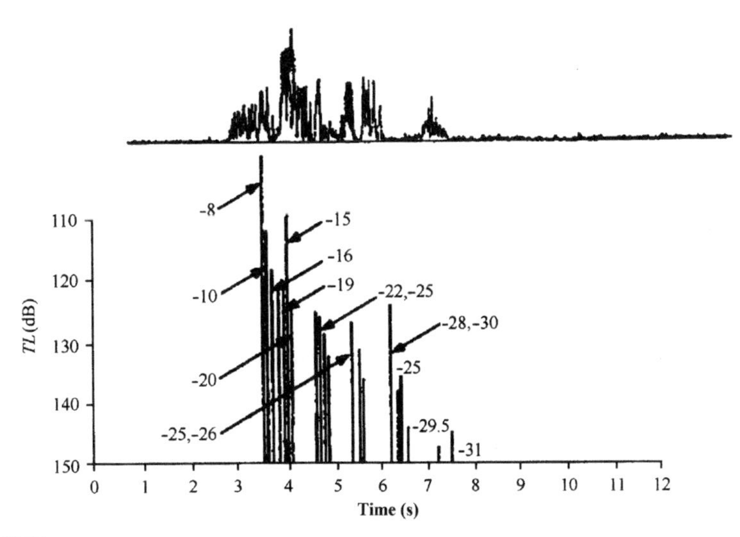

FIGURE 2.30 Observed and calculated data. Eigenrays are identified by grazing angles.

which is just below the source depth and is a typical sound–velocity profile to form a deep-sea sound channeling.

The observed and calculated data for station A at the range of 695 km are shown in Fig. 2.30.

The program identifies the four main ray bundles: (1) the −6.5 to −20 degree rays, (2) the −22 to −25 degree rays, (3) the −26 to −28.3 degree rays, and (4) the −29 to −31 degree rays. These wave packets are separated by about 1 s, and all have at least one bottom reflection; the lower levels of the later arrivals are due to greater bottom loss.

The propagation paths to station C have the ranges about 600 km. Results, as shown in Fig. 2.31, identify the −24.5 degree ray arriving first, distinctly earlier than the others. Then the three main bundles of rays arrive with the angle ranging from −0.3 to −14.7 degree, −23.5 to −27.7 degree, and −30 to −31.1 degree, respectively. The last groups of arrivals are identified as the −34 to −35 degree ray packet, which are bottom bounce rays and have the greatest transmission loss.

The computed eigenrays indeed reflect the observed signal structures both for the overall clumps of energy and the fine structure of spiky arrivals in each clump. The agreement is good, especially considering the variability caused by the ship's drift.

2.3.3.2 *Multipath Structures in Shallow-Water Acoustic Channel Existing Thermocline*

The long-range multipath identifications in deep-sea sound channeling have been introduced where the multipath structures are more stable.

In the case of shallow-water acoustic channels, on account of the effects of both the seawater medium itself and the boundaries, sound propagations become very complicated. In summer and autumn, a thermocline often exists in many shallow-water zones, and thus received signals have complicated structures because of an effect on the multipath propagations. Zhang had explained the multipath structures received in situ where a thermocline appears by means of the ray theory [12].

In the experiments, explosives detonated at 7 and 25 m were used as sources. The sea bottom was rather flat, the water depth was about

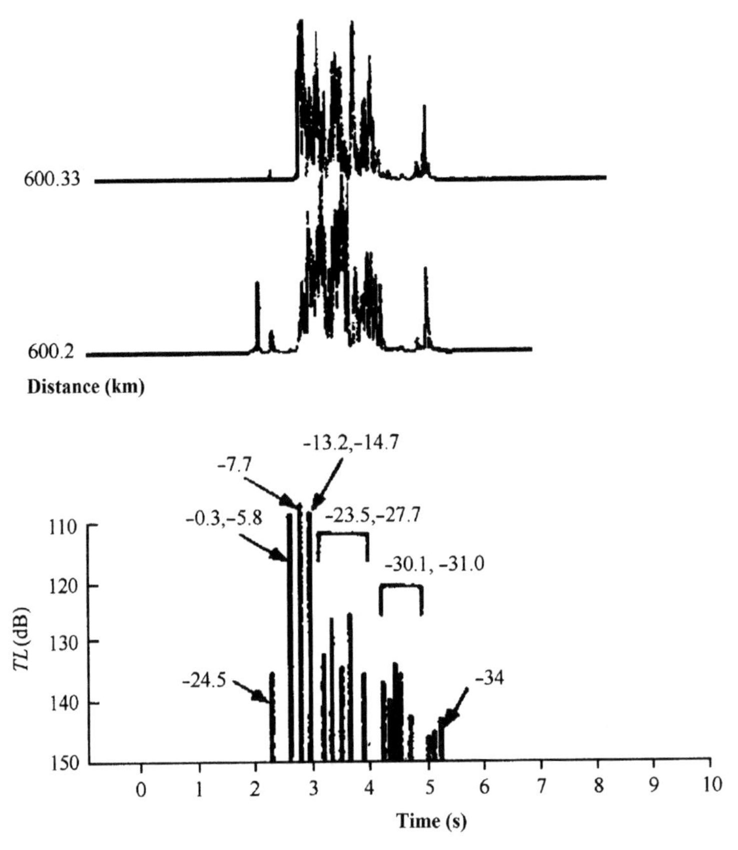

FIGURE 2.31 Observed and calculated data. Eigenrays are identified by grazing angles.

38 m, and the sound velocity profile was plotted in Fig. 2.32. The thermocline was at depths ranging from 11 to 15 m. The signals were received by a 32-element vertical array and a set of 2-element hydrophones. After filtering the original signals by a digital band-pass filter, the waveforms with multipath structure under various frequencies could be obtained.

Provided that the source and hydrophone were both located above the thermocline, the received waveforms are shown in Fig. 2.33A, where the transmission range is 3.5 km, and center frequencies are 3.2, 5.0, and 8.0 kHz, respectively. We can see that the received waveforms for different frequencies consist of a series of pulses (wave packets); the time interval T_1 between two adjacent pulses is about 7.5 ms.

Provided that the source depth was below the thermocline and the hydrophone was above that, the received waveforms are shown in Fig. 2.33B. The received waveforms are also composed of a series of pulses, but the specific structures are different, ie, besides the pulses that appear at about 7.5, 15, and 22.5 ms, respectively, as shown in Fig. 2.33A, some other pulses also appear at about 4, 11.5, and 19 ms, respectively: the time interval between two adjacent pulses $T_2 \cong 4.0$ ms, as shown in Fig. 2.33B.

Under the condition of a typical thermocline (see Fig. 2.34), the time intervals between adjacent pulses in multipath structures can approximately be calculated by using the ray acoustic theory. When the sound source and hydrophone were both located above the thermocline, a train

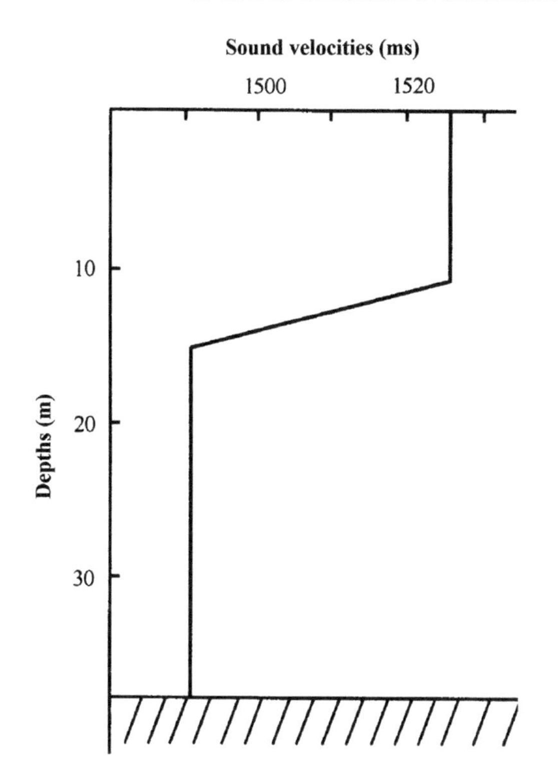

FIGURE 2.32 Sound velocity profile in experimental interval.

of approximately equi-spaced pulses will appear, and the arrival time interval between two adjacent pulses will be given by

$$T_1 \cong \frac{2H\sqrt{c_0^2 - c_1^2}}{c_0 c_1}\left(1 + \frac{2d}{3H}\right) \tag{2.141}$$

When the source depth was below the thermocline, while the hydrophone depth was above the thermocline, the time interval between the adjacent pulses is T_2 or $T_2 - T_1$, where T_2 is given by

$$T_2 \cong \frac{2z\sqrt{c_0^2 - c_1^2}}{c_0 c_1} \tag{2.142}$$

where z is the vertical distance between the hydrophone and the sea bottom.

According to the calculations from Eqs. (2.141) and (2.142), $T_1 = 7.25$ ms and $T_2 = 3.7$ ms, respectively, which are basically consistent with measured values (refer to Fig. 2.33).

The propagations of sound pulses in shallow-water acoustic channels with a thermocline had further been studied by Zhang [13], and the results of theoretical calculations have been compared with the measured data, including waveforms versus depths, ranges, and frequencies (refer to Section 2.3.4).

The waveforms versus depths are shown on the left-hand side of Fig. 2.35, where the transmission range is 1.9 km, the center frequency is 3 kHz, and the bandwidth is 1/3 octave. When the sound source and hydrophone were both located above the thermocline, the received waveform will be composed of a series of wave packets with a time interval to be 7.5 ms, ie, a comb structure appears. Provided that the sound source was located above the thermocline and the hydrophone was located below that, the wave packets are denser, and there is another packet that appears in the medium of 7.5 ms. When the sound source and hydrophone are both below the thermocline, sound energy will center, and the comb structure will disappear. Moreover, there exists a larger time delay (about 30 ms) in comparison with the two waveforms mentioned earlier (refer to Fig. 2.35).

When the sound source and hydrophone are both above the thermocline, the waveforms under different transmission distances are shown in Fig. 2.36, where the transmission distances are 1.9, 3.5, and 5.3 km, respectively, and the center frequency is 3 kHz. We see that the variations of waveform structures versus the distances are not remarkable, and the time intervals between the adjacent wave packets have not been changed. The measured data are consistent with the results obtained by theoretical calculations.

Provided that the source and receiver are both above the thermocline, the waveforms versus

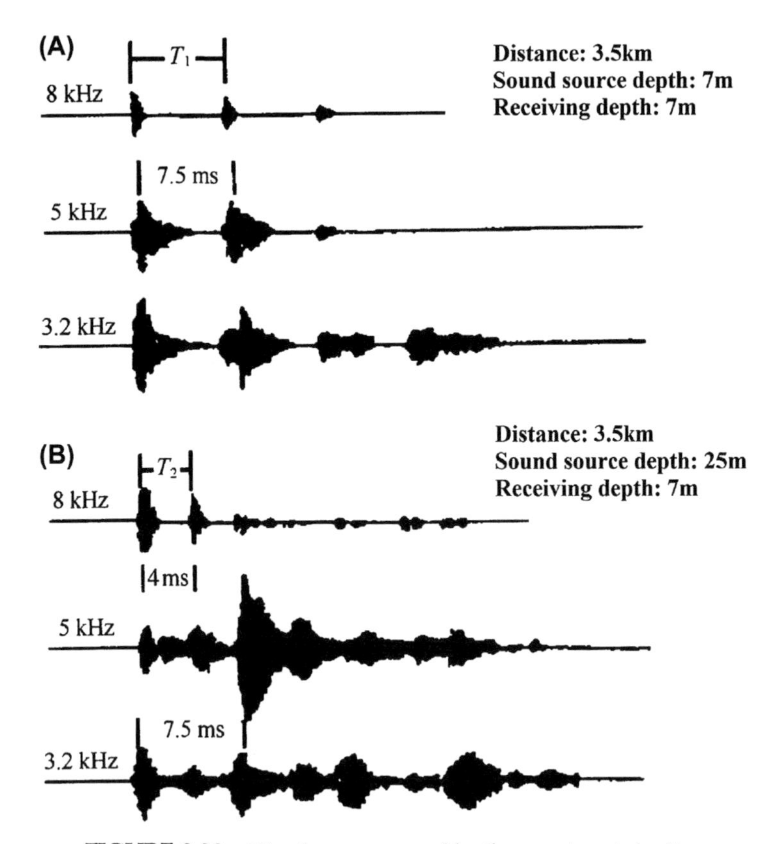

FIGURE 2.33 Waveforms measured by the experiments in situ.

frequencies are shown in Fig. 2.37. Three frequencies of 3, 5, and 7 kHz were employed in the experiments, and the transmission distance $r = 1.9$ km. We see that the waveforms have the similar comb structures for different frequencies, although they are slightly different in amplitudes. Moreover, the wave packets with maximum values will shift to the front parts of multipath structures with increasing frequencies.

It is valuable to note that the multipath structures are stable enough in some transmission circumstances.

The waveforms that are created by five different explosives at the same position with the interval of 5 s are shown in Fig. 2.38, where the depths of sound source and hydrophone were 7 and 8 m, respectively, the bandwidth is 900–1100 Hz, and $r = 2.62$ km. The stability of waveforms can strongly be reflected by the correlation coefficients of the received signals. In the previous example, the waveform correlation coefficients of any two signals are about 0.79–0.88, while the envelop correlation coefficients are up to 0.92–0.98. The results show that the pulse multipath structures are fairly stable. It is quite valuable to combat against the multipath interference, or even use its energy. Of course, the stability would be variable for different transmission circumstances. The experiments in the Xiamen Harbor have demonstrated that the envelop correlation coefficients are smaller than those mentioned earlier (refer to Chapter 4).

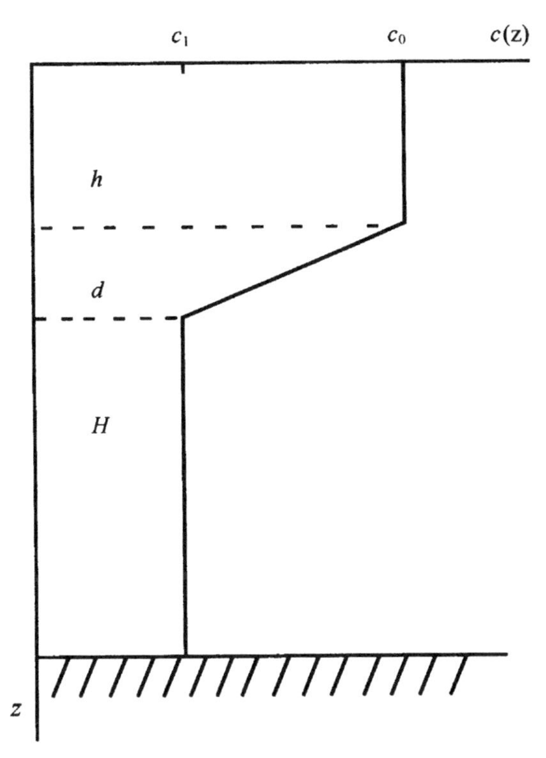

FIGURE 2.34 Simplified model of sound velocity gradient.

2.3.4 Impacts of Multipath Effects on Digital Underwater Acoustic Communications and the Possible Countermeasures to Adapt to the Effects

The multipath propagations are the most essential characteristic for underwater acoustic communication channels.

It is necessary to emphatically point out that the impacts of multipath effects on digital underwater acoustic communications possess duality.

Obviously, the multipath effects may be a principal obstacle to realizing robust digital underwater acoustic communications, in particular for higher data rate ones. However, provided effective signal processing approaches are adopted, the energy of multipath propagations would be used to improve the input SNR of communication receivers.

2.3.4.1 *Unfavorable Impacts of the Multipath Effects*

1. Based on the mechanism causing the multipath effects, we know that it is inefficient to combat against the multipath effects by raising sound power, using a usual band-pass filter, and selecting optimum operating frequencies. That is to say, these approaches cannot improve the input signal-to-multipath interference ratio.

   We also know that if reverberation interference levels are larger than noise levels *NL* in an active sonar, the detections of targets has to operate against the reverberation background. Similarly, if the multipath interference levels are larger than *NL*, communication sonars must also detect the information signals against the multipath interference background, which is possibly corresponding to a maximum communication range provided the efficient signal processing schemes suppressing the multipath interference are not adopted.
2. Multipath effect is a major cause of underwater acoustic channels becoming nonlinear.

   When underwater acoustic communications are carried out in shallow-water acoustic channels, or just beneath the sea surface in deep-sea ones, the multipath structures will be affected by the sound reflected from the sea boundaries. The duration of signal pulse τ_s may be longer than the time differences arriving at a hydrophone between direct-arriving sound pulse and multipath pulse due to sound reflection from them. In the case of deep-sea sound channeling, τ_s is possibly larger than the minimum time difference arriving at a hydrophone between adjacent sound pulses. In these cases, a nonlinear effect appears due to the multipath interference effect (refer to Figs. 1.3, 1.9 and 2.29). Therefore the underwater acoustic channels no longer

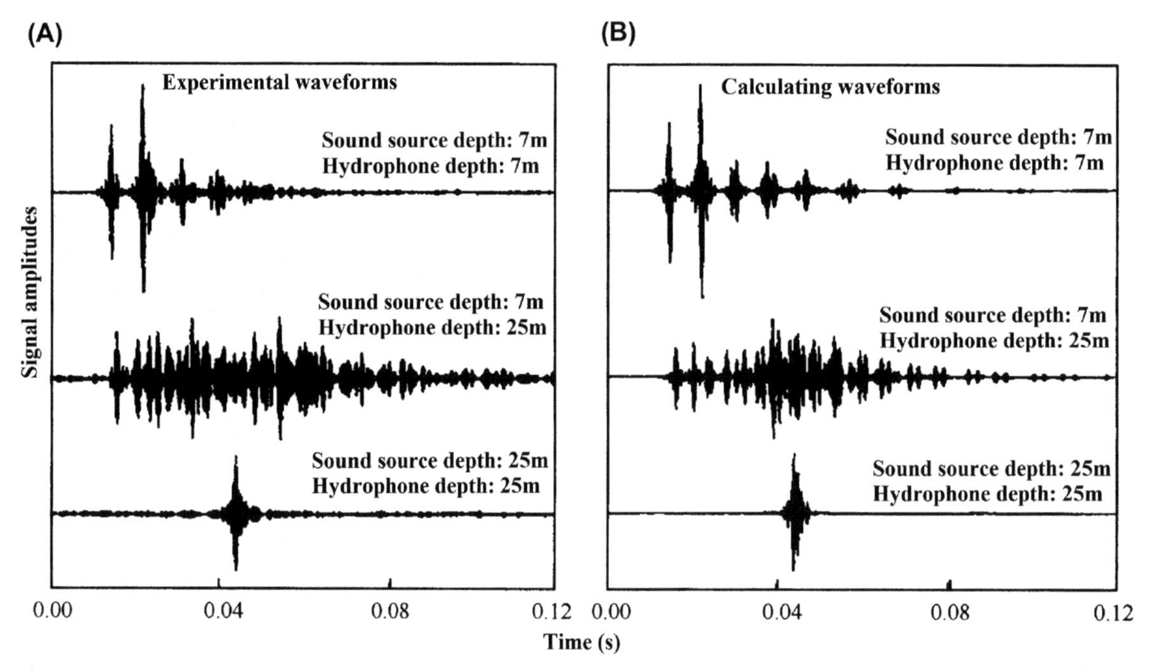

FIGURE 2.35 Wave structures versus depths acquired by experiments and theoretical calculations. (A) Experiment waveforms; (B) Calculating waveforms.

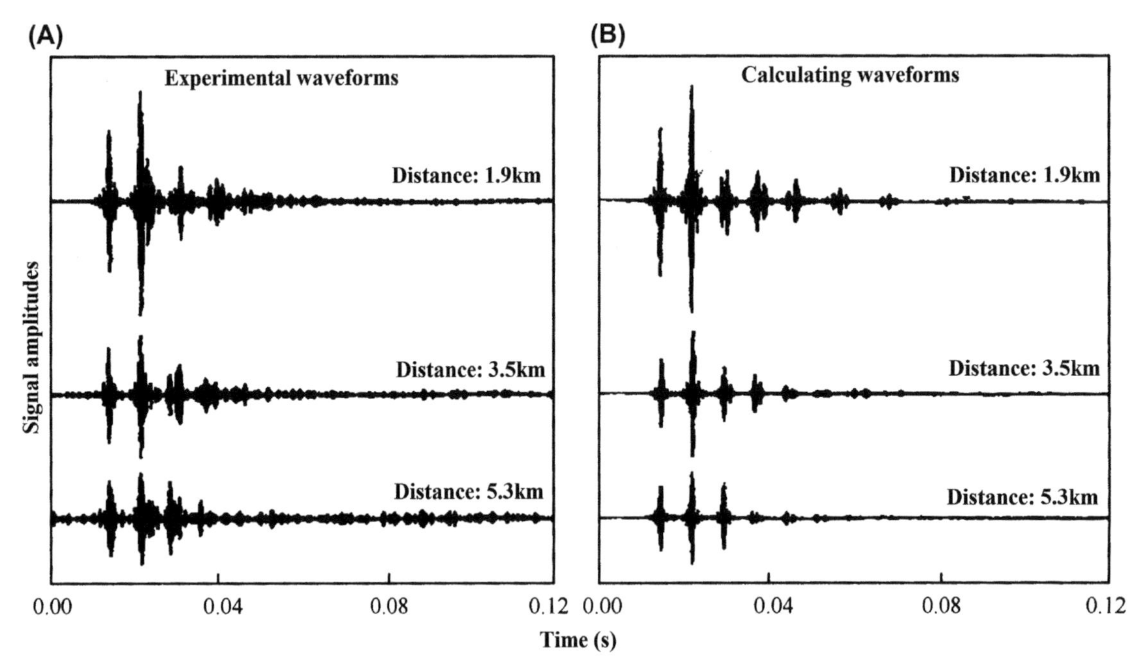

FIGURE 2.36 Wave structures versus distances acquired by experiments and theoretical calculations. (A) Experiment waveforms; (B) Calculating waveforms.

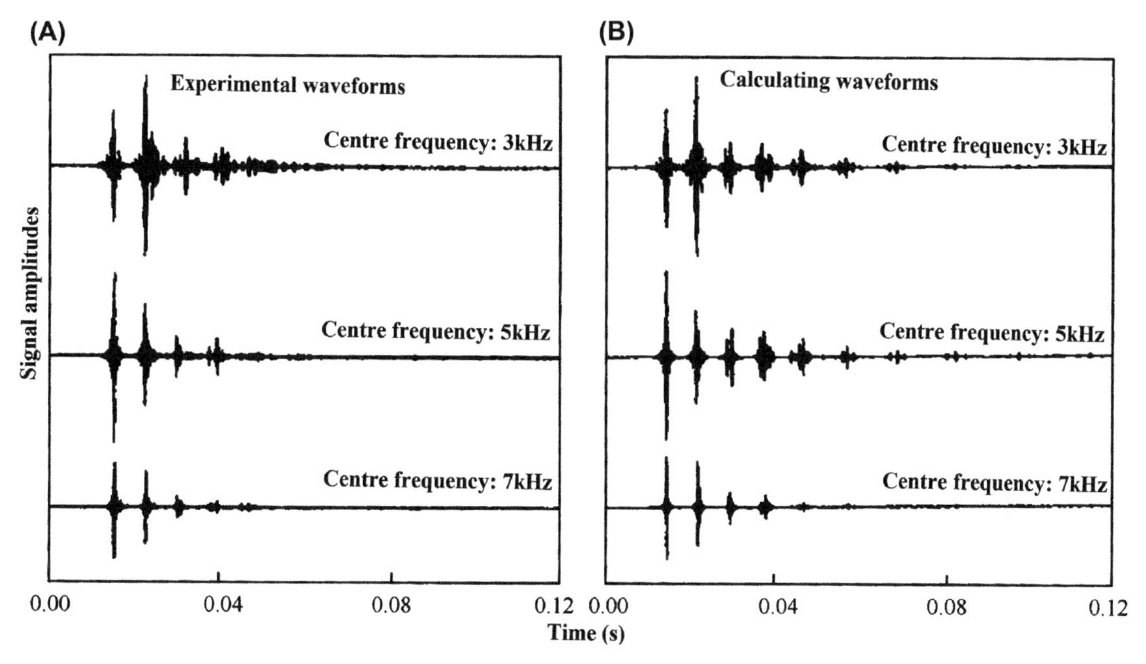

FIGURE 2.37 Wave structures versus frequencies acquired by experiments and theoretical calculations. (A) Experiment waveforms; (B) Calculating waveforms.

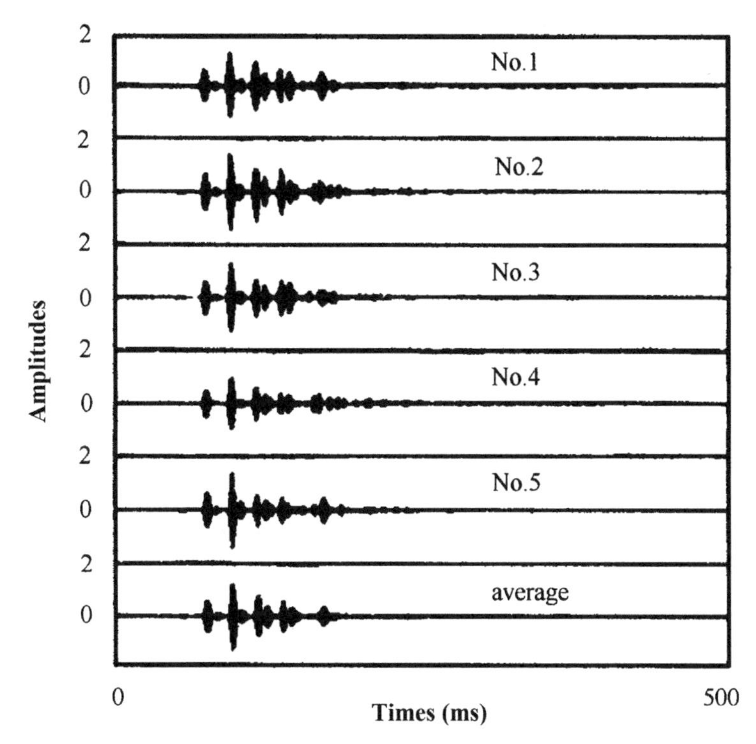

FIGURE 2.38 Received signals of different explosives at the same position.

satisfy the condition of a linear system from the point of view of signal processing. The performances of signal detections by using the matched filter or the cross-correlator will thus remarkably be reduced or even have their applications be restricted.

3. Combating against ISI caused by the multipath effects is a difficult but key task.

   A certain data rate R is expected in an actual underwater acoustic communication sonar. But the total delay spread of the multipath effects T_M may be lengthened up to several hundred milliseconds in some long-range underwater acoustic communications. Provided there is a simple scheme (ie, by selecting the repetitive period of pulse signals $T_0 > T_M$ to combat against ISI), and perhaps $R = 1/T_M$, it is possible that R is only several bps. Once $T_0 < T_M$ is selected to raise R, ISI will appear. In particular, the multipath pulse with maximum amplitude sometimes does not arrive at receiver at the first time. In other words, at the multipath of the first order (refer to Fig. 2.37), the signal decision by usually utilizing amplitude threshold is thus inefficient for the multipath structures with randomly distributing amplitudes.

4. The random spatial-temporal-frequency variability of multipath structures is an essential obstacle to combat against multipath interference or even use its energy.

   The complexity for multipath structures reflects the manifold property of marine communication circumstances. If the structures are invariable or the variable rates are slow enough in comparison with the operating time intervals of underwater acoustic communications, matching with underwater acoustic communication channels would principally be realized, although the complexity in designing corresponding signal processing schemes will increase. For example, the multipath structures in deep-sea acoustic channeling are also complicated while more stable, and excellent performance to match with the channeling has been realized [15].

   If the complexity of multipath structures is accompanied by a rapid and severe random spatial-temporal-frequency variability, it will be very difficult to suppress such a multipath interference, or even use its energy. In particular, in the case of mobile communications, channel parameters and operating conditions will randomly change in the whole mobile process, for example, if the positions of a sound source as well as hydrophone have a remarkable change in depth direction, their multipath structures are also different. In particular, when they both pass through the thermocline, the comb structure will disappear. Moreover, a large relative time delay up to several ten milliseconds appears (refer to Fig. 2.35). How to adapt to rapidly and severely variable multipath structures is also an extremely difficult problem.

5. From the point of view of digital signal processing of underwater acoustic communications, there are several kinds of multipath structures that are particularly difficult to adapt to:
   a. Wave packets in multipath structures have dense distributions. It is thus difficult to separate them and make corresponding signal processing to suppress them.
   b. Multipath structures are composed of a series of wave packets with approximately equal amplitudes, as shown in Fig. 2.39.

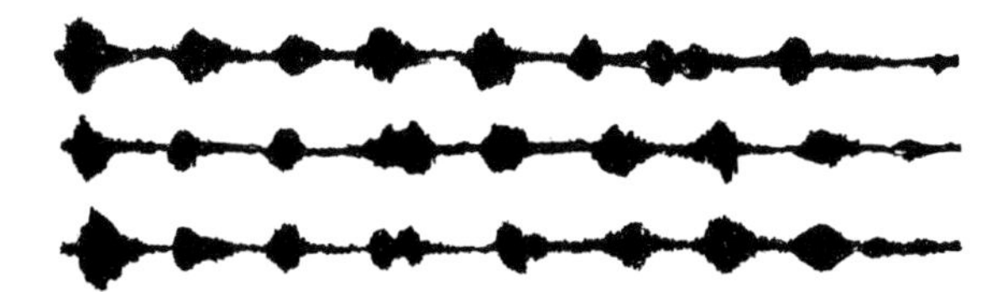

FIGURE 2.39 Multipath structures that have approximately equal amplitudes.

The signals are acquired from the Yellow Sea area with a thermocline [12], where the sound source and hydrophone were both located above the thermocline, and $r = 16$ km. To raise the R, signal pulses are generally transmitted successively; in this case, received signals would be superposed by all wave packets arriving at a hydrophone at the same time with the corresponding numbers (to be eight here) of wave packets, in which only one is expected to be detected.

c. Amplitude distributions of multipath structures are related to operating frequencies. When f is higher, for example above 15 kHz, the wave packet with maximum amplitude will generally appear at the multipath of the first order. But that will shift to the multipath of the higher orders with decreasing frequencies, as shown in Fig. 2.33B. Provided that a receiver has a wider bandwidth, such as in an SS system underwater acoustic communication (ie, in which there exist the multipath structures with remarkably different amplitude distributions), signal detections are quite difficult.

d. Operating frequencies must be selected to be lower for long-range underwater acoustic communications. Therefore, T_M will be lengthened, up to several hundred milliseconds. How to realize the robust communications at expected higher data rates is also an extremely difficult problem.

2.3.4.2 Possible Countermeasures to Adapt to the Unfavorable Impacts Due to the Multipath Effects

2.3.4.2.1 GENERAL APPROACHES

2.3.4.2.1.1 REDUCING DATA RATES R
Provided a specific communication sonar is permitted to operate at a lower R, such as in a text communication, the duration of signal pulse τ_s or pulse repetitive period T_0 may be selected to be larger than T_M. In particular, if an FH system is employed, we can select to set $n\tau_s > T_M$, where n are the numbers of frequency codes that are not repetitively used, and ISI caused by the multipath effects will disappear.

2.3.4.2.1.2 INCREASING OPERATING FREQUENCIES We know that T_M will reduce with increasing operating frequencies. In particular the wave packet with maximum amplitude will also shift to the front part of a multipath structure, or even at the multipath of the first order. Perhaps it has an amplitude greater than that of other wave packets, so the signal detection by using the amplitude threshold decision would be simple and efficient (refer to Section 3.2.1). Of course, increasing f means that the communication ranges will correspondingly be decreased.

2.3.4.2.1.3 SHARPENING THE DIRECTIVITY OF TRANSDUCERS Generally speaking, the positions of transmitting and receiving transducers for underwater acoustic communications are unknown to each other. Therefore, the transducer with omni-directivity in the horizontal direction would generally be selected; however, the vertical directivity may be designed to be sharp. So, the high-order virtual sound sources (refer to Section 2.3.2) may be thought of as being absent, and T_M is thus reduced.

In the case of fixed-point underwater acoustic communications, such as in underwater acoustic networks, the directivity of the transducer must carefully be designed, including adopting the array of hydrophones and adaptive beam forming techniques to obtain a suitable one, which not only will reduce T_M and NL but also will raise SL. Consequently, the required sound power is correspondingly lowered, which is quite available to adapt to low energy consuming conditions. In particular, the sound energy is provided by a better set operated under the water for a long time.

2.3.4.2.1.4 SELECTING AVAILABLE COMMUNICATION CIRCUMSTANCES AND OPERATING STATES Since the multipath structures have a spatial-temporal variability, if experimental seasons and sea areas are permitted to be selected, the performances of underwater acoustic communications will remarkably be improved.

If underwater acoustic communications are carried out in an approximately vertical direction in deep-sea regions, such as the communication between surface ship and deep-water AUV (autonomous underwater vehicle), the multipath interference will disappear by using a transducer with a suitable directivity.

2.3.4.2.2 POSSIBLE SIGNAL PROCESSING SCHEMES THAT MAY ADAPT TO MULTIPATH PROPAGATION CIRCUMSTANCES AT A CERTAIN RANGE

1. In the case of more stable or slowly varying multipath structures, it is efficient to use adaptive channel equalizers [15].
2. In the case of a weak multipath communication condition, such as for a short-range underwater acoustic communication in which high operating frequencies are used, not only T_M will be reduced, but also the wave packet of the first order has an amplitude greater than that of the following ones, and the amplitude decision would be efficient (refer to Section 3.2.1). In this case, it is unnecessary to use the complicated path diversity and valuable to establish civil communication sonars.
3. In the case of long-range, low-frequency underwater acoustic communications, the amplitudes of multipath structures are larger, and T_M is also increased. We propose to use a new adaptive anti-multipath scheme (refer to Chapter 3) to adapt to this communication situation at a certain range since that would correctly detect information signals that are superposed by the multipath interference by being based on the sparse characteristics of multipath structures. This idea is similar to sparse partial equalizers [14]. Of course, specific signal processing methods between them are different.

In addition, inspired by the iterative information exchange in Turbo decoders (referring to Chapter 4), a Turbo equalization technique has been developed. That has been shown to provide significant performance gains, even for severe ISI channels through iterative soft-input/soft-output equalization and decoding.

2.3.4.3 *Feasibility of Using the Multipath Energy*

1. Strictly speaking, $G_c(z) = 0$ is absent in the seawater media, therefore there is not a sound ray traveling along a straight line. The geometric operation range D (refer to Section 2.2) is only several kilometers for a sharp negative $G_c(z)$. The multipath pulses have been converted to the received signals in shadow zones, just as an active sonar receives the target signal that is reflected from sea bottom in the zones. In particular, the multipath energy will be used in long-range underwater acoustic communications where direct sound signal is absent due to the sound refraction effect [14].
2. We have known that multipath structures are caused by sound reflections from the sea boundaries and sound refraction in layered inhomogeneous seawater media, which can be thought of as time diversity signals. For example, the multipath structures shown in Fig. 2.39 are equivalent to eight time diversity signals, which will exist with fluctuations in time-amplitude domains when they travel in random inhomogeneous underwater acoustic channels (refer to Section 2.4). Provided we

can acquire these pulses and then combine them with an optimum model, we expect to obtain an optimum signal detection result under the criterion of maximum output SNR against the background of multipath propagations.

The signal processing scheme mentioned before is referred to as a path diversity or Rake receiver in radio communications. Obviously, the conventional Rake receiver (refer to Chapter 3) does not adapt to complicated and rapidly varying multipath structures encountered in underwater acoustic communication channels. However, if it is modified by an adaptive operating model, it will be expected to adapt to the multipath structures in the channels and obtain the excellent result to use the multipath energy (refer to Chapter 3). It is a key technique to improve the whole performance of underwater acoustic communications, in particular for long-range ones.

3. We can select the condition that satisfies the repetitive period of signal $T_0 > T_M$ for low data rate underwater acoustic communications. In this simple case, multipath pulses are equivalent to time diversity signals. For example, an underwater acoustic telecontrol communication sonar employed in an underwater acoustic releaser (refer to Section 4.2) permits selecting a low data rate equal to 20 bps; therefore, $T_0 = 50$ ms. We know that $T_M < 40$ ms in shallow-water acoustical channels for the operating frequencies to be above 20 kHz. The multipath energy is thus used efficiently (see Fig. 4.30).

Note that the acoustic time-reversal method, which is extended by optical phase conjugation, is a new signal processing system in underwater acoustic communications and close attention should be paid to that at present (see Section 1.3.3 and relative references). In the cases of relatively stable underwater acoustic channels and operating conditions, as communication between fixed points, it may be expected that this system would realize adaptive channel matches under the condition without any prior-known channel parameters, or even to use the multipath energy.

2.3.4.4 *Prediction of Multipath Structures*

Because the multipath structures have a sensitive spatial-temporal variability, to acquire ample data by applying the experiments on site is very difficult. However, we can adopt laboratory simulations to make up for the lack of those at a certain range. It is valuable for designing underwater acoustic communication equipment.

2.3.4.4.1 BELLHOP MODEL BASED ON GAUSSIAN BEAM TRACING METHOD

Simulating multipath structures represented by relative amplitudes for the sea area, as shown in Fig. 2.21, with the flat sea bottom (depth of 1100 m) is plotted in Fig. 2.40A. The sound source and hydrophone are both located at a depth of 100 m, and the transmission distance is 75 km. Those for variable depth (from 500 to 900 m) sea area are shown in Fig. 2.40B, where the hydrophone is located at 50 m, and the transmission distance is 52 km. The upper and lower simulating multipath structures in Fig. 2.40A and B correspond to typical sound velocity profiles in summer and winter, respectively. The total delay spreads of multipath propagations T_M for (a) and (b) both exceed 1 s, but the distributing characteristics of multipath structures between them are essentially different. The most multipath pulses for latter are separated from each other; therefore, it is easier to combat this kind of multipath interference by using adaptive anti-multipath signal processing schemes (see Chapter 3).

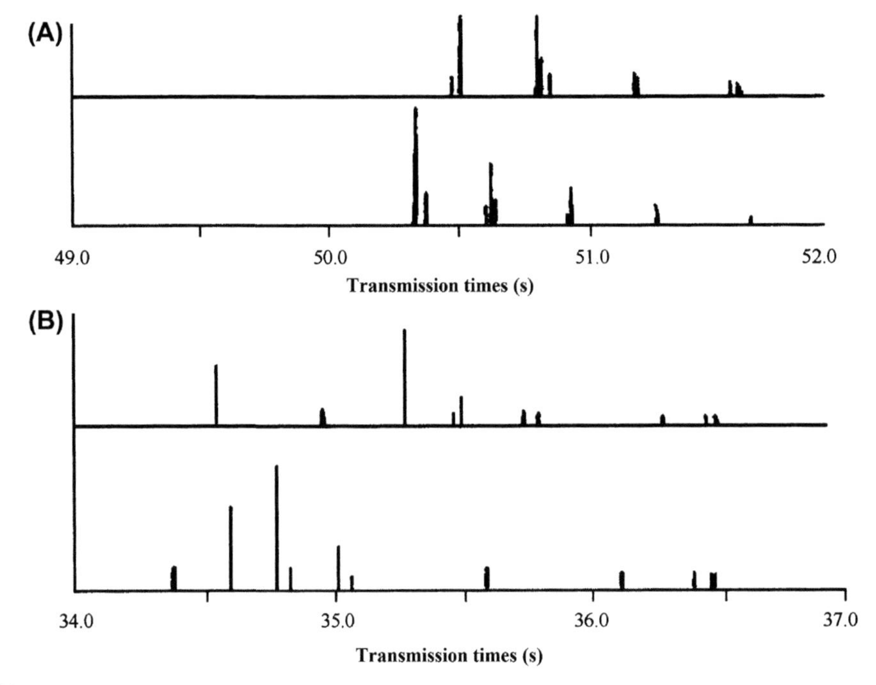

FIGURE 2.40 Simulating multipath structures ($f = 1$ kHz). (A) Correspond to typical sound velocity profiles in summer; (B) In winter.

2.3.4.4.2 NORMAL MODE THEORETICAL METHODS

Since the ray acoustic theory is only suitable for high-frequency underwater acoustic communications, it cannot give frequency-dependent waveforms. So, normal mode theory has gained more attention in recent years.

In some underwater acoustic channels, in particular the deep-sea sound channeling, inhomogeneity may be regarded as "low variable"; therefore WKBZ (Wenzel, Kramers. Brillouin, zhang) normal mode theory can be employed in the channeling, and more accurate and clear solutions will be obtained.

Zhang has calculated the waveform structures of pulses in underwater sound fields by means of a WKBZ normal mode method. The calculated waveforms for band-limited sound signals in the north Pacific Ocean sound channeling are shown in Fig. 2.41. The sound source and hydrophone were both located at the sound channeling axis (686 m), and the propagation distances are 250, 500, 750, and 1000 km, respectively. The values on the left-hand side of Fig. 2.41 represent received distances; the values on the right-hand side are the times corresponding to the positions appearing as peak pulses; the relative times appearing as signal waveforms are signed at the horizontal axis. To make convenient comparisons, the magnitudes of four waveforms in Fig. 2.41 have been multiplied by the factors of $1, \sqrt{2}, \sqrt{3}$, and 2, respectively.

From Fig. 2.41, we see that provided the transmitting and receiving depths are both at the sound channeling axis, the maximum amplitudes for the four transmission distances are all

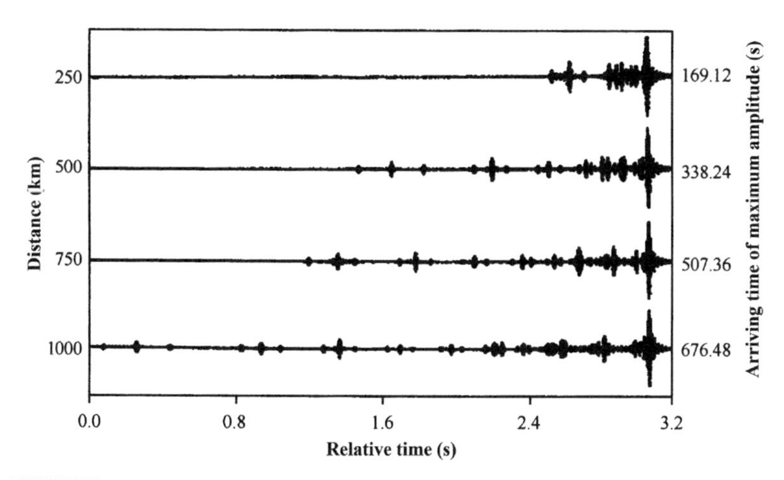

FIGURE 2.41 Multipath structures for band-limited signals at the four distances.

inversely proportional with the square root of the distances, ie, they obey the cylindrical spreading law. Moreover, the number of received pulses will increase with increasing propagation distances. The last arriving signals for the four distances correspond to the sound rays traveling along the sound channeling axis and have maximum amplitudes, while the first arriving signals correspond to the ray traveling along the paths with a maximum deviation from the sound channeling axis and thus have a maximum average sound velocity.

As mentioned before, provided the inhomogeneity of the seawater may be regarded as "low variable," we can utilize the WKBZ normal mode theory to make accurate predictions of multipath structures. In the case of an existing thermocline, sound reflections from that will obviously appear and have a remarkable effect on sound propagations when sound waves pass through it. Zhang [16] has calculated the pulse waveforms by using a beam-displacement ray-mode (BDRM) theory under the condition of an ideal thermocline. Assume that the sea surface is an absolutely soft boundary, there is an ideal negative thermocline in seawater itself; its density is ρ, the sound velocities at the upper and lower layers are c_0 and c_1, respectively, and the sea bottom is regarded as a homogeneous fluid half-space whose density, sound velocity, and absorption coefficients are ρ_b, c_b, and α, respectively. Since the effect of the sea bottom on sound propagations is only exhibited by the reflection coefficient, it is easy to extend that with layered structures. By using the cylindrical coordinate shown next, the sea surface at $z = 0$, and depth $z = h_0 + h_1$. The sound velocity profile is shown in Fig. 2.42.

Because the seawater is regarded as a two-layered media, the lower layer can be thought of as a homogeneous fluid sound channel, and the upper layer is considered as the external boundary of the channel. The sound pressure $P(\omega,r,z)$ (the time factor $e^{-i\omega t}$ has been neglected) at any point in the channel satisfies the Helmholtz equation:

$$\nabla^2 P + k_1^2 P = -\frac{1}{2\pi r}\delta(r)\delta(z - z_0) \qquad (2.143)$$

where $k_1 = \omega / c_1$, ω is the angular frequency. The term on the right-hand side of Eq. (2.143) expresses a broad band explosion sound source located at $r = 0$, and $z = z_0$. If $k_1 r \gg 1$, the sound field excited by this sound source with

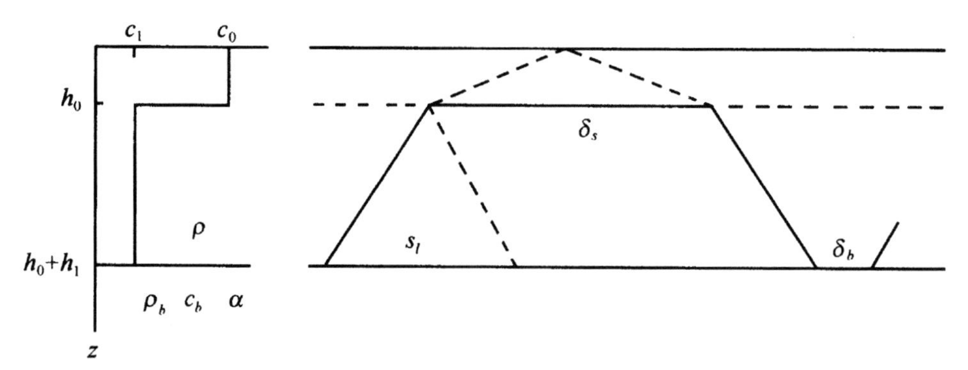

FIGURE 2.42 Shallow-water model with ideal negative thermocline.

unit intensity will be expressed by the sum of the normal modes as

$$P(\omega, r, z_0, z) = \sqrt{\frac{8\pi}{r}} e^{i\frac{\pi}{4}} \sum_l \psi(z_0, v_l)\psi(z, v_l)\sqrt{v_l} e^{iv_l r} \tag{2.144}$$

where ν_l are eigenvalues, and $\psi(z,\nu_l)$ are eigenfunctions. Their calculations are the key programs to realize the predictions of sound fields. According to Eq. (2.143) and boundary conditions, the eigenvalue equation, including the upper layer and sea bottom reflection coefficients V_s and V_b, is given by

$$1 - V_s(v)V_b(v)e^{i2h_1\sqrt{k_1^2 - v^2}} = 0 \tag{2.145}$$

where $V_s = e^{i\psi_s(v)}, V_b = |V_b(v)|e^{i\psi_b(v)}$. Substituting V_s and V_b into Eq. (2.145), we obtain

$$\begin{aligned} 2h_1\sqrt{k_1^2 - v_l^2} + \psi_s(v_l) + \psi_b(v_l) - i\ln|V_b(v_l)| \\ = 2l\pi \quad l = 0, 1, 2, \ldots \end{aligned} \tag{2.146}$$

where $-\ln|V_b(v_l)|$ is the sea bottom reflection loss. Generally, ν_l is a complex number, that is $\nu_l = \mu_l + i\beta_l$. The real part μ_l is the horizontal wave number of the normal mode of lth orders, while the imaginary part β_l is the exponential attenuation coefficient of the normal and satisfies $\beta_l \ll \mu_l$. By separating Eq. (2.146) into real and imaginary parts, we obtain

$$\begin{aligned} 2h_1\sqrt{k_1^2 - \mu_l^2} + \psi_s(\mu_l) + \psi_b(\mu_l) \\ = 2l\pi \quad l = 0, 1, 2, \ldots \end{aligned} \tag{2.147}$$

$$\beta_l = \frac{-\ln|V_b(\mu_l)|}{S_l(\mu_l) + \delta_s(\mu_l) + \delta_b(\mu_l)} \tag{2.148}$$

where $S(\mu_l) = \frac{2h_1\mu_l}{\sqrt{k_1^2 - \mu_l^2}}$ are the cycle distances of eigenrays of the normal modes in the water layer, $\delta_s(\mu_l) = -\frac{\partial\psi_s}{\partial\mu}\Big|_{\mu_l}$ are the beam-displacements of the eigenrays of the normal modes at the thermocline boundary surface, and $\delta_b(\mu_l) = -\frac{\partial\psi_b}{\partial\mu}\Big|_{\mu_l}$ are the beam-displacements of the eigenrays of the normal modes at the sea bottom. β_l can thus be expressed by $S_l(\mu_l)$, $\delta_s(\mu_l)$, $\delta_b(\mu_l)$, and the reflection coefficient of the sea bottom. So, β_l is related to reflection properties of the sea bottom, and it is also controlled by the sound velocity profile of the sea water.

The eigenfunctions of the normal modes for an ideal negative thermocline in a shallow-water sound field can be derived by passing through a series of calculations and are given by

$$\psi(z,\mu_l) = \begin{cases} \dfrac{\sqrt{2}\cos\dfrac{\psi_s}{2}\sin\left(z\sqrt{k_0^2-\mu_l^2}\right)}{\left(k_1^2-\mu_l^2\right)^{1/4}(S_l+\delta_s+\delta_b)^{1/2}\sin\left(h_0\sqrt{k_0^2-\mu_l^2}\right)} & (0 \le z \le h_0) \\ \dfrac{\sqrt{2}\cos\left[(z-h_0)\sqrt{k_1^2-\mu_l^2}+\dfrac{\psi_s}{2}\right]}{\left(k_1^2-\mu_l^2\right)^{1/4}(S_l+\delta_s+\delta_b)^{1/2}} & (h_0 \le z \le h_0+h_1) \end{cases} \tag{2.149}$$

The sound field of the normal mode for a harmonic point source may be obtained by calculating the eigenvalue ν_l and eigenfunction $\Psi(z,\nu_l)$. According to Eq. (2.144), the special variable (r,z_0,z) is regarded as the parameters of $P(\omega,r,z_0,z)$, ie, the frequency response function of sound channels. Assume the transmitting signal is $s(t)$, and its spectrum is $S(\omega)$, the received signal can therefore be expressed by applying a Fourier transform as

$$P(t,r,z_0,z) = \frac{1}{2\pi}\int_{-\infty}^{\infty} S(\omega)P(\omega,r,z_0,z)e^{-i\omega t}\mathrm{d}\omega \tag{2.150}$$

We may understand the effect of sound channels on sound fields and thus predict the characters of received signals by calculating $P(\omega,r,z_0,z)$.

The BDRM theory has been extended to approximate the actual thermocline [13].

Zhang et al. had observed regular multipath structures in a shallow-water channel with a thermocline and had calculated the waveform structures of pulse signals by using the beam-displacement ray modem theory.

A typical shallow-water thermocline is shown in Fig. 2.43, where h_b is the water depth. In this model, the sound speed is rapidly decreased in the thermocline from h_0 to h_1 and is slowly changed from h_1 to h_b.

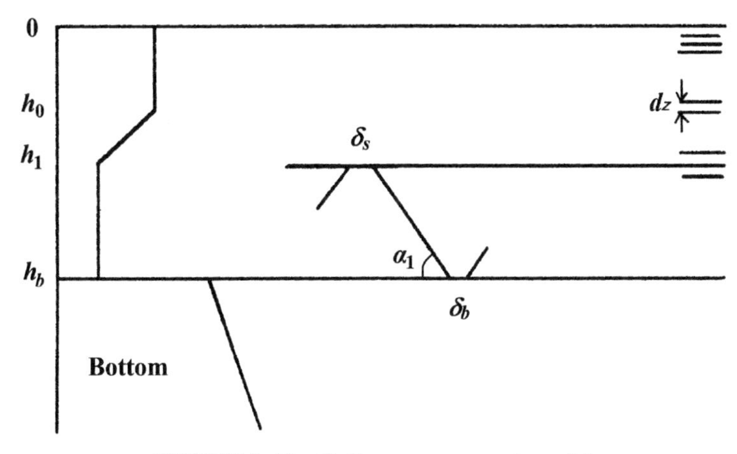

FIGURE 2.43 Shallow-water acoustic model.

By using the Fourier synthesis, the solution relative to a pulsed problem can be represented as

$$p(r, z_S, z; t) = \frac{1}{2\pi}\int_{-\infty}^{\infty} F(\omega)P(r, z_S, z; \omega)e^{-i\omega t}d\omega$$

where $F(\omega)$ is the Fourier transform of the source signal $S(t)$, and $P(r,z_s,z;\omega)$ is the solution of the Helmholtz equation for a harmonic point source. For the stratified model, Fig. 2.43, $P(r,z_s,z,\omega)$ can be expressed as the sum of the normal modes:

$$P(r, z_S, z; \omega) = \sqrt{\frac{8\pi}{r}}e^{i\frac{\pi}{4}}\sum_l \psi_l(z)\psi_l(z_S)\sqrt{\mu_l}e^{i\mu_l r-\beta_l r}$$

where μ_l and $\psi_l(z)$ are the horizontal wavenumber and eigenfunction of the normal mode, respectively. The mode attenuation coefficient β_l is calculated by the BDRM theory:

$$\beta_l = \frac{-\ln|V_s(\mu_l)V_b(\mu_l)|}{S(\mu_l)+\delta_s(\mu_l)+\delta_b(\mu_l)}$$

where $S(\mu_l)$, $\delta_s(\mu_l)$, and $\delta_b(\mu_l)$ are the cycle distance in the water layer, and the upper and bottom beam-displacements, respectively. The eigenfunction in the area from h_1 to h_b can be expressed as

$$\psi(z, \mu_l) = \sqrt{\frac{2}{S(\mu_l)+\delta_s(\mu_l)+\delta_b(\mu_l)}} \times \frac{\sin\left(-\int_{h_{1l}}^{z}\sqrt{\mu_l^2-k^2(z)}dz+\frac{\pi}{2}-\frac{\varphi_s}{2}\right)}{\left\{BE^{\frac{4}{3}}-DE^{\frac{2}{3}}\left[k^2(z)-\mu_l^2\right]+16\left[k^2(z)-\mu_l^2\right]^2\right\}^{\frac{1}{2}}}, h_l < z < h_b$$

where $B = 2.152$, $D = 1.619$, $E = \left|\frac{dk^2(z)}{dz}\right|$, and φ_s is the upper boundary reflection phase.

The eigenfunction in the area from 0 to h_1 can be calculated by the finite difference. According to the boundary condition at h_1, we can get the φ_s.

The signal waveforms calculated by means of BDRM theory have been compared with the experiments in situ, as shown in Figs. 2.35–2.37. The figures on the left-hand side of them represent experimental waveforms, while the figures on the right-hand side of them are obtained by applying theoretical calculations. We can see that the calculated waveforms are quite consistent with the experimental ones. It means that the predictions of multipath structures in underwater acoustic channels would be possible provided that the relative parameters have been understood in detail.

2.4 FLUCTUATION OF TRANSMITTED SOUND IN UNDERWATER ACOUSTIC COMMUNICATION CHANNELS [17]

We have discussed the basic laws of transmitted sound in the underwater acoustic communication channels that are regarded as time-invariant and layered media, including both the sea surface and sea bottom that are thought of as flat horizontal boundaries.

The fluctuations of transmitted sound in randomly inhomogeneous underwater acoustic communication channels, including rough boundaries, will be analyzed in this section, which will further tend to actual underwater acoustic communication channels.

The randomness of the channels is mostly caused by the following:

1. There are the blobs of inhomogeneity generated by turbulence in the body of the seawater, which are sometimes called thermal microstructures. They have different temperatures, or the index of refraction, and different scales randomly distributed in the body. When sound waves travel through this kind of media, the phenomena of random scattering and refraction will appear.
2. Random disturbances due to internal waves in the sea [1]. Because the sound field fluctuations caused by them have sufficiently long time scales, such as from several minutes to several hours, they are generally larger than the operating time intervals of underwater acoustic communications, and thus they will not be introduced here.
3. Sea surface as the upper boundary of underwater acoustic channels is randomly surging; moreover the sea bottom is also rough. The sound scattering from them is also a major reason causing the fluctuations of sound fields.

Obviously, the wandering of directional transducers is one of reasons causing signal fluctuations, in particular in civil underwater acoustic communications in which mechanical stabilization is generally not used.

Because of the existences of both random inhomogeneity in the body of the seawater and rough boundaries, a sound scattering effect appears. That is to say, sound energy will again be distributed in all directions. There are three objects that are valuable to be investigated for this effect:

1. A regular component of a sound field is reduced due to the sound scattering. The sound scattering in an inhomogeneous water body is a reason to cause the sound transmission loss *TL* as mention earlier; moreover, randomly rough boundaries will also cause sound energy to be lowered in the specific direction. If a rough boundary is regarded as a plane with equivalent reflection coefficient, it will also be lowered.
2. Because the sound mean field (regular component) is superposed by random sound scattering fields, the fluctuations of signal field appear, and coherence characteristics will correspondingly be weakened.
3. Because of sound backward scattering, the interference background of reverberation appears. It is a kind of interference that is difficult to overcome for active sonars. Similarly, forward scattering from the sea boundaries will cause violently fluctuating multipath structures.

How to adapt to the fluctuations of sound signals, including the severe signal fading due to the multipath interference mentioned in Section 2.3, is a very difficult problem in designing signal processing systems employed in digital underwater acoustic communications.

The basic law of the fluctuations of sound signals traveling in underwater acoustic communication channels will first be discussed in this section. Then, their impacts on digital underwater acoustic communications will be analyzed. Some efficient countermeasures to adapt to the fluctuations will also be included.

2.4.1 Fluctuation of Transmitted Sound Caused by the Random Inhomogeneity in the Body of the Seawater [6]

We have discussed sound propagations in the model of layered inhomogeneous and time-invariant channels. In fact, the index of refraction is a random variable in actual seawater medium. To describe its random characteristics, the mean square of the index of refraction is first introduced as

$$\overline{\mu^2} = \overline{\left(\frac{\Delta c}{c_0}\right)^2} \tag{2.151}$$

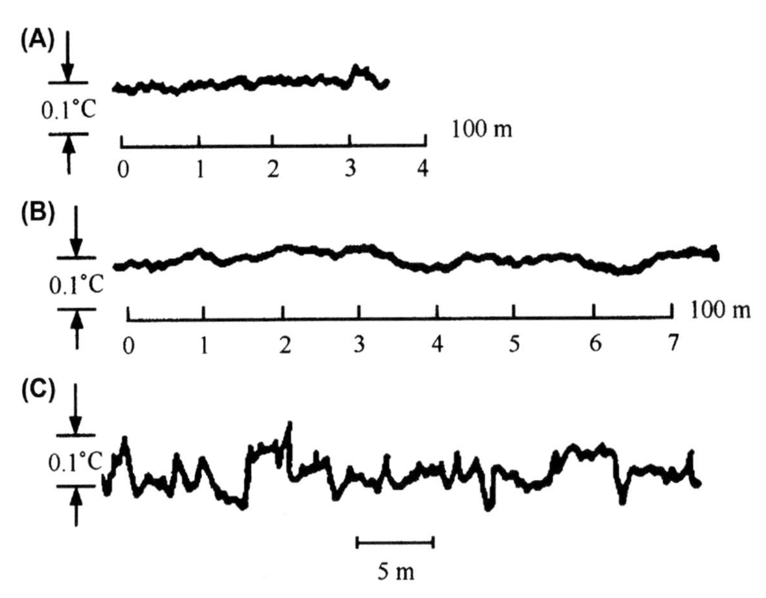

FIGURE 2.44 Records of the thermal microstructure.

where c_0 is the mean sound velocity, and $\overline{(\Delta c)^2}$ is the mean square deviation of sound velocity.

Experiments in situ have proven that the fluctuation of the index of refraction is mostly caused by the fluctuation of the temperature of the seawater, which is sometimes called thermal microstructure.

Fig. 2.44A and B show the records of the temperature fluctuation in a surface mixed layer with the depths to be about 5 m and at a main thermocline where the depth is 50 m, respectively, which were acquired by means of a sensitive submarine-mounted thermopile carried through the sea by the submarine's motion [5]. $\overline{\mu^2} = 8 \times 10^{-2}$ and the mean scale of inhomogeneity $a = 50$ cm in the mixed layer have been measured. The larger and longer fluctuations at the thermocline may associate with the internal waves. The other observations of the thermal microstructure made use of a recording thermometer with more rapid response. The record by using that is shown in Fig. 2.44C, where $\overline{\mu^2} = 5 \times 10^{-9}$, and $a \approx 60$ cm.

In general, the index of refraction μ in the seawater is a spatial-temporal random process:

$$\mu = \mu(\xi, \eta, \zeta, t)$$

where (ξ,η,ζ) are the coordinates of scattering points. Besides requiring one to find $\overline{\mu^2}$, the statistical characteristics of the spatial-temporal correlation must be analyzed as viewed from the signal processing since the characteristic of signal fluctuating fields would be one of major parameters for designing a digital communication sonar.

The general expression of spatial-temporal correlation function with respect to μ is given by

$$B(\mathbf{r}, t) = \overline{\mu(\xi_1, \eta_1, \zeta_1, t_1)\mu(\xi_2, \eta_2, \zeta_2, t_2)}$$

Analyzing a generalized nonstationary random process is a very complicated and difficult problem. Provided μ can approximately be regarded to be a stationary in time process, then $B(\mathbf{r},t)$ only depends on time difference τ:

$$B(\mathbf{r}, t) = \overline{\mu(\xi_1, \eta_1, \zeta_1, t)\mu(\xi_2, \eta_2, \zeta_2, t - \tau)}$$

If the spatial statistical characteristics of μ are still homogeneous, $B(\mathbf{r},t)$ is only related to the distance ρ between two received points:

$$B(\rho,t) = \overline{\mu(\xi_1 - \xi_2, \eta_1 - \eta_2, \zeta_1 - \zeta_2, t)\mu(\xi_1 - \xi_2, \eta_1 - \eta_2, \zeta_1 - \zeta_2, t - \tau)}$$

Under some specific operating conditions of underwater acoustic communications that are only necessary to consider spatial correlativity or time correlativity, respectively, such as a single transmitting-receiving transducer commonly employed in a communication sonar, in such a case, we only require analyzing the time correlativity and the corresponding time autocorrelation function:

$$B(t) = \overline{\mu(t)\mu(t-\tau)}$$

Similarly, the spatial correlativity is only considered in some specific communication fields, such as in spatial diversity techniques. In such cases, $B(\rho,t)$ reduces to the spatial correlation function:

$$B(\rho) = \overline{\mu(\xi_1 - \xi_2, \eta_1 - \eta_2, \zeta_1 - \zeta_2, t)^2}$$

The two types of correlation coefficients are usually used for theoretical investigations. One is exponential:

$$R(\rho) = e^{-|\rho|/a} \tag{2.152}$$

The other is Gaussian:

$$R(\rho) = e^{-\rho^2/a^2} \tag{2.153}$$

Of the two, the latter is the more physically reasonable since the former involves discontinuous changes in temperature fluctuations.

2.4.1.1 *Sound Scattering in a Weak Randomly Inhomogeneous Medium [17]*

Besides the sound velocity in the seawater being a random variable, its density is also a spatially random function. Since the fluctuation of the sound velocity is much larger than that of the density, the sound scattering caused by the fluctuation of the density can therefore be neglected.

Sound pressure p satisfies the wave equation:

$$\frac{1}{c^2}\cdot\frac{\partial^2 p}{\partial t^2} - \nabla^2 p = 0 \tag{2.154}$$

It is difficult to find the general solution of Eq. (2.154). Provided the seawater may be considered as a weakly random inhomogeneous medium, we can use an infinitesimal disturbance method to find that.

When the random fluctuation of the sound velocity Δc is much less than the mean sound velocity c_0, that is $c = c_0 + \Delta c$, and $\Delta c \ll c_0$, Eq. (2.154) becomes

$$\nabla^2 p - \frac{1}{c_0^2}\left(1 - \frac{2\Delta c}{c_0}\right)\frac{\partial^2 p}{\partial t^2} = 0 \tag{2.155}$$

Correspondingly,

$$p = p_0 + p_1 \quad p_1 \ll p_0 \tag{2.156}$$

where p_0 and p_1 are the definite and random parts of p, respectively.

Substituting Eq. (2.156) into Eq. (2.155), we get

$$\nabla^2 p_0 + \nabla^2 p_1 - \frac{1}{c_0^2}\left(1 - \frac{2\Delta c}{c_0}\right)\left(\frac{\partial^2 p_0}{\partial t^2} + \frac{\partial^2 p_1}{\partial t^2}\right) = 0$$

$$\nabla^2 p_0 - \frac{1}{c_0^2}\frac{\partial^2 p_0}{\partial t^2} = 0$$

p_0 satisfies the following wave equation:

The scattering wave of one order approximation p_1 thus satisfies the following wave equation:

$$\nabla^2 p_1 - \frac{1}{c_0^2}\frac{\partial^2 p_1}{\partial t^2} = -\frac{2\Delta c}{c_0^3}\frac{\partial^2 (p_0 + p_1)}{\partial t^2}$$

The factor $\frac{\partial^2 p_1}{\partial t^2}$ on the right-hand side of this expression may be neglected according to the assumption of infinitesimal disturbance. Therefore,

$$\nabla^2 p_1 - \frac{1}{c_0^2}\frac{\partial^2 p_1}{\partial t^2} = -\frac{2\Delta c}{c_0^3}\frac{\partial^2 p_0}{\partial t^2} \quad (2.157)$$

Introduce a sign Q and let $4\pi Q = (2\Delta c/c_0^3)(\partial^2 p_0/\partial t^2)$, and Eq. (2.157) becomes

$$\nabla^2 p_1 - \frac{1}{c_0^2}\frac{\partial^2 p_1}{\partial t^2} = -4\pi Q \quad (2.158)$$

It means that individual inhomogeneous medium elements have become the secondary scattering sources generating scattering waves p_1 under the action of the primary wave p_0, and the Q is thus called source intensity. Provided p_0 is replaced by the incident plane wave traveling along x axis,

$$p_0 = A_0 e^{j(\omega t - kx)} \quad k = \omega/c_0 \quad (2.159)$$

We obtain

$$\nabla^2 p_1 + k^2 p_1 = \frac{2k^2\Delta c}{c_0} A_0 e^{j(\omega t - kx)} \quad (2.160)$$

This is a nonhomogeneous wave equation, and its solution is

$$p_1 = -\frac{A_0}{4\pi}\int_V \left(\frac{2k^2\Delta c}{c_0}e^{-jk\xi}\right)\frac{e^{-jkr}}{r}\mathrm{d}V \quad (2.161)$$

Eq. (2.161) can also be written as

$$p_1 = \frac{k^2 A_0}{2\pi}\int_V \mu(\xi,\eta,\zeta,t)\frac{e^{-jk(\xi+r)}}{r}\mathrm{d}V \quad (2.162)$$

We can see that the scattering sound pressure p_1 in Eqs. (2.161) and (2.162) is expressed by the volume integrals of the fluctuation of the sound velocity and the deviation of the index of refraction, respectively. The limits of integral would include the whole volume of an inhomogeneous medium generating the sound scattering wave p_1. Assume the inhomogeneity of the medium

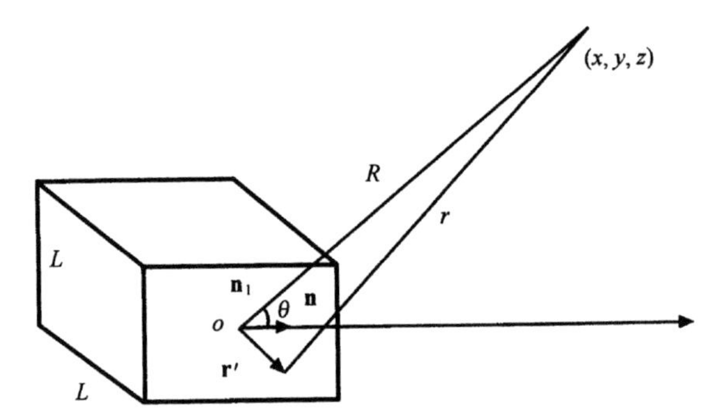

FIGURE 2.45 Scattering diagram generated by an inhomogeneous quadrate.

is restricted in a quadrate with the sides L sufficiently large, as shown in Fig. 2.45, where the center of the quadrate is at the original point of coordinate O, R at the distance from original point O to the observing point (x,y,z), n is the direction of the incident sound wave, n_1 is the direction of the scattering wave, θ is the scattering angle, r' is the vector radius of the scattering point (ξ,η,ζ), and r is the distance from scattering point to the observing point. Therefore,

$$R = \sqrt{x^2 + y^2 + z^2}$$

$$r = \sqrt{(x-\xi)^2 + (y-\eta)^2 + (z-\zeta)^2}$$

When L $\gg a$ and $r \ll L$, we have

$$\overline{|p_1|^2} = \frac{A_0^2 k^4 \overline{\mu^2} V}{4\pi^2 R^2}\int_V R(\rho)e^{-jk(n-n_1)\bullet(r_1'-r_2')}\mathrm{d}V$$

where $(r_1' - r_2')$ is the vector radius between two scattering points.

If we know that the spatial correlative coefficient $R(\rho)$, scattering sound intensity $\overline{|p_1|^2}$ will be obtained.

In the case of $R(\rho) = e^{-\rho/a}$, we have

$$\overline{|p_1|^2} = \frac{2A_0^2 k^4 a^3 \overline{\mu^2} V}{\pi R^2 \left(1 + 4k^2 a^2 \sin^2\frac{\theta}{2}\right)^2} \quad (2.163)$$

If $R(\rho) = e^{-\rho 2/a2}$, we will get

$$\overline{|p_1|^2} = \frac{A_0^2 k^4 a^3 \overline{\mu^2} V}{4\sqrt{\pi}R^2} e^{-k^2 a^2 \sin^2\frac{\theta}{2}} \tag{2.164}$$

We can see that by examining Eqs. (2.163) and (2.164), the more violent fluctuation of the temperature and the higher the operating frequencies, the more intense the scattering sound field will be. Moreover, the farther the distance between a received point and scattering volume, the weaker the scattering signal becomes. In general, the sound scattering caused by the blobs of the inhomogeneity possesses directivity. In the case of the microscale blobs of the inhomogeneity, that is $ka \ll 1$, and $k^2a^2 \sin(\theta/2) \cong 0$, $\overline{|p_1|^2}$ is thus independent of θ. In other words, the sound scattering is approximately isotropic. In the case of the blobs of the inhomogeneity with huge scales (ie, $ka \gg 1$), the sound scattering has a sharp directivity.

2.4.1.2 Fluctuations of Sound Signals in Weakly Random Inhomogeneous Media

We know that the total sound field is the sum of the initial incident wave and the scattering wave due to the presence of the temperature random inhomogeneity, thus the sound field is also a randomly fluctuating field. Now, we will discuss the relations between fluctuations of media and that of sound signals by means of the method of Petov infinitesimal disturbance.

Assume a plane wave,

$$p_0 = A_0 e^{j(\omega t - kx)} \tag{2.165}$$

that travels from the left half-space to the right half-space in the direction of the x axis, as shown in Fig. 2.46. The medium is homogeneous in the field of $x < 0$ and is inhomogeneous in the field of $x > 0$. The observing point is in the right half-space. Let the total sound pressure be

$$p = A(r)e^{j[\omega t - S(r)]} \tag{2.166}$$

where $A(r)$ and $S(r)$ are the amplitude and phase of the sound pressure at the observing point, which are unknown functions. Writing the sound pressure by another expression,

$$p = A_0 e^{j[\omega t - \Psi(r)]} \tag{2.167}$$

Let Eq. (2.166) equal Eq. (2.167), and we get

$$\Psi(r) = S(r) + j\ln[A(r)/A_0] \tag{2.168}$$

The real and imaginary parts of the function $\Psi(r)$ represent the phase and the logarithmic value of amplitude ratio of the sound pressure, respectively. Provided $\Psi(r)$ may be found, the fluctuations in the phase and amplitude of sound pressure can thus be determined.

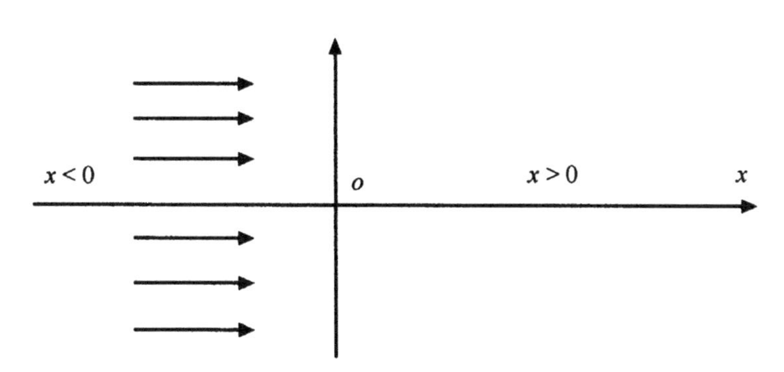

FIGURE 2.46 Plane wave traveling to right half-space with random inhomogeneity.

Substituting Eq. (2.167) into the wave Eq. (2.154), we obtain

$$(\nabla \Psi)^2 + j\nabla^2 \Psi = (\omega/c)^2 = n^2 k^2 \quad (2.169)$$

where $n = c_0/c$, and $k = \omega/c_0$. Taking a zero order approximation, that is $\Psi = \Psi_0$, it satisfies the wave equation at $n = 1$ for a homogeneous media. Therefore,

$$(\nabla \Psi_0)^2 + j\nabla^2 \Psi_0 = k^2 \quad (2.170)$$

Let $\Psi = \Psi_0 + \Psi'$, Ψ' be an infinitesimal disturbance quantity, which is caused by the fluctuation of media. Moreover, considering $n = 1 + \mu$, we may find the following expression by referring to Eqs. (2.169) and (2.170):

$$2(\nabla \Psi_0 \cdot \nabla \Psi') + j\nabla^2 \Psi' = 2\mu k^2 + \left[\mu^2 k^2 - (\nabla \Psi')^2\right]$$

If $\frac{1}{k}\nabla \Psi'$ and μ have the same order of magnitude, that is $\frac{1}{k}|\Delta \Psi'| \ll 1$. The terms in the bracket in this expression will therefore be neglected, and this expression can approximately be expressed as

$$2(\nabla \Psi_0 \cdot \nabla \Psi') + j\nabla^2 \Psi' = 2\mu k^2 \quad (2.171)$$

Introduce a new function W and let $\Psi' = e^{jkx} W$. By substituting $\Psi_0 = kx$ and Ψ' into Eq. (2.171), we get

$$\nabla^2 W + k^2 W = -j2\mu k^2 e^{-jkx} \quad (2.172)$$

Eq. (2.172), just like Eq. (2.160), is a nonhomogeneous wave equation, and its solution is

$$W = \frac{jk^2}{2\pi} \int_V \frac{1}{r} e^{-jk(\xi + r)} \mu(\xi, \eta, \zeta, t) dV$$

We can find

$$\Psi' = -\frac{jk^2}{2\pi} \int_V \frac{1}{r} e^{-jk[r-(x-\xi)]} \mu(\xi, \eta, \zeta, t) dV \quad (2.173)$$

where r is the distance between scattering point and observing point. Since

$$\Psi'(r) = \Psi(r) - \Psi_0 = S(r) - \Psi_0 + j\ln\left(\frac{A(r)}{A_0}\right) \quad (2.174)$$

the real and imaginary parts in Eqs. (2.173) and (2.174) will equal each other. Therefore,

$$\Delta\varphi = S(r) - \Psi_0 = \frac{k^2}{2\pi} \int_V \frac{\sin k[r-(x-\xi)]}{r} \mu(\xi, \eta, \zeta, t) dV \quad (2.175)$$

$$E = \ln\left(\frac{A(r)}{A_0}\right) = \frac{k^2}{2\pi} \int_V \frac{\cos k[r-(x-\xi)]}{r} \mu(\xi, \eta, \zeta, t) dV \quad (2.176)$$

where $\Delta\varphi$ is the phase fluctuation, though amplitude $A(r) = A_0 + \Delta A$. Provided the values of the fluctuation in amplitude are small enough, that is $\Delta A \ll A_0$, we have $E = \ln[A(r)/A_0] \approx \Delta A/A_0$ that expresses the fluctuation of relative amplitudes.

We will further simplify both Eqs. (2.175) and (2.176) next.

Assume the incident wave is a plane wave beam. Since the operating frequencies f employed in digital underwater acoustic communications are commonly higher, the scales of the inhomogeneous blobs in the seawater are also greater, and the condition of $ka \gg 1$ is generally satisfied. For example, take $f = 10$ kHz and $a = 60$ cm, and thus $ka = 25$. Therefore the sound scattering due to the temperature inhomogeneity has a sharp directivity. For an incident plane wave with a limited beam, the inhomogeneous volume that causes the sound scatting and signal fluctuations at receiving point $M(L,0,0)$, in fact, may be confined in a cone, as shown in Fig. 2.47.

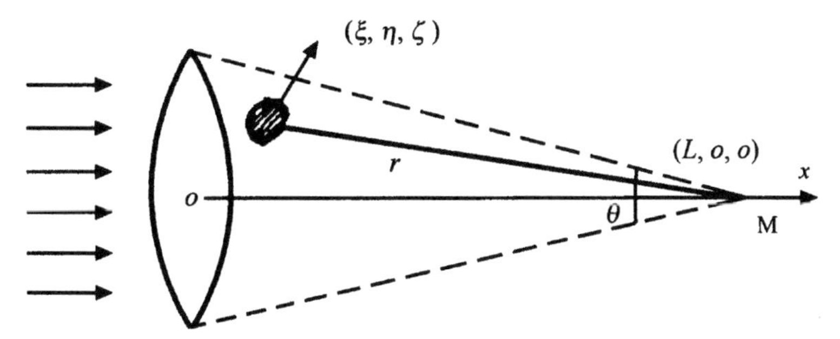

FIGURE 2.47 Conic scatting volume formed by directional sound scattering.

Now, the integral variable ξ is in the integral between limits 0 and L. The limits of integral variables both η and ζ are smaller than that of ξ. So, Eqs. (2.175) and (2.176) may be simplified as

$$\Delta\varphi = \frac{k^2}{2\pi}\int_0^x \times \iint_{-\infty}^{\infty} \frac{\sin\frac{k^2(\eta^2+\zeta^2)}{2(x-\zeta)}}{x-\zeta}\mu(\xi,\eta,\zeta,t)\mathrm{d}\xi\mathrm{d}\eta\mathrm{d}\zeta \tag{2.177}$$

$$E \equiv \frac{\Delta A}{A_0} = \frac{k^2}{2\pi}\int_0^x \times \iint_{-\infty}^{\infty} \frac{\cos\frac{k^2(\eta^2+\xi^2)}{2(x-\xi)}}{x-\xi} \times \mu(\xi,\eta,\zeta,t)\mathrm{d}\xi\mathrm{d}\eta\mathrm{d}\zeta \tag{2.178}$$

Since μ is a random function, both $\Delta\varphi$ and E are also random ones. Field measurements have demonstrated that the fluctuation in temperature is approximately homogeneous in space and approximately stationary in time in general sea areas. Therefore μ, and thus $\Delta\varphi$ and E, are all the random processes that are homogeneous in space and stationary in the time domain. Provided the random functions describing the processes are known, we may obtain some statistical characteristics of both $\Delta\varphi$ and E.

2.4.1.2.1 FLUCTUATING MEAN SQUARE VALUES OF THE PHASE AND AMPLITUDE OF SOUND SIGNALS

The fluctuating intensity of sound signals caused by the fluctuation of the seawater temperature and its variable laws will first be discussed.

The mean square values of amplitude $\overline{E^2}$ will first be found; the calculation of $\overline{\Delta\varphi^2}$ is similar to that of $\overline{E^2}$.

We have assumed that the fluctuation of μ is caused by that of the temperature and whose fluctuating characters may approximately be considered as isotropy. Therefore,

$$\overline{\mu(\xi_1,\eta_1,\zeta_1,t)\mu(\xi_2,\eta_2,\zeta_2,t)} = \overline{\mu^2}R(\rho) \tag{2.179}$$

where $R(\rho)$ is the spatial correlation coefficient of the index of refraction, and $\rho = \sqrt{(\xi_1-\xi_2)^2+(\eta_1-\eta_2)^2+(\zeta_1-\zeta_2)^2}$ is the distance between two receiving points.

According to Eq. (2.178), we obtain

$$\overline{E^2} = \frac{k^4\overline{\mu^2}}{4\pi^2}\int_0^x\int_0^x\iint_{-\infty}^{\infty}\iint \times \frac{1}{(x-\xi_1)(x-\xi_2)}\cos\frac{k(\eta_1^2+\xi_1^2)}{2(x-\xi_1)} \times \cos\frac{k(\eta_2^2+\xi_2^2)}{2(x-\xi_2)}R(\rho)\mathrm{d}\xi_1\mathrm{d}\xi_2\mathrm{d}\eta_1\mathrm{d}\eta_2\mathrm{d}\zeta_1\mathrm{d}\zeta_2 \tag{2.180}$$

In the case of $R(\rho) = e^{-\rho^2/a^2}$, by applying the complicated calculations of sextuple integrals, we can find the mean square values of both $\Delta\varphi$ and E, which are related to wave parameters $D = \frac{4L}{ka^2}$.

When D has a moderate value, we will get

$$\overline{(\Delta\varphi)^2} = \frac{\sqrt{\pi}}{2}\overline{\mu^2}k^2aL\left(1 + \frac{1}{D}\arctan D\right) \quad (2.181)$$

$$\overline{E^2} = \frac{\sqrt{\pi}}{2}\overline{\mu^2}k^2aL\left(1 - \frac{1}{D}\arctan D\right) \quad (2.182)$$

When $D \gg 1$, the secondary terms on the right-hand side of Eqs. (2.181) and (2.182) will be neglected, so

$$\overline{(\Delta\varphi)^2} = \overline{E^2} = \frac{\sqrt{\pi}}{2}\overline{\mu^2}k^2aL \quad (2.183)$$

The condition of $D \gg 1$ is always satisfied for digital underwater acoustic communications. We can see that the intensity of fluctuation of both $\Delta\varphi$ and E depends not only upon μ and a, but they also are proportional to the transmission distances and the square of the operating frequencies.

2.4.1.2.2 CORRELATIVITY OF FLUCTUATIONS

The fluctuating intensity $\overline{E^2}$ and $\overline{(\Delta\varphi)^2}$ have been discussed. To further describe the statistical characteristics of fluctuations of sound signals, we must resort to correlation functions.

2.4.1.2.2.1 TIME AUTO-CORRELATION FUNCTION To determine the time auto-correlation characteristics of fluctuations of sound signals, direct measurement in situ is certainly an available approach. However, there are a number of factors to affect the fluctuations of sound signals. How to acquire fluctuating data caused alone by the fluctuation of the temperature of the seawater is quite difficult. Therefore analyzing the relations between fluctuation of sound signals and that of the temperature of the seawater is significant because the measurements in situ for the latter are easier than those of the former.

The fluctuation in the temperature of the seawater medium in general sea areas may be considered as having the characteristics of being statistically homogeneous in space and stationary in time domain, so the correlation characteristics of μ may be described by means of a spatial-temporal correlation function:

$$\overline{\mu(\xi_1, \eta_1, \zeta_1, t)\mu(\xi_2, \eta_2, \zeta_2, t)} = \overline{\mu^2}R'(\xi_1 - \xi_2, \eta_1 - \eta_2, \zeta_1 - \zeta_2, t_1 - t_2) \quad (2.184)$$

That is to say, the spatial-temporal correlation coefficient R' is only determined by the difference of spatial coordinates and the space of time, which is independent of specific spatial positions and specific moments.

Let us further assume that R' can be separated into the multiplication of two functions: one is only dependent upon spatial coordinates, and the other is only dependent upon the space of time:

$$R'(\xi_1 - \xi_2, \eta_1 - \eta_2, \zeta_1 - \zeta_2, t_1 - t_2) = R(\xi_1 - \xi_2, \eta_1 - \eta_2, \zeta_1 - \zeta_2)M(t_1 - t_2) \quad (2.185)$$

Provided the media are isotropic, we get

$$R'(\xi_1 - \xi_2, \eta_1 - \eta_2, \zeta_1 - \zeta_2, t_1 - t_2) = R(\rho)M(\tau) \quad (2.186)$$

where $\tau = (t_1 - t_2)$. We should note that this assumption will not satisfy if the seawater medium has strong turbulence. In this case, the spatial-temporal variability of the seawater medium is related and thus cannot simply be separated.

According to Eq. (2.168), the temporal auto-correlation function of the fluctuation in amplitude at the same received point $(L,0,0)$ and different moments of t_1 and t_2 can be expressed as

$$B_E(\tau) = \overline{E(L,0,0,t_1)E(L,0,0,t_2)}$$

Consider Eqs. (2.185), (2.186), and (2.178); moreover, the factor $M(\tau)$ is independent of integral variables, and the temporal auto-correlation function of the fluctuation in amplitude is thus given by

$$B_E(\tau) = \frac{k^4\overline{\mu^2}}{4\pi^2}M(\tau)\int_0^x\int_0^x\int\int_{-\infty}^{+\infty}\int\int \times \frac{1}{(x-\xi_1)(x-\xi_2)}\cos\frac{k(\eta_1^2+\xi_1^2)}{2(x-\xi_1)}\cos\frac{k(\eta_2^2+\xi_2^2)}{2(x-\xi_2)}R(\rho)\mathrm{d}\xi_1\mathrm{d}\xi_2\mathrm{d}\eta_1\mathrm{d}\eta_2\mathrm{d}\zeta_1\mathrm{d}\zeta_2 \tag{2.187}$$

By comparing Eq. (2.177) with Eq. (2.180), we get $B_E(\tau) = M(\tau)\cdot\overline{E^2}$. Therefore,

$$R_E(\tau) = M(\tau) \tag{2.188}$$

The conclusion mentioned is also available for the fluctuation in the phase of sound signals. It is a simple but important conclusion, since that means the characteristics of temporal auto-correlation for the fluctuation of sound signals may approximately be considered to be consistent with that for the fluctuation of seawater medium under the same specific conditions. Therefore, the measurements for temporal auto-correlation characteristics with respect to the fluctuation of sound signals may approximately be replaced by those of the fluctuation of the sea water medium. Obviously, the measurements for the latter are easily carried out.

2.4.1.2.2.2 SPATIAL CORRELATION CHARACTERISTICS

2.4.1.2.2.2.1 Longitudinal Correlation Function Assume an incident plane wave is traveling along the x direction, two receivers are located at $(L_1,0,0)$ and $(L_2,0,0)$ in this direction, and the distance between two points $\Delta L = L_2 - L_1$, as shown in Fig. 2.48. Based on Eqs. (2.177) and (2.178), we can obtain the spatial correlation functions of the sound amplitude and phase between them.

Provided the wave parameter $D \gg 1$ (ie, the distance between a receiver and sound source is sufficiently larger than the scale of the inhomogeneous blobs a), and the spatial correlation

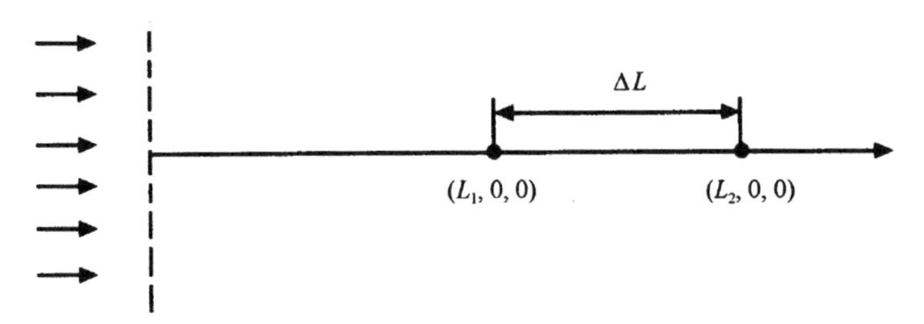

FIGURE 2.48 Positions of receiving points for finding longitudinal correlation function.

coefficient of the index of sound refraction is Gaussian, we can obtain

$$\overline{E_1 \cdot E_2} = \overline{\Delta\varphi_1 \cdot \Delta\varphi_2} = \frac{1}{2}\overline{\mu^2}k^2 L_1 \int_{-\infty}^{\infty} \frac{e^{-\xi^2/a^2}}{1 + 4 \cdot \frac{(\Delta L)^2 - 2\xi\Delta L + \xi^2}{k^2 a^4}} d\xi \tag{2.189}$$

If ΔL is less, then $ka \gg 1$ can usually be satisfied, and Eq. (2.189) will be simplified as

$$\overline{E_1 \cdot E_2} = \overline{\Delta\varphi_1 \cdot \Delta\varphi_2} = \frac{1}{2}\overline{\mu^2}k^2 L_1 \int_{-\infty}^{\infty} e^{-\xi^2/k^2a^2} d\xi = \frac{\sqrt{\pi}}{2}\overline{\mu^2}k^2 a L_1 \tag{2.190}$$

This expression is perfectly consistent with Eq. (2.193) that expresses the mean square values of the amplitude and phase fluctuations. It means that the fluctuations in both sound amplitude and phase are perfectly correlative provided the distance between any two receiving points is nearer in the sound transmitting direction. Thus the correlation functions expressed by Eq. (2.190) are independent of ΔL.

Provided $\Delta L \gg a$, according to Eq. (2.189), we get

$$\overline{E_1 \cdot E_2} = \overline{\Delta\varphi_1 \cdot \Delta\varphi_2} = \frac{\sqrt{\pi}\overline{\mu^2}k^2 a L_1}{2\left[1 + \left(\frac{2\Delta L}{ka^2}\right)^2\right]} \tag{2.191}$$

If Eq. (2.191) is expressed by means of correlation coefficients,

$$R_E = \frac{\overline{E_1 \cdot E_2}}{\overline{E^2}}, \quad R_{\Delta\varphi} = \frac{\overline{\Delta\varphi_1 \cdot \Delta\varphi_2}}{\overline{(\Delta\varphi)^2}}$$

the longitudinal correlation coefficients under the condition of $\Delta L \gg a$ are given by

$$R_E = R_{\Delta\varphi} = \frac{1}{1 + \left(\frac{2\Delta L}{ka^2}\right)^2} \tag{2.192}$$

The longitudinal spatial correlation coefficients R_E and $R_{\Delta\varphi}$ will decrease with increasing ΔL. Letting ΔL correspond to $R_E = R_{\Delta\varphi} = e^{-1}$ is defined as the longitudinal correlation radius ΔL_0, and according to Eq. (2.192), we get

$$\Delta L_0 \approx 0.6ka^2$$

Since $ka \gg 1$, then $\Delta L_0 \gg a$. So, the longitudinal correlation radii of sound signal fluctuations are much larger than those of the fluctuation of the index of refraction. For example, take $f = 10$ kHz and $a = 75$ cm, and it follows that $\Delta L_0 = 14$ m. The longitudinal correlativity of the fluctuation of sound signals is strong enough when they travel in a weakly random inhomogeneous medium. This valuable performance may be employed in anti-noise signal processing (refer to Section 2.5).

2.4.1.2.2.2.2 Transversal Correlation Function Two receiving points are placed at $(L,0,0)$ and $(L,\Delta y,0)$, as shown in Fig. 2.49, in

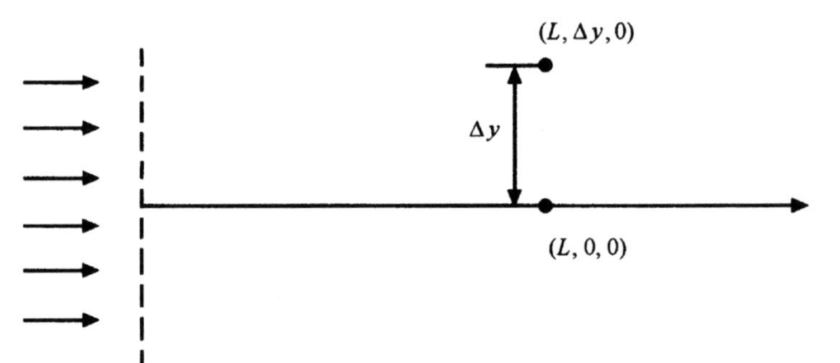

FIGURE 2.49 Positions of receiving points for finding transversal correlation function.

which Δy is perpendicular to the direction of the traveling sound wave.

In the case of $L \gg a$, $ka \gg 1$ and the correlation coefficient of the index of refraction being Gaussian, by applying complicated mathematical operations, we can obtain transversal correlation coefficients of the phase and amplitude $R_{\Delta\varphi}$ and R_E, respectively.

Provided $D \ll 1$, we have

$$R_{\Delta\varphi} = e^{-(\Delta y/a)^2}$$

$$R_E = e^{-\left(\frac{\Delta y}{a}\right)^2}\left[1 - 2\left(\frac{\Delta y}{a}\right)^2 + \frac{1}{2}\left(\frac{\Delta y}{a}\right)^4\right]$$

Provided $D \gg 1$,

$$R_{\Delta\varphi} = \frac{e^{-\left(\frac{\Delta y}{a}\right)^2} + \frac{1}{D}\left[\frac{\pi}{2} - S_i\left(\frac{1}{D}\frac{\Delta y^2}{a^2}\right)\right]}{1 + \frac{1}{D}\arctan D} \tag{2.193}$$

$$R_E = \frac{e^{-\left(\frac{\Delta y}{a}\right)^2} - \frac{1}{D}\left[\frac{\pi}{2} - S_i\left(\frac{1}{D}\frac{\Delta y^2}{a^2}\right)\right]}{1 - \frac{1}{D}\arctan D} \tag{2.194}$$

where S_i is the integral hyperbolic sine.

$R_{\Delta\varphi}$ and R_E versus $\Delta y/a$ at $D \ll 1$ are plotted in Fig. 2.50A, and the phase correlation coefficient $R_{\Delta\varphi}$ (curve 2), just as that of the fluctuation of the refraction index, belongs to the Gaussian mode. The correlation coefficient of amplitude (curve 2) does not fit, but its correlation radius has the same order of magnitude with that of the fluctuation of the refraction index.

$R_{\Delta\varphi}$ and R_E versus $\Delta y/a$ at $D = 10$ are plotted in Fig. 2.50B. Curve 2 describes the correlation coefficient of fluctuation of refraction index. We see that there are similar variable laws among curve 1 (R_E), curve 2 (the correlation coefficient of the refraction index), and curve 3 ($R_{\Delta\varphi}$). In other words, the transversal correlation radii

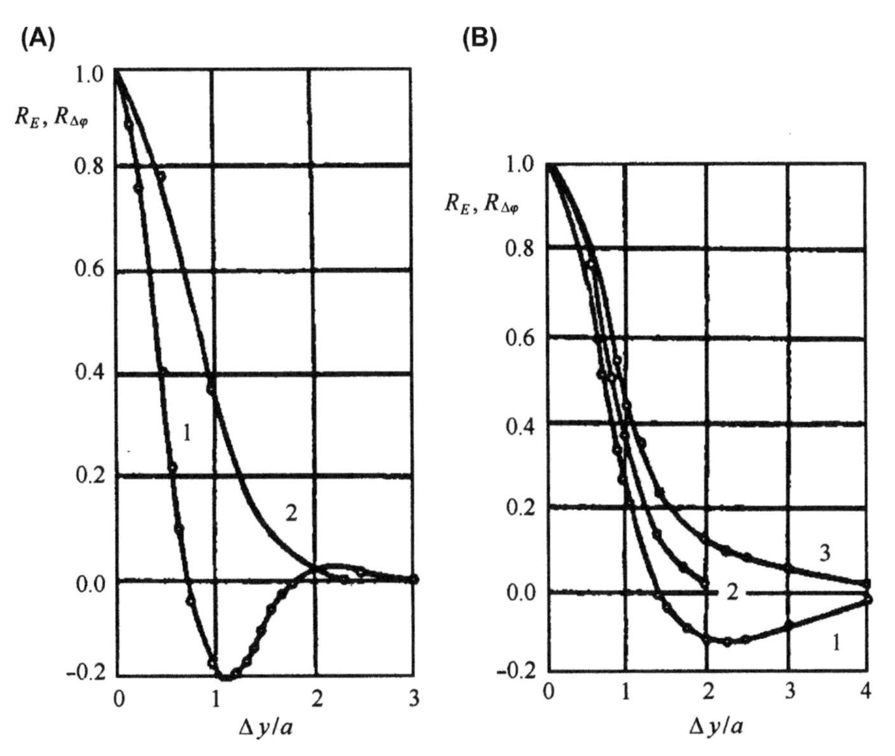

FIGURE 2.50 Diagrams of transversal correlation coefficients: (A) $D \ll 1$; (B) $D = 10$.

for $R_{\Delta\varphi}$ and R_E have the same order of magnitude as that of the fluctuation of the refraction index. We should note that the longitudinal correlation radii are much greater than transversal ones.

The fluctuating characteristics of both the phase and amplitude calculated by using the infinitesimal disturbance method for a plane sound wave traveling in a weakly inhomogeneous media have been introduced. The fluctuating characteristics for a general spherical sound wave have further been calculated by Mintzer et al. under the condition of $L \gg a^2/\lambda$ [7]. They obtained

$$\overline{E^2} = \overline{\Delta\varphi^2} = \overline{\mu^2}k^2L_nL \tag{2.195}$$

where L_n is the integration scale of the inhomogeneity of media as

$$L_n = \left[B_\mu(0)\right]^{-1}\int_0^\infty B_\mu(\rho)\mathrm{d}\rho$$

where $B_\mu(\rho)$ is the correlation function of μ for a stationary homogeneous field.

In the case of a Gaussian correlation coefficient, they obtained corresponding formulas available for arbitrary distances as follows:

$$\overline{\Delta\varphi^2} = 0.5\sqrt{\pi}\overline{\mu^2}k^2aL\{1 + 0.5[F(1.1;3/2;iD/4) + F(1.1;3/2;-iD/4)]\} \tag{2.196}$$

$$\overline{E^2} = 0.5\sqrt{\pi}\overline{\mu^2}k^2aL\{1 - 0.5[F(1.1;3/2;iD/4) + F(1.1;3/2;-iD/4)]\} \tag{2.197}$$

where $F(\alpha,\beta,\gamma,z)$ is a hypergeometric function.

2.4.2 Sound Scattering Forms Random Boundaries and the Fluctuation of Transmitted Sound in Underwater Acoustic Communication Channels

2.4.2.1 Sound Scattering From the Sea Surface

We have discussed the sound scattering and fluctuation of transmitted sound in weakly random inhomogeneous media in the previous section. In fact, the sound scattering from rough boundaries of the sea is another major reason to cause the fluctuations of sound signals.

2.4.2.1.1 ECKART THEORY [18]

Let us consider a simplified model to exhibit the relations between a mean reflection coefficient of system synthesis $\langle V\rangle$ and statistical characteristics of the sea surface. Let the fluctuating sea wave heights $\upsilon = \upsilon(x)$ and $\langle \upsilon\rangle = 0$ (refer to Fig. 2.51).

Take $\upsilon = 0$ as a reference boundary, and the fluctuating path difference of the reflected sound waves caused by the fluctuation of υ equals $2\upsilon \sin \chi$; then the additional fluctuation in phase is given by

$$K = 2k_0\upsilon \sin \chi \tag{2.198}$$

where $k_0 = \omega/c$. K is called the Rayleigh parameter that is a criterion describing the roughness of the sea surface [3]. When K tends to zero, which means that the additional fluctuation in phase also equals zero, the sound reflection from such a sea surface can be thought of as a specular

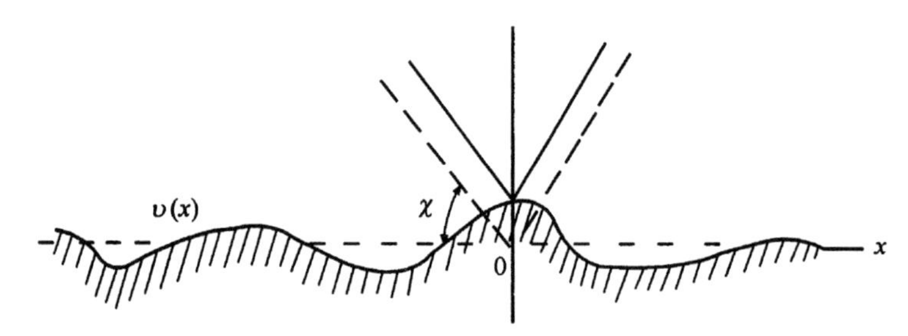

FIGURE 2.51 Model of fluctuating sea surface.

one. The sound scattering effect is more violent with increasing K.

Provided the Rayleigh reflection coefficient may be considered as not being affected by the fluctuation of the sea surface (that is corresponding to take a Kirchhoff approximation), the reflection field can be expressed as

$$p = V_R p_0 e^{i\phi_0 - 2ik_0 v \sin\chi}$$

where $p_0 e^{i\phi_0}$ is related to an incident field, and V_R is the Rayleigh reflection coefficient for a flat boundary. It is generally permitted to let $p_0 e^{i\phi_0} = 1$, so the random reflection coefficient is given by

$$V = V_R e^{-2ik_0 v \sin\chi} \tag{2.199}$$

Let the probability density function (PDF) for a random surface be $W(v)$, and we have

$$\langle V \rangle = V_R \int_{-\infty}^{+\infty} \exp(-2ik_0 v \sin\chi) W(v) dv \tag{2.200}$$

For a Gaussian surface, we obtain

$$W(v) = \sigma^{-1}(2\pi)^{-1/2} \exp\left(\frac{-v^2}{2\sigma^2}\right) \tag{2.201}$$

where σ is rms wave height.

Substituting Eq. (2.201) in Eq. (2.200) gives

$$\langle V \rangle = V_R \exp\left(-2k_0^2\sigma^2 \sin^2\chi\right) \tag{2.202}$$

This is the famous Eckart result.

The mean reflection coefficient given by Eq. (2.202) is agreeable to the experimental results for small Rayleigh parameters, while there are some deviations under the condition of $k\sigma \sin\chi > 1$.

By examining Eq. (2.200), the $\langle V \rangle$ and $W(v)$ are a Fourier transform pair, so we may acquire the statistical characteristics of the random surface $W(v)$ according to the measured results of the $\langle V \rangle$:

$$W(v) = \frac{1}{\pi V_R} \int_{-\infty}^{\infty} \langle V \rangle \exp(2ik_0 v \sin\chi) d(k \sin\chi)$$

2.4.2.1.2 THEORY OF INFINITESIMAL DISTURBANCE

The method of infinitesimal disturbance to solve scattering field generated by a statistically rough surface is that by converting this surface into a smooth one on which "virtual" sound sources are distributed, as a result it will be changed into a problem to solve the smooth surface with nonhomogeneous boundary conditions.

Let the heights of the fluctuation relative to $z = 0$ be $\xi(x,y)$, and the absolute soft boundary condition is given by

$$\psi[x, y, \zeta(x, y)] = 0 \tag{2.203}$$

If the height of the fluctuation is small in comparison with the wavelength of the sound wave and the surface is flat enough, Eq. (2.203) will be converted into the following condition at $z = 0$:

$$\psi + \zeta \frac{\partial \psi}{\partial z} = 0 \tag{2.204}$$

Let sound field be expressed by

$$\psi = \langle \psi \rangle + \psi_p \tag{2.205}$$

where $\langle \psi \rangle$ is a mean field, and ψ_p is a scattering field that satisfies

$$\langle \psi_p \rangle = 0 \tag{2.206}$$

By substituting Eq. (2.205) into Eq. (2.204) and then finding its mean value, we get

$$\langle \psi \rangle + \left\langle \zeta \frac{\partial \psi_p}{\partial z} \right\rangle = 0 \tag{2.207}$$

According to Eqs. (2.204), (2.205), and (2.207), we obtain

$$\psi_p + \zeta \frac{\partial}{\partial z} \langle \psi \rangle + \left(\zeta \frac{\partial \psi_p}{\partial z} - \left\langle \zeta \frac{\partial \psi_p}{\partial z} \right\rangle \right) = 0$$

The third term on the left-hand side of this expression is a high-order infinitesimal magnitude and can be neglected; we can obtain the second boundary condition at $z = 0$ as

$$\psi_p + \zeta \frac{\partial}{\partial z} \langle \psi \rangle = 0 \tag{2.208}$$

According to the Green formula, we have

$$\psi_p(x,y,z) = -\frac{1}{2\pi}\iint_{-\infty}^{\infty}\psi_p(x_1,y_1,0)\frac{\partial}{\partial z}\left(\frac{\exp(ikR)}{R}\right)dx_1dy_1 \quad (2.209)$$

where $R = \sqrt{(x-x_1)^2+(y-y_1)^2+z^2}$. Substituting Eq. (2.208) into Eq. (2.209), we get

$$\psi_p(x,y,z) = \frac{1}{2\pi}\iint_{-\infty}^{\infty}\zeta(x_1,y_1)\left[\frac{\partial}{\partial z}\langle\psi\rangle\right]_{z=0}\times\frac{\partial}{\partial z}\left(\frac{\exp(ikR)}{R}\right)dx_1dy_1 \quad (2.210)$$

Substituting Eq. (2.210) into Eq. (2.207), we can obtain

$$\langle\psi\rangle + \frac{\sigma^2}{2\pi}\iint_{-\infty}^{\infty}B_s(x-x_1,y-y_1)\times\left[\frac{\partial}{\partial z}\langle\psi\rangle\frac{\partial^2}{\partial z^2}\left(\frac{\exp(ikR)}{R}\right)\right]_{z=0}dx_1dy_1 = 0 \quad (2.211)$$

where the auto-correlation function of the rough surface is introduced as

$$\sigma^2 B_s(x-x_1,y-y_1) = \frac{1}{\sigma^2}\langle\zeta(x,y)\zeta(x_1,y_1)\rangle \quad (2.212)$$

where $\sigma^2 = \langle v^2\rangle$.

Consider that the surface is statistically homogeneous and isotropic; the auto-correlation function is only relative to the distance ρ, ie, $B_s = B_s(\rho)$, where $\rho = \sqrt{(x-x_1)^2+(y-y_1)^2}$. The mean field may be expressed by the superposition of the incident wave and the specular reflection wave:

$$\langle\psi\rangle = \exp[i(\alpha_0 x+\beta_0 y-\gamma_0 z)] + V\exp[i(\alpha_0 x+\beta_0 y+\gamma_0 z)] \quad (2.213)$$

where V is the mean reflection coefficient, and (α_0, β_0, γ_0) express the direction cosine of wave number. Obviously, $\langle\psi\rangle$ satisfies the wave equation and the boundary condition of the isotropic surface. By applying relative operations, we then can obtain the simplified expression of V under the following two extreme conditions:

1. By satisfying the following,

$$k\rho_0 \sin^2\chi \gg 1 \quad (2.214)$$

we get

$$V = -1 + 2(k\sigma\sin\chi)^2 \quad (2.215)$$

That is to say, if Eq. (2.214) is satisfied, V will reduce to the Eckart result under the condition of small grazing angles, as shown in Eq. (2.202).

2. By satisfying

$$k\rho_0 \ll 1$$

and the Neumann sea wave spectrum, we get

$$|V| = 1 - 0.56(fH_{cp})^{3/2}H_{cp}^{1/10}\sin\chi \quad (2.216)$$

where f is the sound frequency in kHz, and $H_{cp} = \sqrt{2\pi}\sigma$ is the mean height of sea wave in meters.

3. Experimental researches on the spatial-temporal correlativity of surface scattering fields.

The theoretical calculations for the spatial-temporal correlativity of the surface scattering field will not be discussed here. Some experimental results that are available for designing signal processing systems employed in digital underwater acoustic communications will simply be introduced.

2.4.2.1.2.1 TEMPORAL CORRELATIVITY The temporal correlativity of the fluctuation in amplitude for sound pulses reflected from the sea surface had been experimentally researched by Qulin. Operating frequencies are 4, 7, 11, 15, and 36 kHz, respectively. The duration of the pulses is 3 ms, and repeated frequencies are 3–8 Hz. A sound source and a hydrophone were both located on a slope sea bottom where the depth is 80 m. The distance between the source and hydrophone might be changed from 200 to 1500 m.

Provided the Rayleigh parameter K is less than 1, the temporal correlation coefficients of fluctuations with respect to sea wave and sound amplitude reflected from the sea surface have a similar mode, as shown in Fig. 2.52A and B, respectively; moreover, they have the same vibrating periods and approximately equal correlation radii. When K is increased to be larger than 1, the quasi-period behavior of the fluctuation in the amplitude will disappear; moreover, the correlation radii will also be reduced, as shown in Fig. 2.52C, where $K = 2.2$.

Since the temporal correlation radii of the fluctuation in the amplitude will be reduced with increasing K, the width of the spectrum of sound signal will correspondingly be spread.

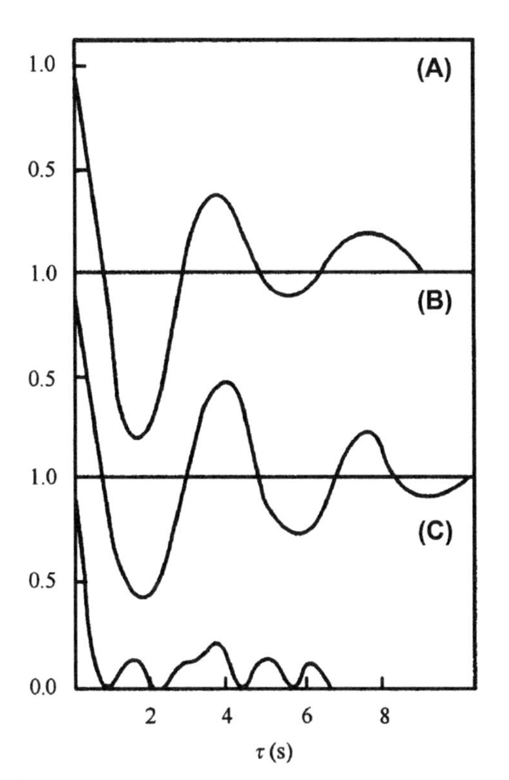

FIGURE 2.52 Temporal correlation coefficient versus K. (A) Sea wave; (B) $K = 0.37$; (c) $K = 2.2$.

Experimental results had proven that the spreading width is only several Hz. It can generally be neglected for some digital underwater acoustic communications.

2.4.2.1.2.2 SPATIAL CORRELATIVITY The spatial correlativity has been measured in some open sea areas by using floating ships. The experimental conditions and operating states were as follows: single frequency pulses with the duration of 10 ms and the carrier frequencies of 1, 2, 3.5, and 5 kHz were transmitted, respectively, at the depth of 150 m, and a receiving transducer array that consisted of three hydrophones mounted normally to each other was located at the depth of 75 m. During transmitting, frequencies were 3.5 kHz and 5 kHz, and the corresponding horizontal distances were about 0.8–0.9 km (grazing angle $\chi \cong 15$ degree). When the frequencies 1 and 2 kHz were used, the distances were about 1.0–1.2 km ($\chi \cong 11.5$ degree). Since rms height of a sea wave is 0.23 m, then K is 0.4, 0.8, 1.7, and 2.4 corresponding to the signal frequencies of 1, 2, 3.5, and 5 kHz, respectively.

The spatial correlation coefficients of the signal fluctuation in the amplitude reflected from the sea surface at the maximum space of two hydrophones that are 6 m in vertical, longitudinal, and transversal directions are shown in Fig. 2.53, where curve 1, 2, 3, and 4 correspond to the frequencies of 1, 2, 3.5, and 5 kHz, respectively. We see that the spatial correlation radii of the fluctuation in the amplitude at these directions will decrease with increasing carrier frequencies or K; moreover, the vertical spatial correlation radii (according to correlation coefficients to be reduced to 0.4–0.5) are less than those in the horizontal direction for these frequencies. While the correlation radii in the longitudinal and transversal directions approximately equal each other, the correlativity of the transversal direction will be lowered with a rate that is slower than that of the longitudinal one.

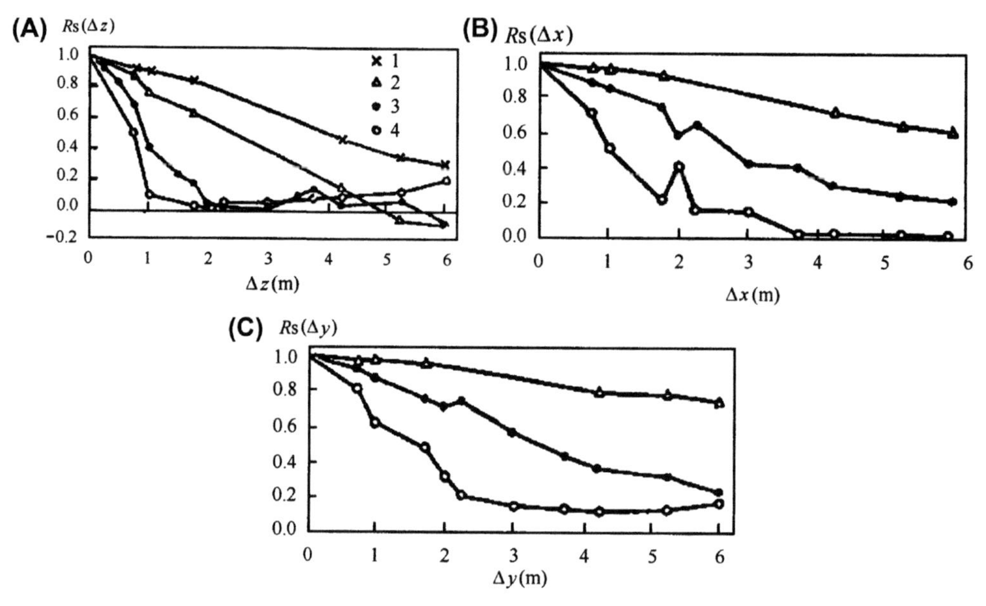

FIGURE 2.53 Vertical (A) longitudinal (B), transversal (C) correlation coefficients of the signal fluctuation in the amplitude reflected from the sea surface.

2.4.2.2 *Acoustic Properties of the Sea Bottom*

The sea bottom is a reflecting and scattering boundary with some characteristics similar to those of the sea surface. However, its effects on underwater acoustic communications are more complicated because of its diverse and multilayered composition.

There are two causes of the reflection of sound from the sea bottom being much more complex than that from the sea surface. First, the bottom is more variable in its acoustic properties because it may vary in composition from hard rock to soft mud. Second, it is often layered with a density and a sound velocity that can change gradually or abruptly with depth. For these reasons, the sound reflection behaviors of the sea bottom are more difficult to be predicted than those of the sea surface.

Now, the sea bottom is simplified to a fluid half-space, so the basic behaviors of reflection and transmission for a plane sound wave incident at different angles on this plane boundary will first be analyzed. Then some experimental results for those acoustic properties will be introduced.

2.4.2.2.1 REFLECTION AND TRANSMISSION OF SOUND WAVE INCIDENT ON AN INTERFACE THAT CONSISTS OF TWO FLUIDS

Assume densities and sound velocities in the upper and lower fluid media are ρ, ρ_1 and c, c_1, respectively, and the incident angle is θ, as shown in Fig. 2.54.

By means of a simple derivation, the reflection coefficient V and transmission coefficient W can be found as follows:

$$V = \frac{m\cos\theta - n\cos\theta_1}{m\cos\theta + n\cos\theta_1} = \frac{m\cos\theta - \sqrt{n^2 - \sin^2\theta}}{m\cos\theta + \sqrt{n^2 - \sin^2\theta}}$$

$$W = \frac{2m\cos\theta}{m\cos\theta + \sqrt{n^2 - \sin^2\theta}} \tag{2.217}$$

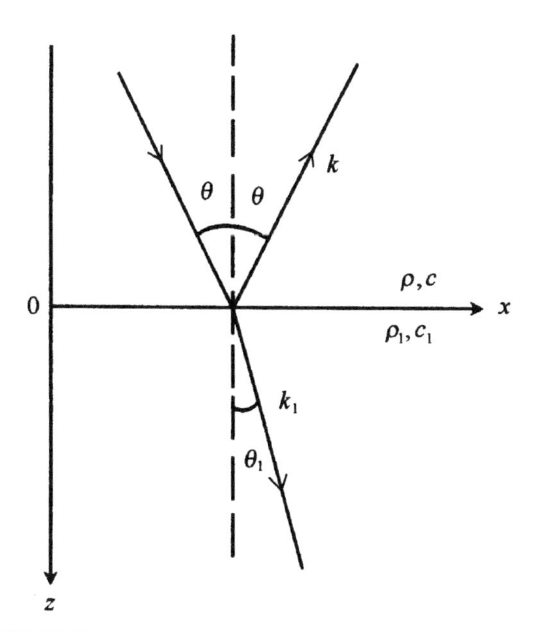

FIGURE 2.54 Parameters using to derive reflection and transmission coefficients.

where $n = c/c_1$ is the index of the refraction, and $m = \rho_1/\rho$. V and W have some significant properties:

1. When $\theta \to \pi/2$, then $V \to -1$ and $W \to 0$, which are independent of the media parameters.
2. Provided that θ satisfies following equation,

$$m\cos\theta - \sqrt{n^2 - \sin^2\theta} = 0, \quad \text{that is } \sin\theta = \sqrt{\frac{m^2 - n^2}{m^2 - 1}} \tag{2.218}$$

then, $V = 0$, and the interface has a total transmission property.

3. Let n be a real number, $n < 1$, and $\sin\theta > n$. In such cases, the reflection coefficient V becomes

$$V = \frac{m\cos\theta - i\sqrt{\sin^2\theta - n^2}}{m\cos\theta + i\sqrt{\sin^2\theta - n^2}} \tag{2.219}$$

or

$$V = \exp(i\varphi), \quad \varphi = -2\arctan\frac{\sqrt{\sin^2\theta - n^2}}{m\cos\theta} \tag{2.220}$$

We see that the module of reflection coefficient $|V| = 1$, which means that the phenomenon of total reflection occurs. The phase difference between reflected and incident waves equals φ.

For a sandy bottom ($m = 1.95$, $n = 0.85$) and its sound absorption effect to be neglected, the most upper curve in Fig. 2.55A and the most lower curve in Fig. 2.55B represent the modules and arguments of reflection coefficients versus grazing angles, respectively.

For actual absorptive media, n is a complex number satisfying $n = n(1 + i\alpha)$, where $\alpha > 0$. The reflection coefficient is

$$V = |V|\exp(i\varphi), \quad |V| < 1$$

The modules and arguments of the reflection coefficients versus grazing angles for different α are plotted in Fig. 2.45A and B, respectively.

When $n > 1$, the sound velocity in the sea bottom is less than that in the seawater, and the total reflection will be absent. For a muddy bottom ($m = 1.56$, $n = 1.008$) and different α, the modules and arguments of the reflection coefficients versus grazing angles are plotted in Fig. 2.56A and B, respectively.

Besides considering the sound fields that are attenuated caused by the reflection loss, we must pay attention to the phase shift due to the sound reflection from the sea boundaries since that is one of the reasons to bring about the phase blur in shallow-water acoustic channels.

2.4.2.2.2 EXPERIMENTAL RESEARCHES ON THE ACOUSTIC PROPERTIES OF THE SEA BOTTOM

We have known that the laws of the sound reflection and scattering from the sea bottom are quite complicated and variable. Theoretical analyses are very difficult. Therefore, the experimental researches on the acoustic properties of the sea bottom are particularly valuable.

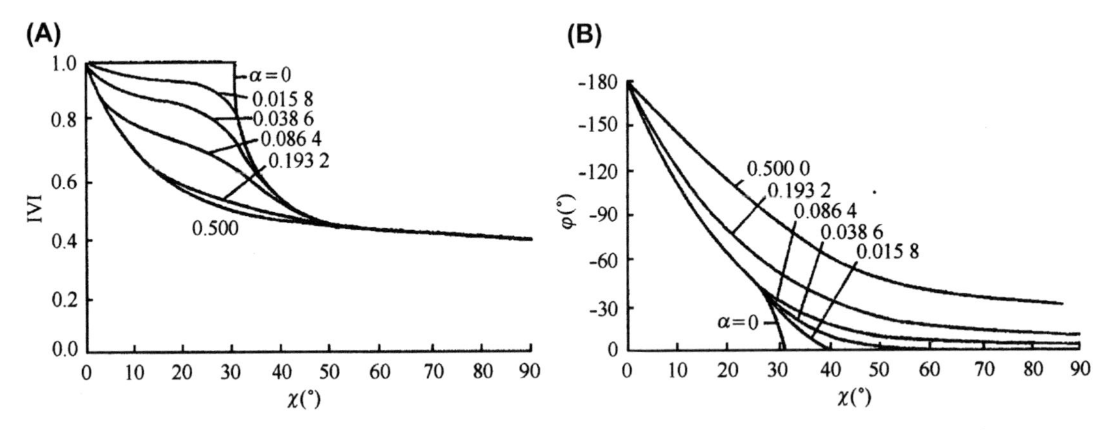

FIGURE 2.55 Module (A) and argument (B) of the reflection coefficients versus grazing angles.

Generally speaking, we can only measure the roughly mean values of the reflection coefficients for some types of the sea bottom, such as for deep-sea bottom and shallow sea bottom with different types (sand, mud, etc.) and smoothness.

Fig. 2.57 shows some measured results for the reflection coefficients of the sea bottom, and Fig. 2.57 (A) correspond to a silt bottom where the depth is 230 m; (B) corresponds to a sand bottom having the depth of 185 m; (C) is a mud bottom, and the depth is 1100 m. The sound velocities in the sea bottom are 3–10% greater than those in the seawater. The black points represent measured data [1].

Sometimes by means of bottom reflection loss,

$$BL = -20\lg|V|$$

to describe the properties of sound reflection from the sea bottom.

According to a lot of field measurements at deep-sea locations, a series of curves of the bottom loss as a function of grazing angles for various frequencies are shown in Fig. 2.58 [13]. The virtual curves are obtained by using an extrapolation method. The curves are average

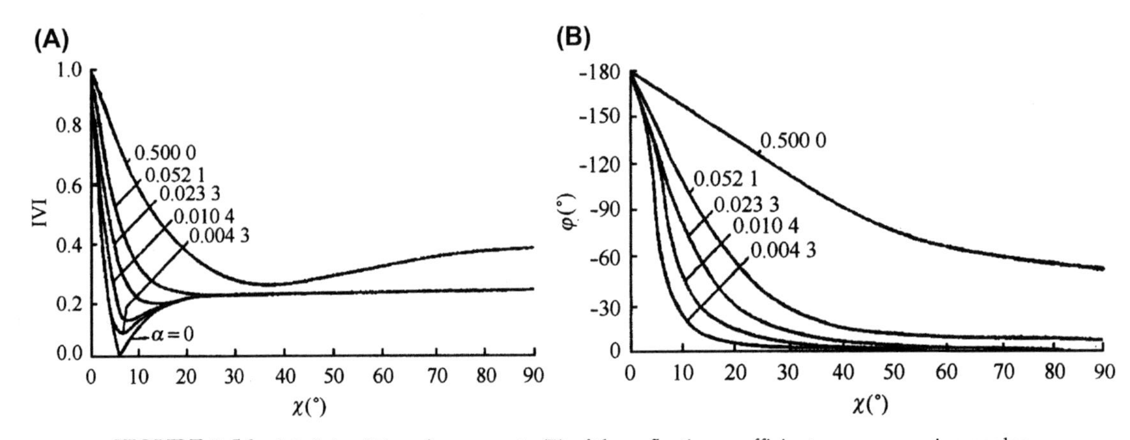

FIGURE 2.56 Modules (A) and arguments (B) of the reflection coefficients versus grazing angles.

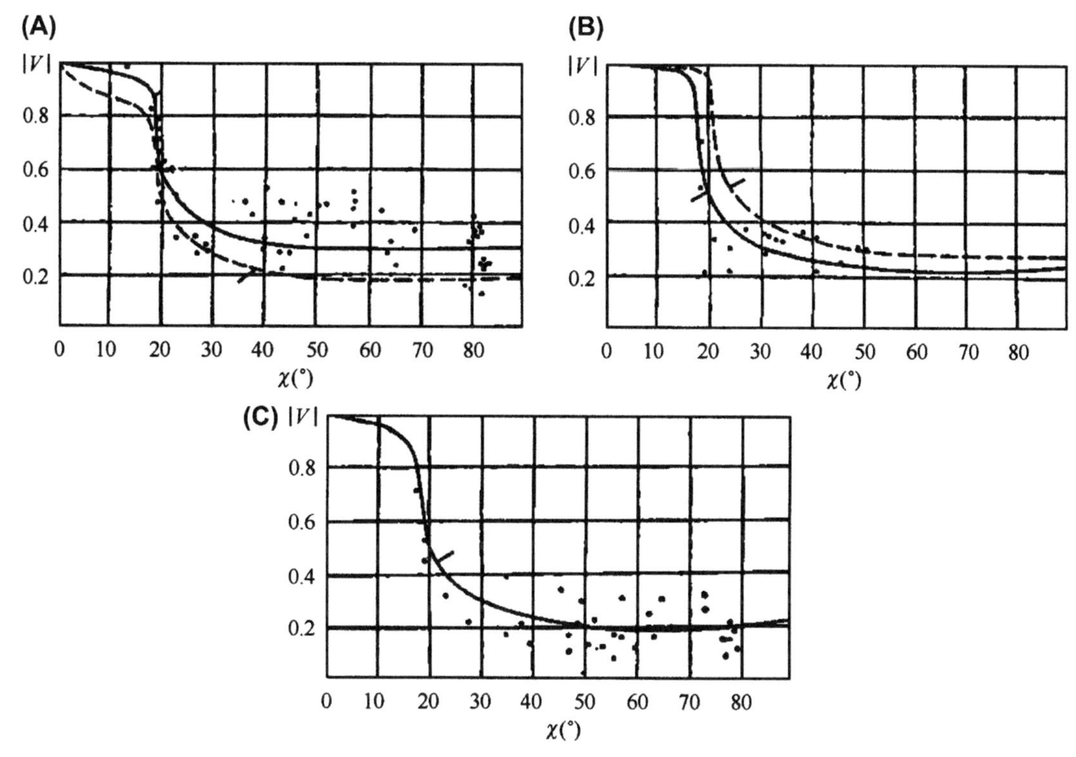

FIGURE 2.57 Experimental results for reflection coefficients of the sea bottom. (A) A silt bottom, D = 230 m; (B) A sand bottom, D = 185 m; (c) A mud bottom, D = 1100 m.

losses for abyssal bottoms and not necessarily representative of any one location; wide variations should be expected for any particular set of measurements.

The bottom losses shown in Fig. 2.58 have the three basic characteristics:

1. There is a "demarcating grazing angle" χ^*. When $\chi < \chi^*$, *BL* is less; when $\chi > \chi^*$, *BL* is larger. χ^* is thus a characteristic parameter of *BL*.
2. At small grazing angles ($\chi < \chi^*$), *BL* will increase with increasing χ.
3. At large grazing angles ($\chi > \chi^*$), *BL* is basically independent of χ, and it may generally be considered a constant.

Based on that, a simplified model of three parameters has been applied to describe the basic characters of *BL* versus χ [1]:

$$-\ln|V(\chi)| = \begin{cases} Q\chi & 0 < \chi < \chi^* \\ -\ln|V_0| = \text{const} & \chi^* < \chi < \dfrac{\pi}{2} \end{cases} \tag{2.221}$$

where Q, χ^*, and $-\ln|V_0|$ are the three parameters, as shown in Fig. 2.59, that may be used to analyze the structures of the sound fields in shallow-water channels, etc.

BL for some types of bottoms has also been measured by Marsh [19]; the results are presented in Table 2.4. These are the examples of

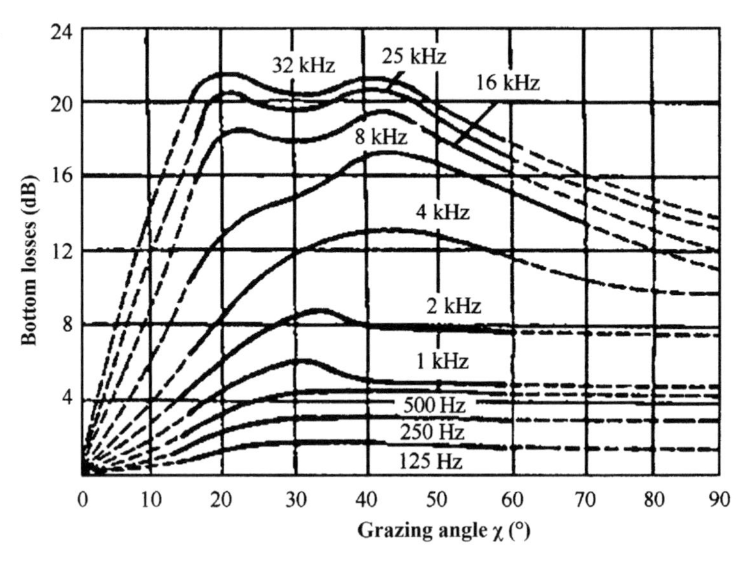

FIGURE 2.58 Bottom losses versus grazing angle for various frequencies.

the range of the reflection losses to be expected for bottoms characterized by words such as "sand" or "mud" that denote the particle size of the sedimentary bottoms to which they refer.

Some experimental results for spatial-temporal correlativity of sound signals traveling over underwater acoustic channels, in particular at longer range communications, will simply be introduced next. In these cases, the integral effects of both random inhomogeneity in the body of the seawater and the roughness of the sea boundaries on the fluctuations of sound signals are considered.

The time-variant characteristics of underwater acoustic channels have been analyzed by using a pulse-to-pulse correlation method [20].

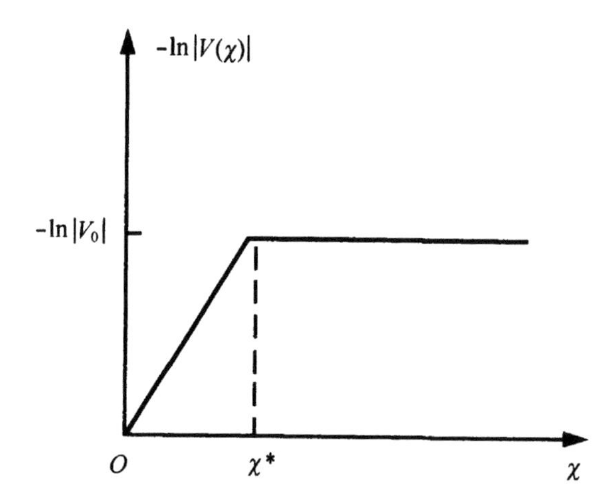

FIGURE 2.59 Model of three parameters for bottom loss.

TABLE 2.4 Measured Reflection Losses for Different Bottom Types

24 kHz, 10 degree Grazing Angle, 17 Stations			
Mud	16 dB		
Mud–sand	10		
Sand–mud	6		
Sand	4		
Stony	4		
Normal Incidence 7 Stations			
	4 kHz	**7.5 kHz**	**16 kHz**
Sandy silt (3 stations)	14	14	13
Fine sand (1 stations)	7	3	6
Coarse sand (1 stations)	7	8	8
Medium sand with rock (1 stations)	8	6	10
Rock with some sand (1 stations)	5	4	10

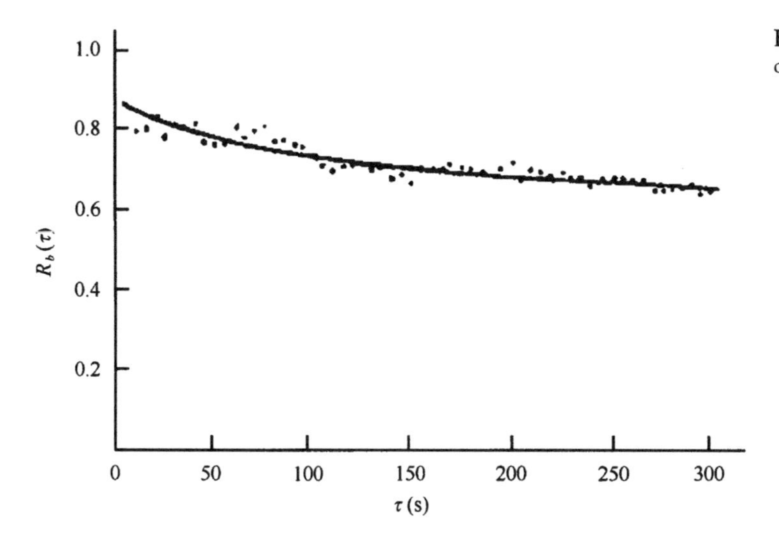

FIGURE 2.60 Pulse-to-pulse correlative coefficients versus t.

The pulse-to-pulse correlation coefficients $R(\tau)$ versus τ in shallow-water acoustic channels at the distance of 45 km are shown in Fig. 2.60. We can see that the steady part of $R(\tau)$ is about 0.7 for a single frequency pulse signal with the duration of 5 ms, which generally corresponds to the bandwidth of 200 Hz. The spreading waveform after passing through the channels is shown in Fig. 2.61.

Time-variant character measured by using pulse-to-pulse correlation method for a single frequency (1.6 kHz) pulse signal with duration of 10 ms at 7.5 km is plotted in Fig. 2.62. The corresponding spreading waveform is shown in Fig. 2.63.

By observing Figs. 2.60 and 2.62, we should note that the pulse-to-pulse correlation coefficients are roughly reduced by obeying the exponential law, but they will tend to a steady value when τ is increased to a certain value. So, in spite of the transmission characteristics of the channels being very complicated, a steady mean part of the coefficients always exists. That would be a major basis to correctly detect arriving signals by using corresponding signal processing schemes.

Fig. 2.64 shows the time correlativity for shallow-water acoustic channels. The time correlation coefficients at 16 and 31 km are shown in Fig. 2.64A and B, respectively. The center frequency is 700 Hz, and the bandwidths are 50, 100, and 200 Hz, respectively.

The spatial correlation coefficients $R(\rho)$ versus distance ρ acquired by two measured results in the same channel mentioned earlier are shown

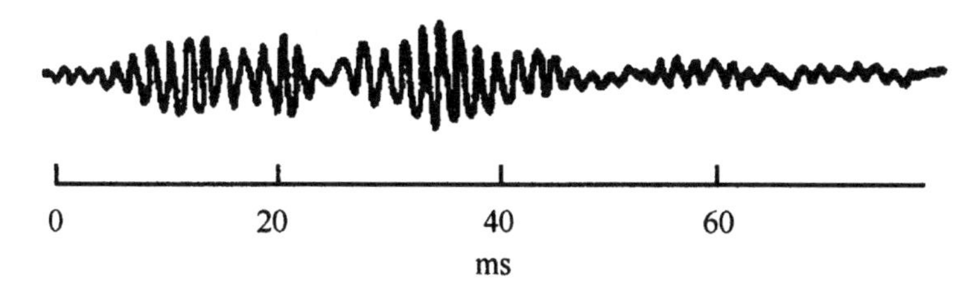

FIGURE 2.61 Received waveform for a pulse signal.

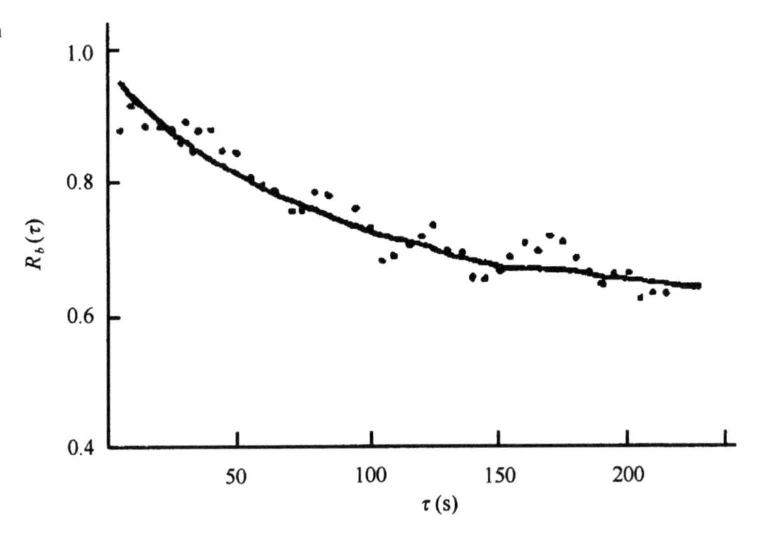

FIGURE 2.62 Time variability for a 7.5 km transmission channel.

in Fig. 2.65, where the center frequency is 400 Hz and the transmission range is 35 km.

We have performed the experimental investigation with respect to pulse-to-pulse correlativity combining with the development of a sound releaser (refer to Chapter 4) in the Xiamen Harbor, and typical measured results are shown in Fig. 2.66. The real and virtual curves represent the pulse-to-pulse correlation coefficients at the ranges of 1.8 and 3.7 km, respectively. The center frequency is 20 kHz, which was employed in the telecontrol communication of the releaser, and the duration of the pulse is 5 ms, with sea state 3 appearing during the experimental interval.

According to the analyses mentioned before, we can see that the spatial-temporal correlativity of underwater acoustic communication channels closely depends on the properties of the channels, operating frequencies, communication ranges, etc. Therefore, it is necessary to carry out corresponding predictions or measurements in situ for them, as these data are valuable for designing a communication sonar in which correlation detection schemes are employed.

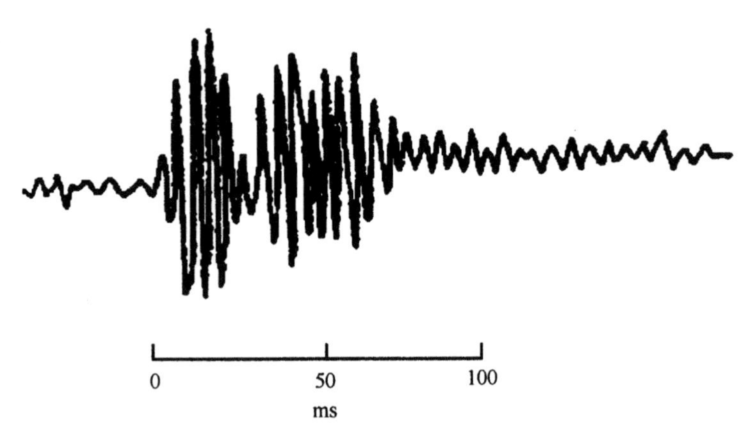

FIGURE 2.63 Spreading signal received due to multipath propagations.

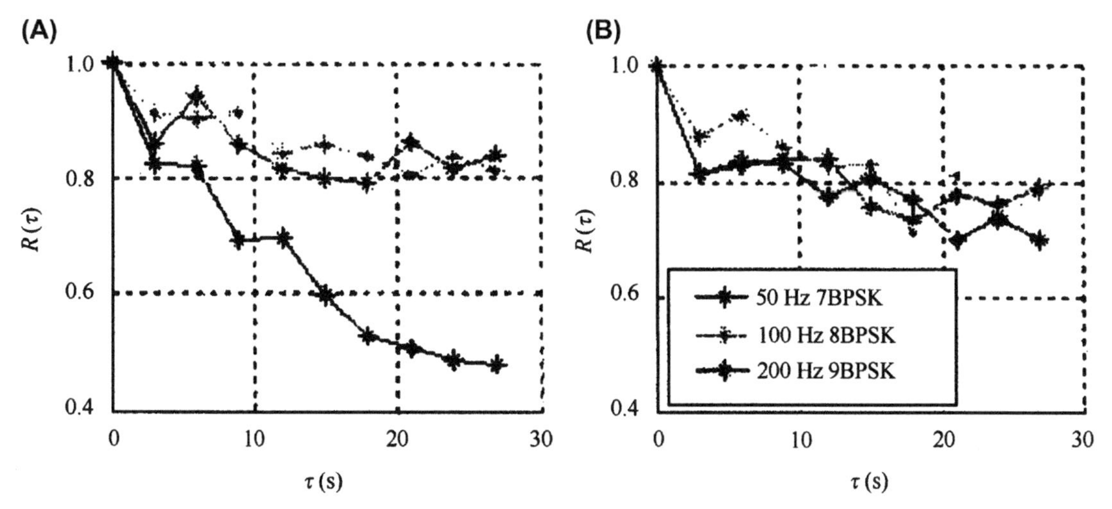

FIGURE 2.64 Time correlativity for shallow-water acoustic channels. (A) 16 km; (B) 31 km.

2.4.3 Impacts of the Fluctuations of Transmitted Sound on Digital Underwater Acoustic Communications and the Possible Countermeasures to Adapt to the Fluctuations

2.4.3.1 *Impacts of the Fluctuations of Transmitted Sound on the Communications*

2.4.3.1.1 FLUCTUATION IN THE AMPLITUDE OF THE SOUND SIGNALS

The fluctuation in the amplitude of sound signals will cause some signal code elements to remarkably fade, which is equivalent to input SNR being reduced, or even some code elements will be lost if their amplitudes are thus less than the threshold of amplitude decision. Therefore BER will rise, or the communications could even be inefficient.

There are the two major reasons to cause the fluctuation in the amplitude that must be paid attention to.

1. Random inhomogeneity in the body of the seawater

   According to Eq. (2.183), we can see that the mean square value of the fluctuation in the amplitude of sound signals traveling over

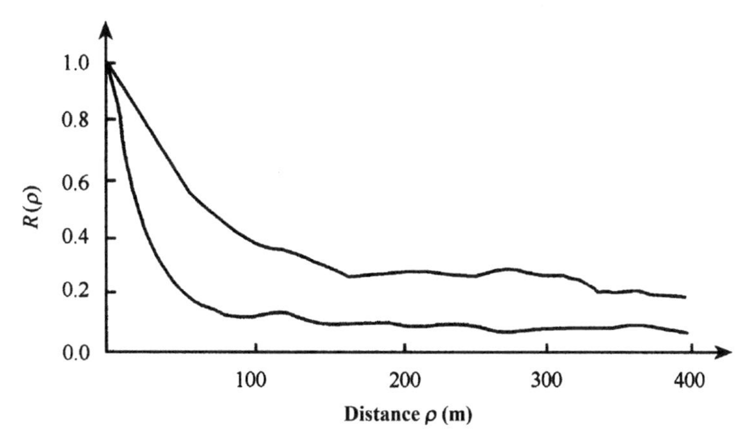

FIGURE 2.65 Spatial correlativity for underwater acoustic channels.

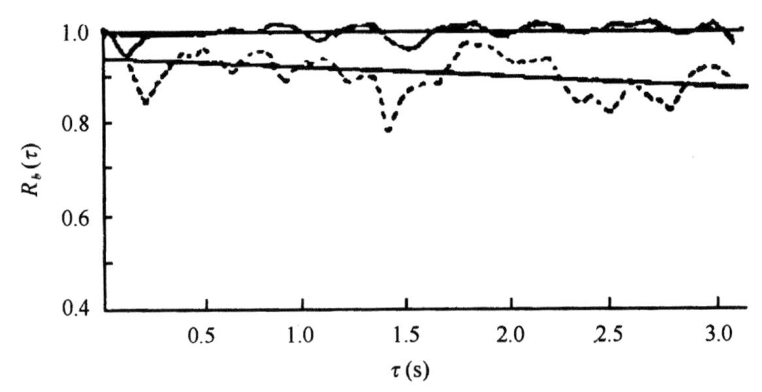

FIGURE 2.66 Pulse-to-pulse correlativity in Xiamen Harbor shallow-water area.

weak, randomly inhomogeneous media under the condition of the wave parameter $D \gg 1$ is proportional to $\overline{\mu^2}$, f^2, and L. Therefore the two operating situations in which the fluctuation in the amplitude must particularly be paid attention to are as follows.

One is relative to long-range digital underwater acoustic communication. In such a case, large transmission loss TL will encounter. In addition, the fluctuation in the amplitude has an accumulating effect with increasing communication ranges. Therefore the input SNR will remarkably be reduced. Perhaps we have to detect the information signals under the condition of weak SNR.

The other is with respect to high-frequency underwater acoustic communications. Since TL will remarkably rise with increasing operating frequency, the fluctuation in amplitude will also be more violent. Therefore BER will also rise. It means that high operating frequencies are only adaptively employed in short-range underwater acoustic communications.

2. Sound reflection and scattering from the sea boundaries

Sound interference effect, which is caused by the superposition between direct and reflected sound waves from the sea boundaries, will lead to severe amplitude fading.

The sound scattering from the sea boundaries causes not only the attenuation of the regular component of the sound field, but it also brings about the fluctuation of the sound field due to the superposition between the regular component of the sound field and scattering sound waves; the correlativity of the sound field will weaken too.

We have also known that the received signals at long-range underwater acoustic communications basically come from multipath propagations. Obviously, those signals have a behavior of random fluctuation in amplitude. Moreover, the behavior remarkably depends on specific communication channels, operating frequencies, incident angles, etc. Therefore, how to efficiently convert the energy of multipath propagations to useful signals to be detected is more difficult.

2.4.3.1.2 FLUCTUATION IN THE PHASE OF SOUND SIGNALS

The applications of coherent detection schemes, such as PSK and SS-DS (refer to Chapter 3), that are based on the phase of the sound signal to be detected correctly are restricted. They are even inefficient in digital underwater acoustic communications due to the fluctuation in the phase of sound signals.

Similarly, accurately tracing synchronizing signals, which is a key technique in digital communications, by adopting coherent detection schemes must be considered with caution.

There are still two main reasons to produce the fluctuation in the phase that must be paid attention to.

1. Fluctuation in the phase caused by the inhomogeneity in the body of the seawater
   The fluctuations in the phase and amplitude have the similar laws under some specific conditions, as expressed in Eq. (2.183). Therefore the impacts of the fluctuation in the phase due to the inhomogeneity on long-range or high-frequency digital underwater acoustic communications will be remarkably violent.
2. Fluctuation and sudden change in the phase due to the sound reflection from the sea boundaries
   The additional phase shifts of the sound wave reflected from the sea surface are determined by Rayleigh parameter K, as expressed in Eq. (2.198). We note that the additional phase shifts are proportional to the rms wave heights, operating frequencies, and $\sin\chi$. That is to say, the shifts are a random variable. Provided $K \to 0$, the sea surface can be considered to be acoustically smooth, and in such a case, the phase shift tends to 180 degrees.
   The phase variations of sound reflection from the sea bottom are more complex (refer to Figs. 2.55 and 2.57) than those from the sea surface. The reflected phase shifts from the former depend on the properties of the sea bottom, grazing angles, operating frequencies, etc. Obviously, they are also a random variable.
   Obviously, how to adapt to the random rapid fluctuation in the phase in digital underwater acoustic communications is a complicated and difficult problem. Therefore, the application of coherence detection schemes to communication sonars will be limited.

2.4.3.1.3 FLUCTUATION IN THE FREQUENCY OF SOUND SIGNAL

In the case of fitted-point communications, experimental results had proven that the frequency shifts of sound signals caused by the inhomogeneity in the body of the seawater or the reflections from the sea boundaries are only several Hz, which may generally be neglected in some digital underwater acoustic communications, as they are much less than the bandwidth for a communication sonar.

However, greater Doppler frequency shifts will be encountered for high-speed mobile communications. Because there is not the adaptability of the frequency shifts for the usual signal processing systems (such as matched filter, OFDM) employed in underwater acoustic communications, a wider bandwidth has to be arranged in the systems. To lower the disadvantageous impacts by using a broad bandwidth, such as NL, will raise with increasing that, so some effective approaches to adapt to the greater frequency shifts must be adopted. Of course, it is also a complicated problem to correct the great Doppler frequency shifts for high-speed mobile underwater acoustic communications. We know that a similar problem has been encountered in active sonars.

2.4.3.1.4 SPATIAL-TEMPORAL-FREQUENCY VARIABILITY OF THE FLUCTUATIONS OF SOUND SIGNALS

We know that the fluctuations of sound signals are randomly variable in spatial-temporal-frequency domains. Obviously, how to adapt to such a communication circumstance is quite difficult.

However, the spatial-temporal correlativity of sound signals in underwater acoustic communication channels is much better than that of the marine noise in corresponding frequency ranges (refer to Section 2.5). It is a basic condition to employ spatial-temporal correlation signal processing schemes in digital underwater acoustic communications, and excellent signal detection results will be obtained (refer to Chapter 4).

2.4.3.2 *Possible Countermeasures to Adapt to the Fluctuations of Sound Signals*

2.4.3.2.1 GENERAL MEASURES TO REDUCE THE FLUCTUATIONS

1. Appropriately raising sound source level *SL* will raise the total input SNR and satisfy the preset requirement for BER at the output terminal. In fact, a signal fading in amplitude is equivalent to such a *SL* having a random, fluctuating character.

   The limitations by raising *SL* have been discussed with respect to the sound transmission loss *TL* in Section 2.2.
2. It is an efficient method to lower operating frequencies, including *TL* that will reduce with lowering the operating frequencies, which is equivalent to raising *SL*.

   The fluctuations in the phase and amplitude of sound signals due to the inhomogeneity in the body of the seawater will weaken.

   The regular components of reflected sound signals from the sea boundaries will correspondingly be improved.

   Doppler frequency shift will proportionally be decreased.

   Tracing the signals will be easier.

   However, to obtain these advantages, some prices have to be paid; for example, lowering operating frequencies means that the communication data rates will possibly be reduced. In particular the multipath effects will be violent. So, an optimum operating frequency for a specific communication sonar must be selected in detail, which is relative to its whole performance.
3. To adapt to the randomly spatial-temporal-frequency variable fluctuations of received signals, tolerant designing tactics would be adopted.

   Because of the complicated and variable peculiarity of the fluctuations of sound signals in underwater acoustic communication channels, it is difficult to design a suitable signal processing system employed in digital communication sonars with robust performances.

   To improve the robustness of communication sonars, we must pay attention to avoid the possible "threshold effect": provided the input SNR is lowered to a certain value, the performances of digital communications will severely be reduced. Therefore, presetting the relative parameters of communication sonars, including $(DT)_i$, *SL*, *TL*, and *NL* must give adequate room for adapting to unexpected large fluctuations of signals. Of course, in such a designing tactic, some specifications of communication sonars, such as maximum communication ranges and data rates, will correspondingly be reduced.
4. Selecting a suitable modulation scheme is one of the basic approaches to adapt to the random fluctuations of sound signals.

   Obviously, some signal processing schemes, such as ASK and QAM, are unsuitable to employ in digital underwater acoustic communications, which have violent fluctuations of signals. Generally speaking, in such cases, especially for long-range underwater acoustic communications where multipath pulses are converted to the signals being processed, using noncoherent detection schemes gives a higher robustness than those of coherent detection ones. For example, it would be better to select an FM scheme since it is less sensitive to the fluctuations of signals.

2.4.3.2.2 SOME SPECIFIC TECHNIQUES FOR ADAPTING TO THE SIGNAL FLUCTUATIONS

1. Utilizing the usual amplitude equalizer to overcome the fluctuation in the amplitude of sound signals.

   Primary experiments have demonstrated that it is efficient to combat the medium fluctuation in the amplitude, as shown in

Fig. 2.67. The input and output of the amplitude equalizer system are plotted at its upper and lower parts, respectively. Input amplitudes between two examining signals have a magnitude difference of about 20 dB. Their output amplitudes approximately equal after applying this equalizer. The tracing time is 0.25 ms. Experiments have also shown that it is inefficient to use that to overcome violent signal fluctuations.

2. Adopting suitable channel coding schemes to adapt to signal fluctuations is efficient, in particular for low data rate underwater acoustic communications. For example, a decision model of binomial distribution law was employed in the telecontrol communication in an underwater acoustic releaser (refer to Chapter 4) with data rates to be 20 bps. We can transmit M pulse signals and take M_0 as the threshold numbers to decide telecontrol instructions, ie, allow $(M - M_0)$ pulses to be lost due to the fluctuations of signals. The robustness of the telecontrol communication had remarkably been improved.
3. Utilizing suitable diversity techniques to adapt to the signal fluctuations is also efficient.

   Since the fading of sound signals in underwater acoustic channels has the property of frequency selection, the frequency diversity is thus efficient (refer to Chapter 3) to combat that.

   Similarly, the fluctuations of sound signals depend on the positions of both transmitting and receiving transducers, so the spatial diversity is also efficient.
4. Adaptive signal processing techniques are basic approaches to combat the fluctuations of signals. Following are some examples.
   It is expected that by means of adaptive paths diversity (Rake receiver) one can realize the optimum signal detection against the background of randomly fluctuating multipath propagations.

   Adaptive total time sampling processing for Pseudorandom signal sequences could replace conventional synchronizing signal detection schemes and has a prospect to solve the difficult problem accurately tracing that in the channels with violent fluctuation in the time domain.

   The adaptive correction of Doppler frequency shift is necessary for some high-speed mobile communications by using a relative corrective network based on the measurements in situ at some typical frequency codes.

   These adaptive signal processing techniques combating against the fluctuations of transmitted sound in underwater acoustic communication channels will be discussed in detail in Chapters 3 and 4.

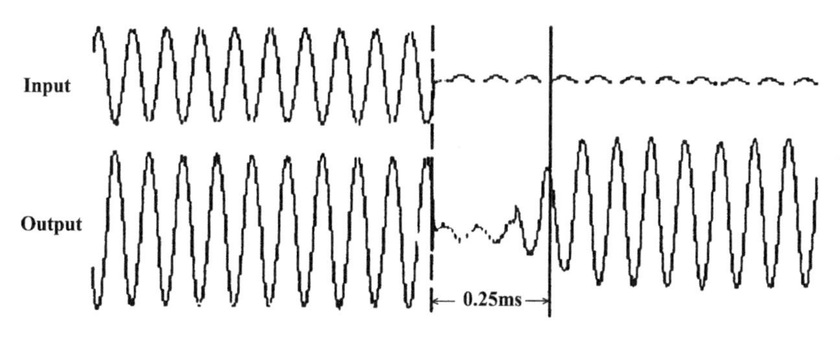

FIGURE 2.67 Input and output of amplitude equalizer.

2.5 NOISE IN THE SEA

Noise in the sea, including ambient noise and the radiated noise of ships, is an interference background field in underwater acoustic channels. It is necessary to predict SNR in underwater acoustic engineering to estimate corresponding communication specifications, such as transmission ranges and data rates. In particular, the spatial-temporal correlativity of the noise in the sea must emphatically be analyzed. Therefore, researching noise fields, just as the signal field, is an important subject.

The noise in the sea is generally considered an additive interference and described by the noise levels *NL*. As viewed from anti-noise, it is necessary to fully utilize the differences of the spatial-temporal correlativity between noise and signal, which are important parameters to develop a new signal processing systems employed in communication sonars.

First, the wave equation without sources will be extended to that with sources in fluid media in this section. According to this equation, any inhomogeneous pressure fields in the media, including vibrating small air bubbles, turbulence, etc., will originate sound waves. Then, *NL* and statistical characteristics for both the ambient noise and the self-noise of ships will be introduced, respectively. Finally, the impacts of the noise in the sea on digital underwater acoustic communications and corresponding countermeasures to combat against that will be discussed.

2.5.1 Wave Equation With Sources in Fluid Media [21]

2.5.1.1 Basic Physical Behaviors of Noise Sources

Any processes generating a nonstationary pressure field, including the vibration and pulsation at fluid boundaries or nonstationary forces acting on fluid, turbulence, temperature fluctuation, etc., will excite sound waves. Every noise source may be described according to its main physical mechanism.

Every basic physical mechanism generating sound pressure fields may correspond to the orders of multipoles occupying superiority in mathematics; a major simple sound source consists of the fluctuation of volume or mass that belongs to the pole of zero order or monopole, such as pulsing air bubbles, cavitations, etc. Sound waves radiated by the monopole are nondirectional, which is described by the scalar quantity F1, and correspond to the spherical vibration of zero order in mathematics. Pulsing force and the vibration of a rigid body will originate even-pole sound sources with the directional pattern of direction cosine, which is expressed by $\frac{\partial F_2}{\partial x}$ in spatial coordinates, ie, the gradient of scalar quantity, and they belong to the spherical vibration of one order in mathematics. As a double even-pole, or four pole, the turbulence in fluid media is described by $\frac{\partial^2 F_3}{\partial x_i \partial x_j}$, which is the spherical vibration of two orders. The sources of the monopole and even-poles only appear at fluid boundaries; while those of the double even-poles can appear in a fluid body itself and in free turbulence areas.

The three kinds of multipole noise sources with the lowest orders, which are well known, are shown in Fig. 2.68.

2.5.1.2 Wave Equation With Sources in Fluid Media

We know that the wave equation with respect to sound pressure in elementary acoustics (including the supposition without sources) is given by

$$\nabla^2 p - \frac{1}{c_0^2}\frac{\partial^2 p}{\partial t^2} = 0$$

We except that the wave equation with sources is

$$\nabla^2 p - \frac{1}{c_0^2}\frac{\partial^2 p}{\partial t^2} = Q \qquad (2.222)$$

where Q is called a source function, including noise sources generated by fluctuating mass sources, fluctuating force sources, and shear stress sources caused be turbulence, etc.

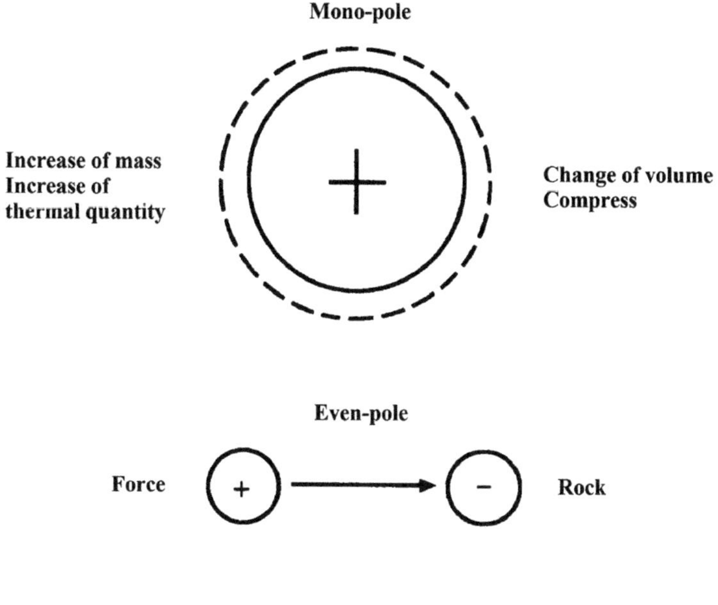

FIGURE 2.68 Three types of sound sources.

The effects of the fluctuating mass and force as noise sources on a wave equation had been described by Stokes and Rayleigh long ago. By considering the fluctuating mass sources, the equation of continuity is given by

$$\frac{\partial \rho}{\partial t} + \frac{\partial (\rho v_i)}{\partial x_i} = q \tag{2.223}$$

Eq. (2.223) is expressed by tensors. q is the ratio generating a new mass in unit volume.

Consider the existence of the force sources, and motion equation is given by

$$\frac{\partial (\rho v_i)}{\partial t} + \frac{\partial (\rho v_i v_j)}{\partial x_j} = -\frac{\partial p}{\partial x_i} + f_i \tag{2.224}$$

where f_i is the pure force that is acted on by external mechanical force on unit volume of fluid (not including the viscous force).

Lighthill then discovered that the shear stresses of turbulence have also an effect on sound sources. He pointed out the jet noise of an airplane could not be explained simply by mass and force sources, and he considered the pulsing fluid as those radiated sources.

Consider the shear stresses of the turbulence; the motion equation becomes

$$\frac{\partial (\rho v_i)}{\partial t} = -\frac{\partial p_{ij}}{\partial x_j} - \frac{\partial (\rho v_i v_j)}{\partial x_j} + f_i \tag{2.225}$$

Eq. (2.225) is called the Lighthill motion equation, where stress tensors

$$p_{ij} = p\delta_{ij}, \quad \delta_{ij} = \begin{cases} 0, & i \neq j \\ 1, & i = j \end{cases}$$

where δ_{ij} are unit tensors.

According to Eqs. (2.223) and (2.225), by combining the usual state equation, the wave equation with sources in fluid media may be derived as

$$\nabla^2 p - \frac{1}{c_0^2}\frac{\partial^2 p}{\partial t^2} = -\frac{\partial q}{\partial t} + \nabla \cdot \mathbf{f} - \frac{\partial^2 T_{ij}}{\partial x_i \partial x_j} \tag{2.226}$$

Eq. (2.226) is called the spreading mode Lighthill equation, where

$$T_{ij} = \rho v_i v_j + p_{ij} - c_0^2 \rho \delta_{ij}$$

is called the Reynolds fluctuating stress tensors that include noise sources generated by the fluctuations of velocity, stress, and density. Generally, $T_{ij} \approx \rho v_i v_j$.

The three terms on the right-hand side of Eq. (2.226) express the major radiation sources in fluid media. The first term expresses the nonstationary mass flow passing through fluid boundaries, which is equivalent to the mono-pole, and its source function is described by the scalar quantify function. The second term expresses the divergence of the nonstationary force acting on a certain fluid boundary, which has the behaviors of the even-poles, and its source function is described by the first-order derivative of scalar quantity. These two kinds of sound sources mentioned before had been regarded by Stokes and Rayleigh. The third term indicates the stresses of turbulence in fluid media itself. Lighthill had derived and proved that these noise sources have the behaviors of double even-poles, and they may be expressed by the second-order derivative of scalar quantity $\partial^2 T_{ij}/\partial x_i \partial x_j$. Therefore, we can consider that there are noise sources of the double even-poles that randomly distribute in the fluid media in which turbulence exists.

Since pressure fluctuations appear due to the fluctuations of velocity in turbulence areas, the transducer will be excited and vibrated to originate turbulence noise. Obviously, the turbulence noise travels with the velocity of turbulence and thus is called "pseudo-sound."

The turbulence usually exists in the sea because there is a current with large Reynolds numbers in it. Therefore the noise caused by the turbulence is an important part of the ambient noise in the sea at low frequency ranges.

A typical kind of the turbulence noise is excited by fluid jets, which would be considered for high-speed sailing, but understanding for that is lacking at present.

According to the similarity theorem in fluid mechanics, turbulence noise intensity $I^{\sim}v^8$, where v is the average velocity of the turbulence. We see that I will violently rise with increasing v.

Obviously, to obtain the solution of Eq. (2.226) is very difficult. As an example, we will take the most simple noise model, which is the monopole noise source, to state that.

In such a case, Eq. (2.266) may be simplified as

$$\nabla^2 p_1 - \frac{1}{c_0^2}\frac{\partial^2 p_1}{\partial t^2} = -\dot{q} \tag{2.227}$$

Since material cannot be generated in the internal of fluid media, any pulsation of mass must be generated at the fluid boundaries, ie, there are not any sound sources inside the body of fluid media. Therefore, the wave equation is efficient; moreover, the amplitudes of sound signals are only controlled by the term on the right-hand side of Eq. (2.227). Provided the sizes of sound sources are smaller than the wavelength of the sound wave, the solution of Eq. (2.227) is

$$p_1(r,t) = \frac{\dot{Q}(t - r/c_0)}{4\pi r} = \frac{\dot{Q}(t')}{4\pi r} \tag{2.228}$$

where r is traveling range, and t' is time delay; moreover,

$$Q \equiv \int_s q\mathrm{d}V = \frac{\mathrm{d}}{\mathrm{d}t}(\rho_0 V) \quad (2.229)$$

Since the fluctuation of the density in fluid media may be neglected, Eq. (2.228) can thus be simplified as

$$p_1(r,t) = \frac{\rho_0 \ddot{V}(t')}{4\pi r} \quad (2.230)$$

This is the most general expression of sound radiation by a monopole (as a small pulsing body) noise source, whose amplitude is proportional to the density of fluid medium and the acceleration of volume and is inversely proportional to the traveling range.

The vibrating small air bubbles in the seawater are typical monopole noise sources, whose volume pulsation can be processed as a mass-elasticity system, in which inertia is provided by additional fluid mass, the effect of the adiabatic compression of air is analogous to a spring, and the resistances of motion are caused by the viscosity of fluid, heat conduct loss, and sound energy radiation. Therefore, we may arrange the second-order differential equation of air volume V and find the corresponding $\ddot{V}$; the radiation sound pressure, corresponding sound intensity, and power at resonant state may be obtained according to Eq. (2.230).

Instantaneous sound power can be expressed as [22]

$$W_N = \frac{\rho_0 \omega_0^4 (\Delta V)^2}{8\pi c_0} e^{-\eta\omega_0 t'} \quad (2.231)$$

where ω_0 is the resonant angle frequency of the air bubbles, ΔV is the difference between instantaneous disturbance and equilibrium volumes, $\eta = \frac{r_0}{\omega_0 m_0}$ is loss factor in which r_0 is resistance coefficient, and m_0 is an additional equally vibrating mass.

2.5.2 Ambient Noise in the Sea [5]

In a loose manner of speaking, ambient noise may be said to be the noise of the sea itself. It is that part of the total noise background observed with a nondirectional hydrophone. The ambient noise is an interference background field in underwater acoustic channels, which may be described by its spectrum levels.

2.5.2.1 Deep-Water Ambient Noise

During World War II, many observations of deep-water ambient noise between 500 Hz and 25 kHz indicated a direct connection between sea state or wind force and the level of ambient noise. Based on these observations, the well-known Knudsen spectra were obtained, having sea state or wind force as a parameter, as shown in Fig. 2.69.

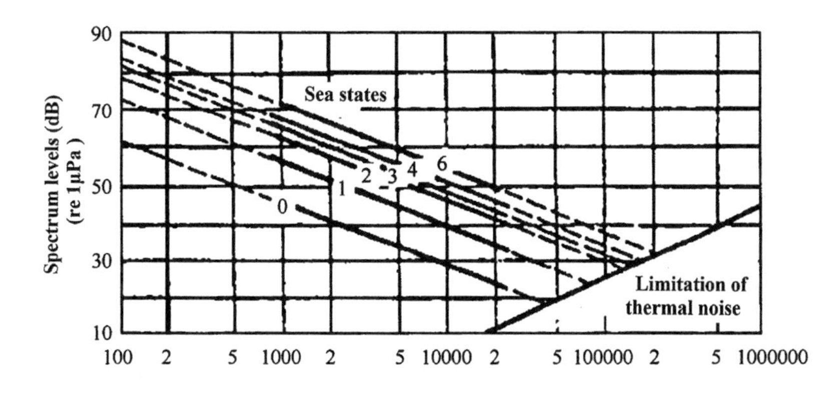

FIGURE 2.69 Kundson spectra of the ambient noise in the deep sea.

Although it is clear that the sea surface must generate the major portion of the ambient noise in this frequency range, the process by which it does so is still uncertain. Perhaps that includes the following:

The most obvious of the processes are breaking whitecaps, which must produce crash noise when breaking occurs.

Another possible source of noise is the flow noise produced by the wind blowing over the rough sea surface.

Another possibility is cavitation, or the collapse of air bubbles formed by turbulent wave action in the air-saturated, near-surface waters.

Still another possible noise-producing process is the wave-generating action of the wind on the surface of the sea.

For prediction purposes, average representative ambient noise spectra for different conditions are required. Such average working curves are shown in Fig. 2.70 for different conditions of shipping and wind speed. The ambient noise spectrum at any location at any time is approximated by selecting the appropriate shipping and wind curves and fairing them together at intermediate frequencies where more than one source is important. The "heavy-shipping" curve is used for locations near the shipping lanes of the North Atlantic; the "light-shipping" curve is appropriate for locations remote from ship traffic.

The deep-sea ambient noise had been thought of as omni-directivity. A lot of experiments in situ have proven that the noise has directivities relative to the following three factors:

1. spatial-distributing behaviors of noise sources,
2. spatial radiation characters of noise sources,
3. propagating conditions in deep-sea sound channeling.

Fig. 2.71 illustrates the distribution of the ambient noise in the vertical direction at frequencies of 112 and 1414 Hz [23]. These are polar plots of the ambient intensity per unit solid angle $N(\theta)$ arriving at a bottomed hydrophone as a function of vertical angle θ. At 112 Hz, more noise appears to arrive at the hydrophone from the horizontal than from the vertical, with a

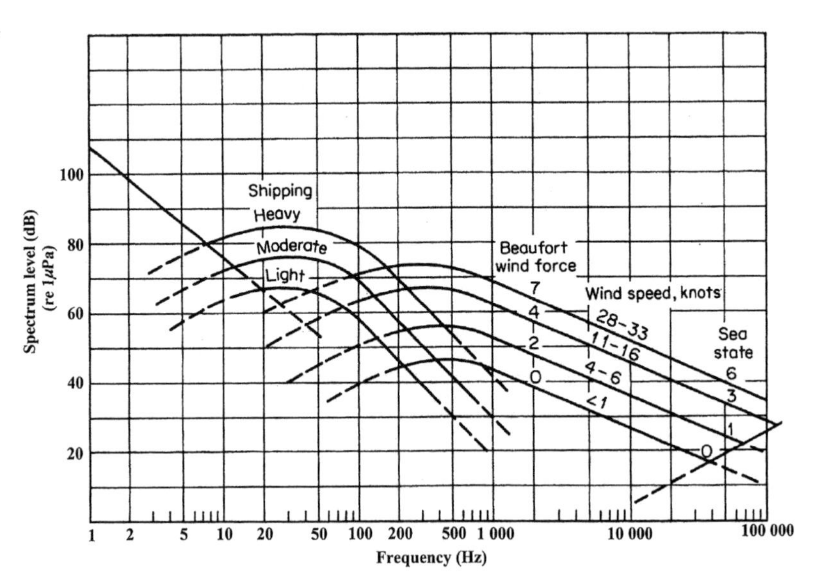

FIGURE 2.70 Average deep-water ambient noise spectra.

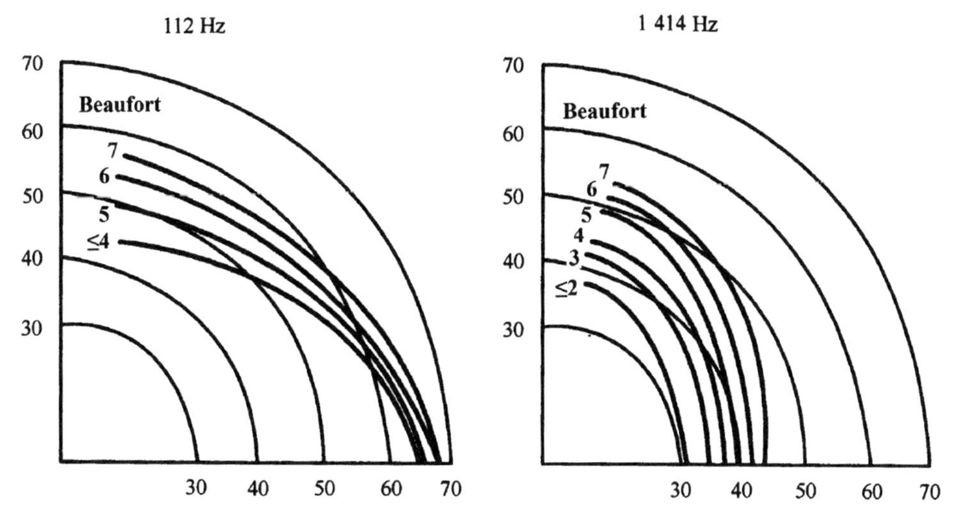

FIGURE 2.71 Distribution of ambient noise in the vertical at a bottomed hydrophone.

difference that diminishes with increasing wind force. At 1414 Hz, the reverse is the case, with more noise arriving from overhead than horizontally, with a difference that increases with increasing wind force.

This directional behavior is consistent with the view that low-frequency noise originates at great distances and arrives at the hydrophone via primarily horizontal paths, whereas high-frequency noise originates at the sea surface more nearly overhead.

2.5.2.2 *Shallow-Water Ambient Noise*

In contrast to the relatively well-defined levels of deep-water ambient noise, the ambient levels in coastal water and in bays and harbors are subject to wide variations. In such locations, the sources of shallow-water noise are highly variable, both from time to time and from place to place.

At a given frequency in shallow-water the noise background is a mixture of three different types of noise: (1) shipping and industrial noise, (2) wind noise, and (3) biological noise. At a particular time and place, the "mix" of these sources will determine the noise level, and because the mix is variable with time, the existing noise level will exhibit considerable variability from time to time and from place to place. As a consequence, only a rough indication can be given of the levels that might be found in bays and harbors and at offshore coastal locations.

A great many measurements of noise inside different bays and harbors were made during World War II. Some examples of the spectra resulting from these measurements are given in Fig. 2.72. Included in the figure is an average line CC showing the average of many determinations between 20 and 200 Hz and is a shaded area giving the location of other measurements at subsonic frequencies.

The typical noise spectrum levels measured in the Xiamen Harbor at the frequency ranges employed in usual civil underwater acoustic communications is shown in Fig. 2.73, which is closely related to the three types of the noise mentioned earlier.

NL is majorly determined by the wind speed in coastal locations. The noise spectrum levels measured in the Yellow Sea at the wind forces 1 to 6 are shown in Fig. 2.74 [1]. We see that

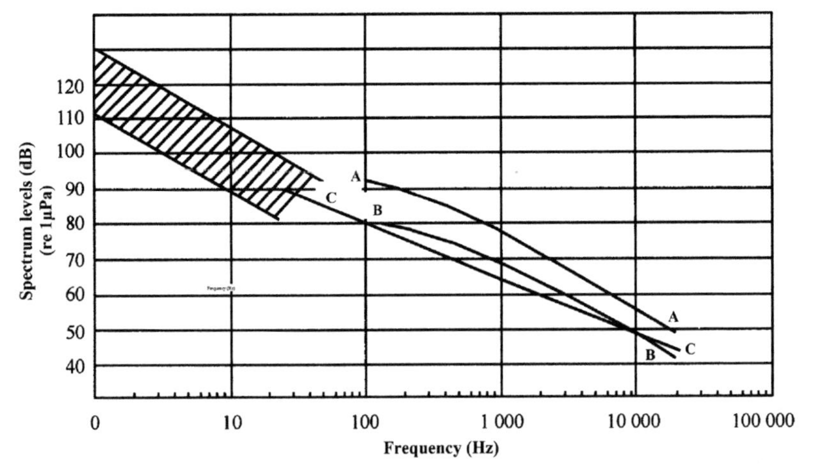

FIGURE 2.72 A summary of noise levels in bays and harbors. AA: A high noise location, a harbor in the daytime. BB: An average noise location. CC: Average of many measurements.

the noise spectrum levels in coastal locations are higher, above 5 dB, than those in deep-water locations.

2.5.2.3 *Intermittent Sources of Ambient Noise*

2.5.2.3.1 BIOLOGICAL SOUNDS

The sounds produced by biological organisms in the sea are many and varied, and they have been extensively studied. Three groups of marine animals are known to make sound: certain kinds of shellfish (Crustacea), certain kinds of true fish, and the marine mammals (Cetacea), such as whales, dolphins, and porpoises. Among the Crustacea, the most important are snapping shrimp, which are ubiquitous inhabitants of shallow tropical and semitropical waters having a bottom of rock, shell, or weed that offers the animals some concealment. These animals make noise by snapping their claws together, as one snaps thumb and forefinger, and thereby produce a broad spectrum of noise between 500 Hz and 20 kHz. Among the fish that should be mentioned are the croakers, which make an intermittent series of tapping noises like that of a woodpecker by means of the contraction of drumming muscles attached to the air bladder; the same principle is involved in beating a

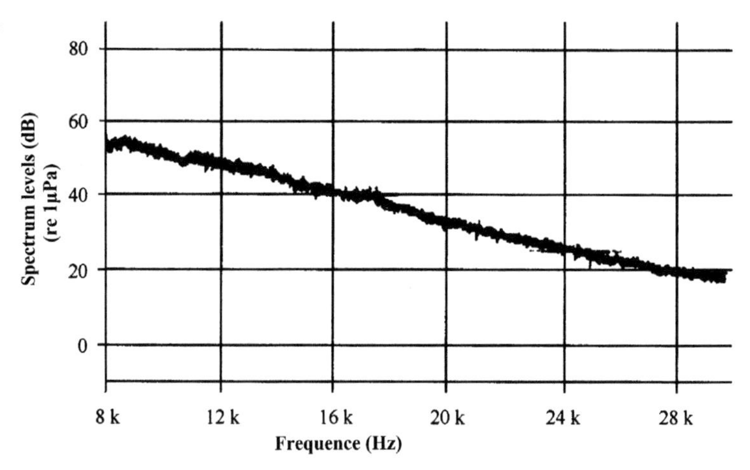

FIGURE 2.73 Spectrum analyses for ambient noise in the Xiamen Harbor.

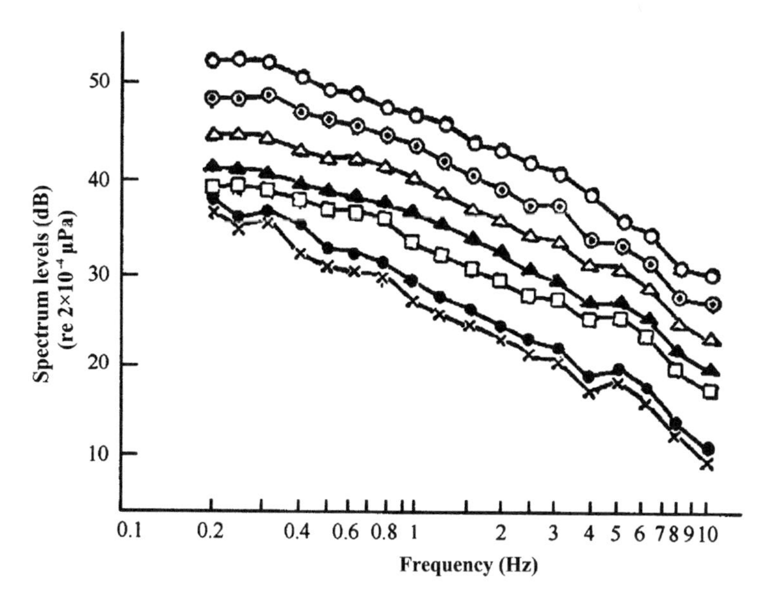

FIGURE 2.74 Spectrum levels of noise in a coastal location versus different wind forces.

drum. The Cetacea include whales and porpoises, which create noise by blowing air through the larynx. Porpoises, for example, produce frequency-modulated whistles that are associated with certain behavior patterns of these animals. Many peculiar chirps, grunts, booms, groans, yelps, and barks heard in the sea are of biological origin.

The experimental researches on the behaviors of the marine biological sounds in the Xiamen Harbor and the Zhou-shan sea area have been carried out by Zhang et al. [24–26].

They had discovered that the probability of exhibiting marine biological evening choruses achieves 86% in the Xiamen Harbor at the time intervals of the astronomical small tide in spring, summer, and autumn. Their continued time are all up to several hours, with noise spectra that mainly distribute in a frequency range from 700 to 1600 Hz, and the major peaks in these seasons are at 80, 1000, and 1250 Hz, respectively. The noise spectrum levels at the intervals at which the evening choruses appear are much higher than those of the ambient noise at sea state 0. For example, an augmentation of 40 dB in a frequency range from 1 to 1.25 kHz has been acquired, as shown in Fig. 2.75, which is the highest noise spectrum level at 1 kHz according to the open reports at that time. This evening chorus with such a high intensity appeared in such a time interval before a strong typhoon would pass through the Xiamen sea area. Weather was particularly hot and suffocating at this interval. We can see that the biological sounds are related to weather conditions, and in particular the high wind noise due to a typhoon will affect the biological sounds. Based on the characters of the noise spectra, the noise sources would be *Johnins belengerii*.

Zhang had also observed the biological sounds in the Zhou-shan sea area. The sounds caused by marine animals had been also discovered, including dawn, noon, and evening choruses. The noise spectrum peaks are below 800 Hz.

2.5.2.3.2 RAIN

Falling rain may be expected to increase ambient noise levels to an extent depending on the rate of rainfall and perhaps on the area over which the rain is falling.

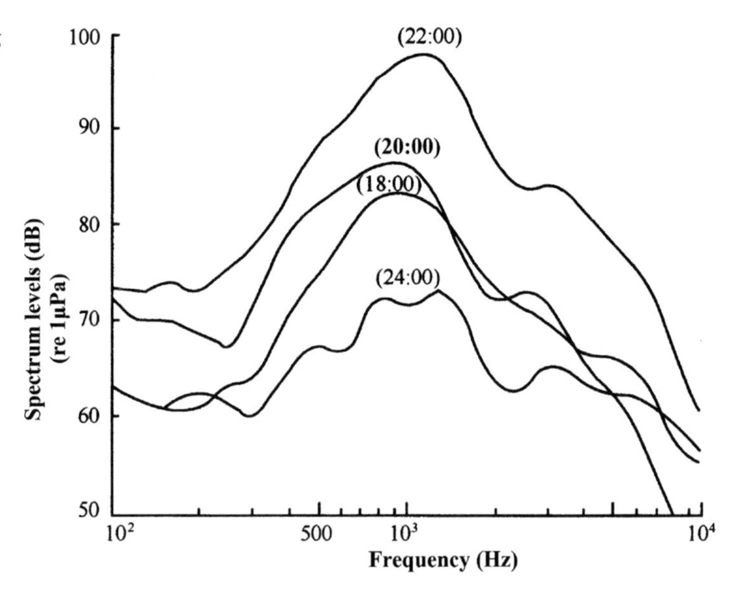

FIGURE 2.75 Spectrum levels of evening choruses caused by biological sounds.

Fig. 2.76 shows three spectra of ambient noise in rain of different intensity, together with the no-rain spectrum for the existing wind speed of 20 to 40 knots [27]. It will be observed that the spectrum of the noise of heavy rain is nearly "white" between 1 and 10 kHz, with a noise increase of 18 dB in a "heavy" rain at 10 kHz over the no-rain spectrum level. In particular, that belongs to an efficient operating frequency band employed in underwater acoustic communications, so we should pay fully attention to that.

2.5.2.4 Statistical Characteristics of the Ambient Noise in the Sea

The purposes researching on the statistical characteristics of the ambient noise in the sea are mainly to find the differences between noise and signal, in particular the differences in spatial-temporal correlativity, which may provide a physical base to design suitable signal processing systems that effectively combat against noise and to select specific relative parameters in the systems.

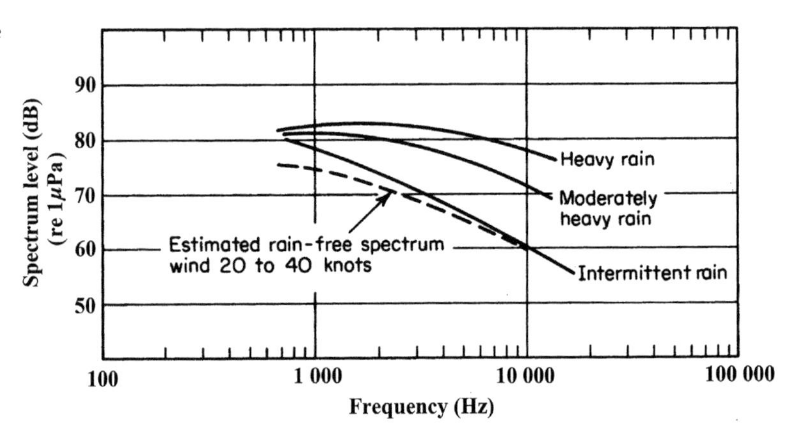

FIGURE 2.76 Ambient noise spectra in rain.

Generally speaking, marine noise belongs to a nonstationary random process. It is very difficult to describe such a process. Using a quasi-stationary scheme is the usual approach to solve that. The total time period being processed in which the random process to be nonstationary may be divided into several sub-periods in which they may be regarded as being stationary. Experiments have demonstrated that the quasi-stationary sub-periods generally are in the range from several seconds to several minutes in which we may analyze them as a stationary random process.

2.5.2.4.1 DISTRIBUTING LAWS OF THE AMBIENT NOISE IN THE SEA

2.5.2.4.1.1 INSTANTANEOUS PDF (PROBABILITY DENSITY FUNCTION) OF THE AMPLITUDE OF THE AMBIENT NOISE Some kinds of ambient noise in the sea, such as molecular motion thermal noise, air bubble pulsating noise, turbulence noise, etc., are consistent with the noise that is originated by a great many sources with random amplitudes and phases. Therefore, the instantaneous PDF of the amplitude for them in a quasi-stationary sub-period will obey Gaussian distributing law based on the law of large numbers.

2.5.2.4.1.2 PDF OF THE ENVELOPE OF THE AMBIENT NOISE Since the instantaneous PDF of the amplitude of the ambient noise in the sea generally obeys Gaussian distributing law, the PDF of their envelope will follow Rayleigh distributing law, as shown in Fig. 2.77A and B, respectively, where A/A_i is the relative amplitude. The relative data were acquired in the Xiamen Harbor and the eastern China Sea, respectively.

2.5.2.4.2 SPATIAL-TEMPORAL CORRELATIVITY OF THE AMBIENT NOISE IN THE SEA

Assume the ambient noise in the sea is a kind of generalized stationary random process in time domain, so its spatial-temporal correlation function would be

$$B_{ik}(\tau) = \overline{n(\mathbf{r}_i, t)n(\mathbf{r}_k, t-\tau)} \tag{2.232}$$

where $\mathbf{r}$ is the position vector from the original point of coordinate to observing point (x,y,z), and $\tau = t_2 - t_1$, which is independent of time reference points.

For particular situations:

If $i = k$, $B_{ik}(\tau)$ will reduce to the temporal correlation function of the noise.

If $\tau = 0$, $B_{ik}(\tau)$ will reduce to the spatial correlation function of the noise.

A normalized spatial-temporal correlation function (ie, a spatial-temporal correlation coefficient) is generally used to express the correlation characteristics of noise fields:

$$R_{ik}(\tau) = \frac{B_{ik}(\tau)}{N} \tag{2.233}$$

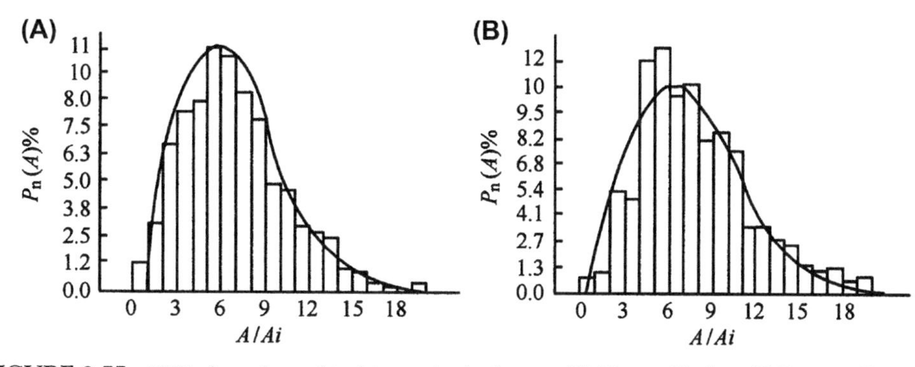

FIGURE 2.77 PDF of envelope of ambient noise in the sea. (A) Xiamen Harbor; (B) Eastern China Sea.

where N is noise power. This coefficient may be used to indicate alone the correlativity of the noise and is independent of its intensity.

The methods for researching the noise fields include both theoretical analyses and actual measurements in situ. We know that the noise fields have remarkable variability, and the theoretical analyzes are thus quite difficult. Therefore, experimental researches on them are important and available.

Some researched results by using the two methods mentioned will simply be introduced next.

It is necessary to establish some simplified models with respect to the ambient noise based on corresponding physical mechanisms in theoretical analyses. There are some models that are more consistent with measured data, and those that are easier to be processed in mathematics will be adopted.

1. Deep-sea distant field model
   In this model, the noise field is formed by the superposition of a great number of plane waves that are independent of each other, and they arrive at received points from the distant deep-sea areas in all directions. The noise field described by this model may be separated to the sum of infinite plane waves in space.
2. Model of isotropic noise field
   This is a simplified spatial distribution model. High-frequency noise (above 50 kHz), such as the thermal noise, approximately obeys this model. Of course, this part of noise is included in general noise fields.
3. Model of the noise field that is produced by the noise sources homogeneously distributed on the sea surface
   In this model, there exist independent noise sources that are distributed homogeneously and continuously on the sea surface, which follow a pattern of power directivity function $g^2(\theta,\omega)$ to radiate the noise into the sea, where θ is the angle drifting from vertical direction. Provided the depth where the hydrophone array is located is greater than the separated spaces between hydrophones, this model may be incorporated into the distant field model.

The results of theoretical calculations for the two typical noise fields in relation to the single-frequency spatial-temporal correlation coefficients of a single-side spectrum will be introduced as follows [28].

1. Isotropic noise field
   In such a case, the spatial-temporal correlation function is not only independent of the selection of the original point of a coordinate, but it also does not change with the rotation of the coordinate system. The spatial-temporal correlation coefficient of the isotropic noise field is thus given by

$$R_{ik}(\tau) = \frac{\sin\frac{\omega}{c}d_{ik}}{\frac{\omega}{c}d_{ik}}\cos\omega\tau \tag{2.234}$$

   where d_{ik} is the distance between two spatial points i and k.
2. The spatial-temporal correlation coefficient for the noise field is formed by the noise sources that are distributed homogeneously on the sea surface. Taking the power directivity function $g^2(\theta,\omega) = \cos^2\theta$, it is consistent with the assumption that the sound source and virtual source caused by the sound reflection from the sea surface will consist of an even-pole noise source. Moreover, it is easier to be processed in mathematics.

For the two hydrophones that are vertically arranged and spaced a distance d_{ik} apart, we can find

$$R_{ik}(\tau) = 2\left[\frac{\sin\omega\left(\frac{1}{c}d_{ik}+\tau\right)}{\frac{\omega}{c}d_{ik}} + \frac{\cos\omega\left(\frac{1}{c}d_{ik}+\tau\right)}{\left(\frac{\omega}{c}d_{ik}\right)^2} - \frac{\cos\omega\tau}{\left(\frac{\omega}{c}d_{ik}\right)^2}\right] \tag{2.235}$$

For the two hydrophones arranged horizontally with a distance of d_{ik},

$$R_{ik}(\tau) = \frac{2J_1\left(\frac{\omega}{c}d_{ik}\right)}{\frac{\omega}{c}d_{ik}}\cos\omega\tau \quad (2.236)$$

Let $\tau = 0$ in Eqs. (2.234)–(2.236), then we can obtain corresponding spatial correlation coefficients versus d_{ik}/λ, as shown in Fig. 2.78. The curves 1 and 2 in this figure correspond to the hydrophones to be arranged in the vertical and horizontal directions, respectively. Curve 3 expresses the spatial correlation coefficient for the isotropic noise field. We can observe that if the space between the two hydrophones d_{ik} exceeds a half-wavelength $\lambda/2$, both the singe-frequency isotropic noise and sea surface distributing noise will approximately be considered as independent noise fields in space.

Some experimental results in situ for the correlation characteristics of the ambient noise in the sea will simply be introduced below.

The measured spatial correlation coefficients for a sea surface distributing noise model in the vertical direction are shown in Fig. 2.79 [10]. The measured data that are indicated by white points are basically agreeable with the prediction curve for the noise whose frequencies range from 600 to 800 Hz at sea state 4. Whereas in the case of smooth sea surface or thc frcquencies ranging from 200 to 400 Hz, the measured data, which are indicated by black points in Fig. 2.79, are remarkably disagreeable with the calculated results. It is expected that the distant noise sources at low frequency range will affect the noise field. There are similar conclusions for the sea surface noise in the horizontal direction. In the case of high speed wind (about 20 m/s) or higher frequency range (above 500 Hz), the measured data for the spatial correlation coefficients are agreeable with theoretical predictions.

In summary, the spatial correlativity for the ambient noise in the sea is weaker, and the correlation radii for both the single-frequency isotropic noise and the sea surface distributing noise are only about a half-wavelength. For example, if the frequency $f = 10$ kHz, the corresponding correlation radius is only 7.5 cm.

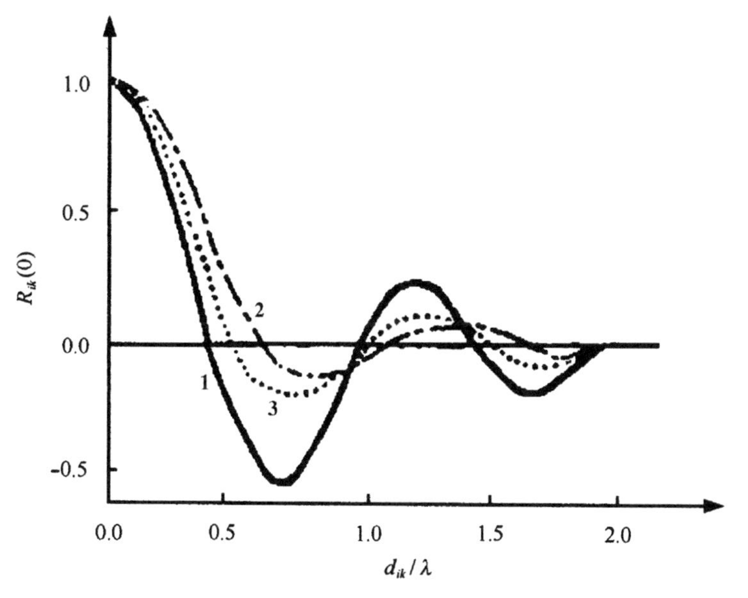

FIGURE 2.78 Single-side, single-frequency spatial correlation coefficients.

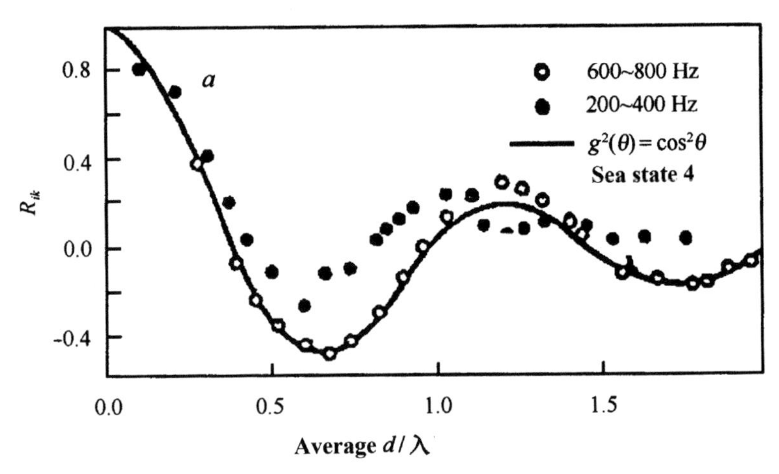

FIGURE 2.79 Measured spatial correlation coefficients.

We have noted that the spatial correlativity of the sound signal is strong (refer to Section 2.4), such as taking $f = 10$ kHz, and the average scale of the inhomogeneity in sea water medium $a = 60$ cm, so the longitudinal correlation radius of the fluctuation in the amplitude of the sound signal is up to 14 m. Of course, the transverse correlation radii are less than that of longitudinal ones, but the former are still equal to the average scale of the inhomogeneity a.

Therefore, provided the spatial correlation detection schemes are employed in digital underwater acoustic communications, the excellent performance combating with the ambient noise can be expected (refer to Chapter 3).

It is complicated to realize spatial correlation detections for common civil underwater acoustic communications, in particular for mobile ones. The schemes of time correlation defections are more suitable for the communications.

By combining the researches on underwater acoustic communications in shallow-water channels (refer to Chapter 3), the time correlativity of the ambient noise has been analyzed [29]. The stationary characteristics of the ambient noise in the channels had first been examined by using the mean value method. It has been demonstrated that the stationary time period is about 1 min.

The time correlation coefficients of the noise envelop at the center frequency of 20 kHz, which corresponds to the operating frequency of the telecontrol communication employed in the underwater acoustic releaser, and that versus τ is shown in Fig. 2.80 by curve 1. We see that the time correlativity of the ambient noise in shallow-water channels is very weak, and the correlative radius is only about 0.5 ms.

Recently, we had performed experimental researches on the instantaneous time correlativity of the ambient noise in shallow-water channels by combining the development of the APNFM (adaptive pseudorandom frequency modulation) signal processing system (refer to Chapter 4). A typical experimental result is shown in Fig. 2.81. The record of the ambient noise is shown in Fig. 2.81A. The calculated result of instantaneous auto-correlativity is shown in Fig. 2.81B. By taking the envelope of the autocorrelation coefficient, we can find the autocorrelation radius is about 0.4 ms. That is to say, the instantaneous auto-correlativity for the ambient noise in the shallow sea channels is also weak enough.

We have noted that the time correlativity of the amplitude fluctuation of the sound signals traveling in the inhomogeneous seawater medium is quite strong; the correlation radii are

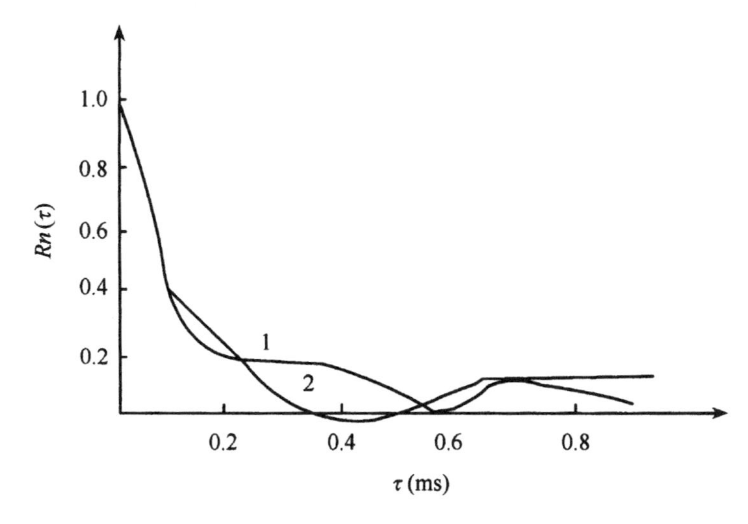

FIGURE 2.80 Time correlation coefficients versus τ for the ambient noise (curve 1) and the self-noise of ships (curve 1).

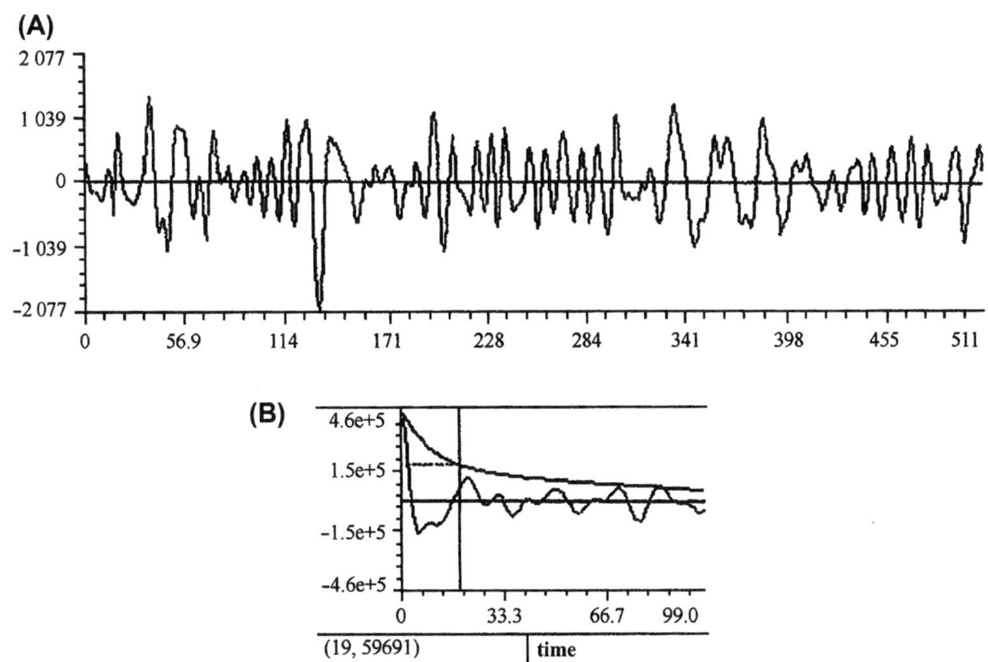

FIGURE 2.81 Analyses of instantaneous correlativity for the ambient noise in the Xiamen Harbor. (A) The ambient noise; (B) The calculated result of instantaneous auto-correlativity.

on the order of several seconds (refer to Fig. 2.64). Therefore it would be efficient to employ the schemes of time correlation detections to combat the ambient noise in digital underwater acoustic communications.

2.5.3 Self-Noise of Ships

As mentioned earlier, the radiated noise of ships is a part of the ambient noise in the sea. But that is of particular importance for passive sonar in which that is regarded as *SL* and for all sonars mounted on ships as *NL*.

This radiated noise that is regarded as background noise for different kinds of sonars is called the self-noise of ships. That is a part of the parameter *NL* in sonar equations.

The *NL* originated by the self-noise depends not only on different radiated sources, but also the propagating paths from these to a hydrophone. In particular, there is little isotropic directivity and correlativity in self-noise fields, so the estimations for parameter *DI* or *GS* in the communication sonar equation are very complicated. Therefore how to estimate the effect of self-noise of ships on a specific communication sonar would carefully be considered.

Because the noise sources between self-noise and radiated noise of ships are the same, the latter will first be introduced.

2.5.3.1 Radiated Noise of Ships

The source of noise on ships, submarines, and torpedoes can be grouped into the three major classes.

1. Machinery noise:
   propulsion machinery (diesel engines, main motors, reduction gears)
   auxiliary machinery (generators, pumps, air-conditioning equipment)
2. Propeller noise:
   cavitation at or near the propeller
   propeller-induced resonant hull excitation
3. Hydrodynamic noise is radiated noise originating in the irregular flow of water past the vessel moving through it and causing noise by a variety of hydrodynamic processes:
   radiated flow noise
   resonant excitation of cavities, plates, and appendages

Of the three major classes of noise just described, machinery noise and propeller noise dominate the spectra of radiated noise under most conditions. The relative importance of the two depends upon frequency, speed, and depth. This is illustrated by Fig. 2.82, which shows the characteristics of the spectrum of submarine noise at two speeds. Fig. 2.82A is a diagrammatic spectrum at a speed when propeller cavitation has just begun to appear. The low-frequency end of the spectrum is dominated by machinery lines, together with the blade-rate lines of the propeller. These lines die away irregularly with increasing frequency and become submerged in the

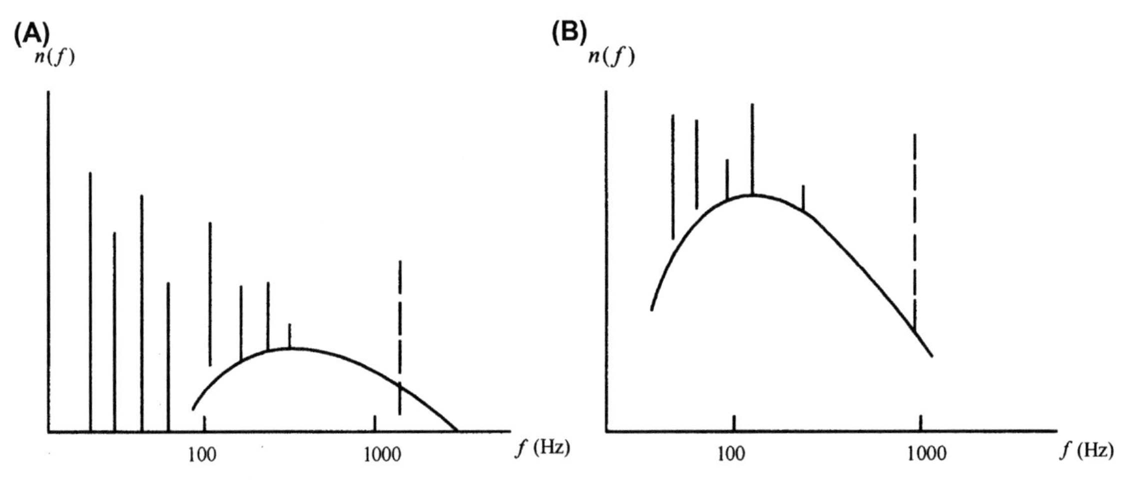

FIGURE 2.82 Diagrammatic spectra of submarine noise at two speeds. (A) At a speed when propeller cavitation has just begun to appear; (B) At a higher speed.

continuous spectrum of propeller noise. Sometimes, as indicated by the dotted line, an isolated high-frequency line or group of lines appears amid the continuous background of propeller noise. These high-frequency lines result from a singing propeller or from particularly noisy reduction gears, if the vessel is so equipped.

At a higher speed, the spectrum of propeller noise increases and shifts to lower frequencies, as shown in Fig. 2.82B. At the same time, some of the line components increase in both level and frequency, whereas others, notably those due to auxiliary machinery running at a constant speed, remain unaffected by an increase in ship speed. Thus, at the higher speeds, the continuous spectrum of propeller cavitation overwhelms many of the line components and increases its dominance over the spectrum. A decrease in depth at a constant speed has, as indicated before, the same general effect on the propeller noise spectrum as an increase in speed at constant depth.

The radiated noise of ships is a kind of random signal, including continuous spectrum, line spectrum, and dynamic spectrum of regular modulation. Based on the analyses of large numbers of experimental data [1], the "average power spectrum" of the radiated noise obtained by passing through a certain confusion in frequency and temporal domains may be described by the power spectrum curve of the Ecs model. The auto-correlation function with the Ecs model can be expressed as

$$B_n(\tau) = e^{-a|\tau|}(\cos \omega_0 \tau + F \sin \omega_0 |\tau|) \quad (2.237)$$

where $a \gg 0$, F is a real number, and its basic shape is shown in Fig. 2.83. Here is the corresponding single-side power spectrum density function:

$$G(\omega) = 2\left[\frac{a + F(\omega + \omega_0)}{a^2 + (\omega + \omega_0)^2} + \frac{a - F(\omega - \omega_0)}{a^2 + (\omega - \omega_0)^2}\right] \quad (2.238)$$

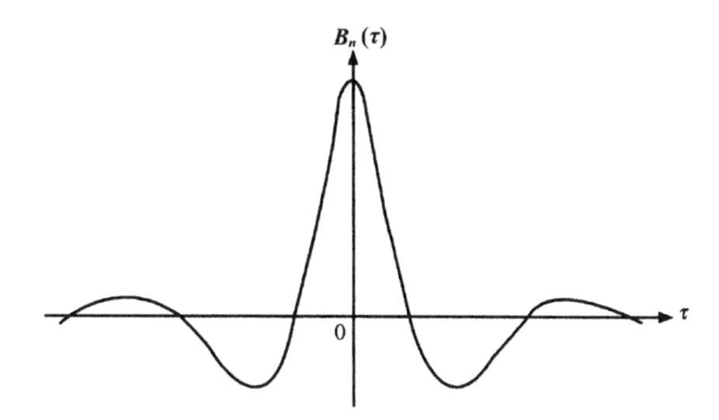

FIGURE 2.83 The basic shape of auto-correlation function of Ecs noise.

The PDF of the amplitude and the spatial-temporal correlativity for the noise of ships, just as for the ambient noise, has been analyzed in developing the underwater acoustic releaser; the center frequency is also 20 kHz.

The PDF of the amplitude of the noise produced by a passenger ship and a naval ship is shown in Fig. 2.84A and B, respectively. They both follow the Rician distribution law at a small ratio of regular component to noise. There exists a regular component in the noise even if its amplitude is small. But for the noise of ships in which the steam turbine is employed, its PDF of the amplitude obeys Rayleigh distribution law, as shown in Fig. 2.85, and corresponding fluctuation rates sometimes are up to 52%. So, the regular component is much small at higher frequencies, such as 20 kHz.

The time correlation characteristics of the noise of ships mentioned before has also been analyzed, and the typical time correlation coefficients versus τ are shown in Fig. 2.80 by curve 2. Although there are slight differences among the different kinds of ships, the time correlativity of the envelope of the noise of ships, just as that of the ambient noise, is weak enough, so the correlation radii are about 0.2 ms.

2.5.3.2 *Self-Noise of Ships [5]*

In the case of the self-noise of ships, just as the radiated noise of ships, machinery noise,

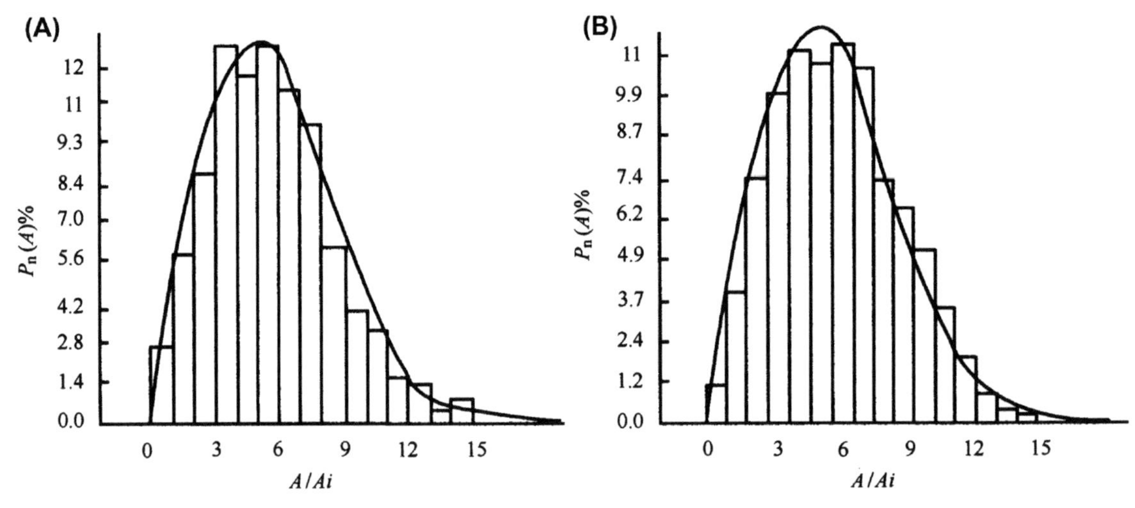

FIGURE 2.84 PDF of amplitudes for the noise of ships. (A) A passenger ship; (B) A naval ship.

propeller noise, and hydrodynamic noise are the three major classes of noise sources. These noise sources radiate sound waves, and then they will travel to a hydrophone along different paths. Therefore, *NL* of self-noise depends remarkably upon the directivity of the hydrophone, its mounting shape, and the location on the ship. Generally speaking, that is placed at a location to be removed as much as possible from the propulsion machinery and the propeller noises of the ships.

Fig. 2.86 is representative of active destroyer sonars operating at frequencies at 10 kHz and below. At very slow speeds, or when lying to, the sonar self-noise level is close to the ambient background level of the sea in the prevailing sea state. At speeds between 15 and 25 knots, the self-noise level increases sharply with speed at the rate of about 1.5 dB/knot, and it represents the dominant contributions of hydrodynamic and propeller noise to the total noise level.

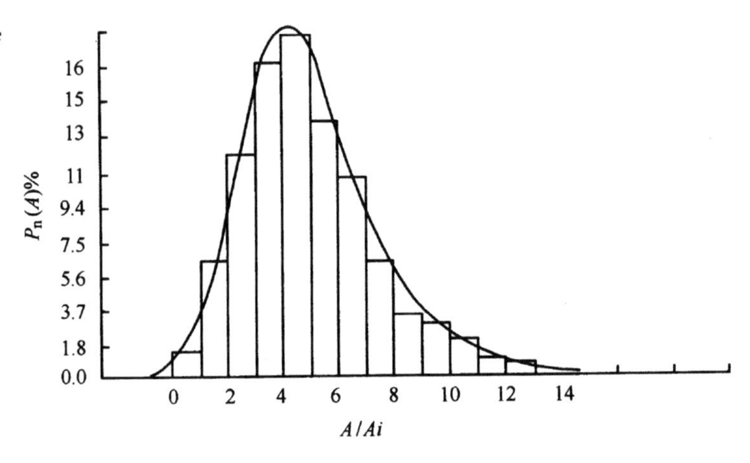

FIGURE 2.85 PDF of amplitudes for the noise of ships using steam turbines.

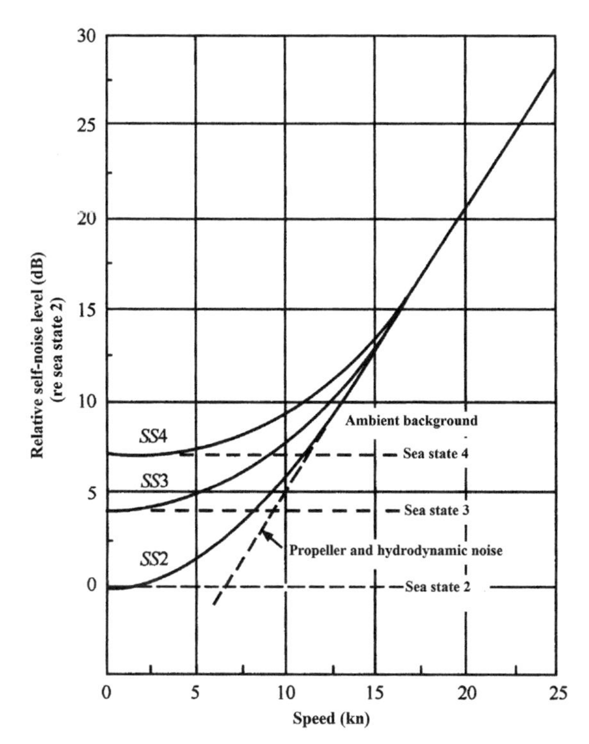

FIGURE 2.86 Self-noise levels versus speed on a destroyer.

Self-noise levels on submarines are illustrated in Fig. 2.87A and B. These show average spectra and the increase of noise with speed. The spectra of Fig. 2.87A are for noisy, average, and quiet installations at a speed of 2 knots. They approximate the levels of deep-water ambient noise at the high-frequency end, but they rise more rapidly with decreasing frequency, probably as a result of machinery noise contributions. The extremely rapid rise with speed suggests the influence of propeller cavitation as the speed increases, as does the fact that the self-noise decreased with increasing depth of submergence. Fig. 2.87B shows the increase of noise level with speed relative to 2 knots.

Of course, the self-noise levels of modern naval ships have considerably been reduced in comparison with those mentioned earlier. How to determine the self-noise levels for a specific communication sonar is quite complicated and difficult, in particular when they depend upon the transmission paths.

2.5.4 Impacts of Background Noise in the Sea on Digital Underwater Acoustic Communications and the Countermeasures to Combat Against the Noise

2.5.4.1 Impacts of the Background Noise on the Underwater Acoustic Communications

The impacts of the background noise on the digital underwater acoustic communications may be analogous with those of the noise in radio channels on radio communications. Moreover, some efficient signal processing schemes to combat against those in radio communications may basically be applied to underwater acoustic communications. Of course, there are some essential differences between them. Therefore, how to adapt to the peculiarity of the background noise in underwater acoustic channels will emphatically be discussed as follows.

1. We know that whether information signals can correctly be detected depends on the input SNR of a receiver. Generally speaking, the transmission loss *TL* is large; *NL* is especially high in underwater acoustic channels. On the other hand, the radiated sound power for communication sonars is also limited. Therefore they have to detect the information signals under the condition of weak input SNR. If SNR is lowered to a certain value for a specific communication sonar, the communication performances will severely be damaged, which is called "threshold effect." How to realize robust digital underwater communication is quite difficult.
2. Since the underwater acoustic channels have the band-limited peculiarity, the bandwidth

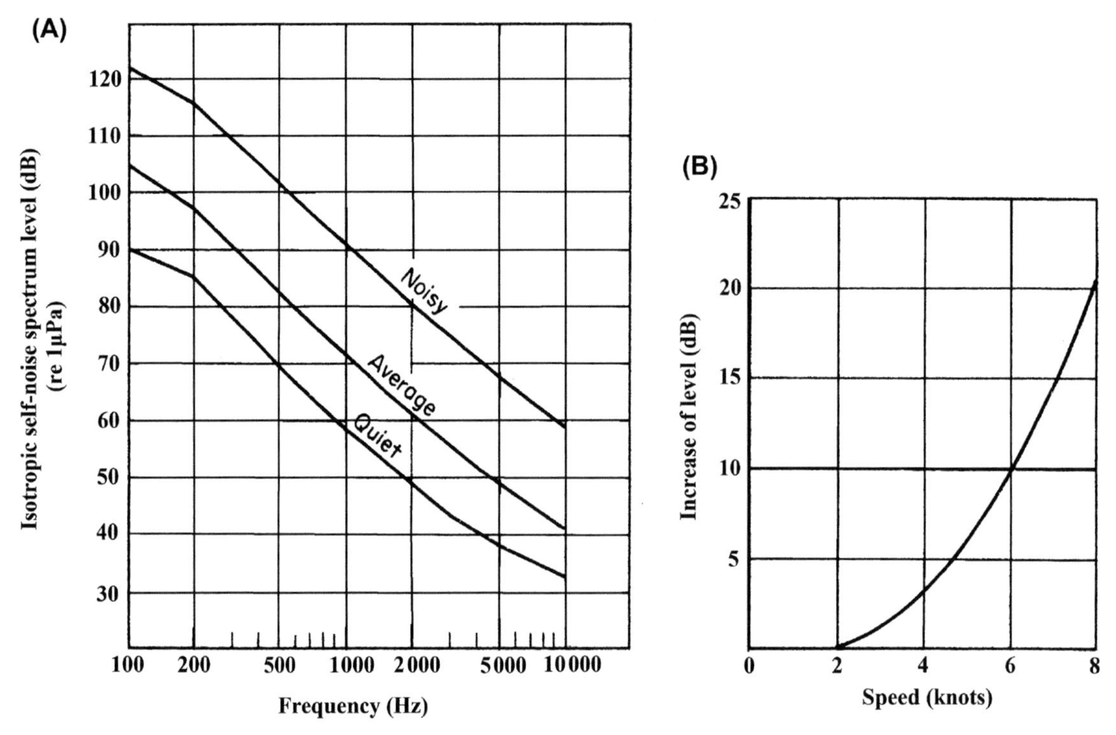

FIGURE 2.87 Average self-noise spectra on submarines. (A) Self-noise spectrum level; (B) Increase of level.

of communication sonars is generally narrower, and the noise spectrum level may be considered to be uniform. According to Eq. (1.18), even if the bandwidth $B = 200$ Hz, *NL* will still rise about 23 dB since *NL* in the channels have relative large values. So, the increase of *NL* due to increasing *B* is one of the reasons to limit high data rate underwater acoustic communications in which wider *B* is necessary.

3. *NL* will rapidly be raised with decreasing the operating frequency in underwater acoustic communications. For example, *NL* will rise 5–6 dB with decreasing the frequencies an octave. By using the approach to decrease operating frequencies to increase communication ranges will thus be restricted. On the contrary, in the case of short range communications, we may raise operating frequency to lower *NL* to make up for the increase of *TL* due to raising the frequency. It is one of the basic reasons to select optimum operating frequencies for a specific communication sonar.
4. Since the self-noise levels of ships will rapidly rise with increasing the speeds of ships, this is a major reason to limit high-speed mobile underwater acoustic communications. Perhaps, to remain at the expected input SNR, the permitted maximum speed of a ship must be restricted, or expected maximum communication ranges have to be decreased to adapt to high-speed mobile communications.
5. We know that *NL* has a severely random spatial-temporal variability, including varied noise propagating conditions. For example, since the propagating conditions in winter

are better than those in summer, *NL* in the former is higher than that in the latter, being up to 7 dB in some sea areas [5]. Therefore, the predictions in relation to *NL* would be carried out, just as those for the sound signal fields.

6. There are some line components in the spectra of the noise of ships; therefore, it is possible to cause the corresponding distortions of signals due to additional modulations.

Based on the effect of the noise on signal detection, there are the two types of interference in receiving theory.

1. Additive interference: input signal is

$$x(t) = s(t) + n(t) \qquad (2.239)$$

where $s(t)$ is the information signal. For linear time-invariant channels, the output is

$$y(t) = s_y(t) + n_y(t) \qquad (2.240)$$

2. Nonadditive interference: one or more parameters of the signal are modulated by the noise, so the signal will change into a random variable. Therefore, this interference sometimes is called modulation noise. Since there is a parasitic modulation in amplitude and phase, the corresponding distortions of the signal appear.

   Generally speaking, the two types of interference exist at the same time; the output is

$$y(t) = s_y(\alpha_1, \alpha_2, \ldots, \alpha_m, t) + n_y(t) \qquad (2.241)$$

where one or more signal parameters of α_1, α_2, …, α_m have been changed into random variables due to nonlinear modulations. We see that provided the operating bandwidth for a communication sonar is selected to fall into the range of appearing line spectra, some parameters of signal, as amplitude and phase, may be additionally modulated. In the cases of long-range digital underwater acoustic communications, in which low operating frequencies have to be employed, nonlinear interference would be paid attention to.

   Of course, the nonlinear interference due to line spectra of the noise of ships may generally be neglected in civil underwater acoustic communications.

2.5.4.2 *Detections of Weak Signals Against the Noise Background*

The randomly spatial-temporal variant laws of background noise in underwater acoustic channels are one of the physical bases for designing signal processing systems employed in communication sonars. Generally speaking, the detection techniques of weak signals in digital radio communications could be applied to digital underwater acoustic communications. Of course, the peculiarity of the background noise in the latter must conscientiously be considered.

2.5.4.2.1 USUAL METHODS TO LOWER *NL*

1. Designing an available band-pass filter is an efficient method to suppress the background noise in the sea. In the case of mobile underwater acoustic communications, Doppler frequency shifts would be considered. For some communication sonars that have to utilize broader bands, such as an SS-FH system (refer to Chapter 3), a tracing filter can be adopted, which will adaptively match every frequency code of a specific FH pattern. Therefore it is possible that *NL* will be reduced up to $1/N$ in comparison with a conventional broad filter, where N is the number of frequency codes forming the pattern.
2. It is an efficient approach to design a suitable directivity of the receiving transducer to obtain a high directivity index *DI*, or the spatial processing gain G_S for a hydrophone array. It is equivalent to lower *NL* to the corresponding numbers of *DI* or G_S in dB.

3. During selecting the optimum operating frequency, the variable laws of *NL* versus operating frequencies would be considered in detail. In particular, in the case of short-range underwater acoustic communications, higher operating frequencies may be chosen, perhaps at which *NL* will be lowered in a value larger than the increase of *TL* by using higher ones.
4. As *NL* has a spatial-temporal variability, it is better to carry out the predictions of *NL* for a specific sea area where underwater acoustic communications will be performed. Perhaps some sea areas with lower background noise may be selected.

Obviously, raising *SL*, using a suitable filter, and selecting the optimum operating frequency are efficient to combat the noise background. The detections of signals against the noise background are only relative to the communication ranges that are far or near, and not relative to the communications that are efficient or inefficient. Under this significance, combating the noise background is thus more efficient than combating the multipath interference background.

2.5.4.2.2 CORRELATION DETECTIONS FOR WEAK SIGNALS AGAINST THE NOISE BACKGROUND

2.5.4.2.2.1 SPATIAL CORRELATION DETECTIONS Since the spatial correlativity of the fluctuation of sound signals is much better than that of background noise in underwater acoustic communication channels, employing the schemes of spatial correlation detections to combat background noise in digital underwater acoustic communications will obtain excellent results. Of course, it is complicated to design and operate a spatial hydrophones array in civil mobile underwater acoustic communications; whereas in fixed-point ones, in particular for data communications in underwater acoustic networks, and auto-observing stations of marine parameters, the robustness of underwater acoustic communications will considerably be improved provided that the spatial correlation detection schemes are used.

2.5.4.2.2.2 TIME CORRELATION DETECTIONS We have known that the time correlativity of background noise in the sea, as well as its spatial correlativity, is quite weak; eg, the correlative radius is generally less than 1 ms. On the contrary, the time correlativity of the fluctuation of sound signals in underwater acoustic channels is better, and the correlation radii generally are on the order of several seconds.

The correlativity will be subject to some lose when sound signals are reflected from the sea surface. Experiments have demonstrated that the correlation radium of the amplitude of the reflected sound signal is roughly consistent with that of the sea wave to be about 1 s at generally small grazing angles. Therefore even if the signals received will come from multipath pulses in long-range underwater acoustic communications, their correlativity is still much better than that of the background noise in the sea.

We see that it is efficient to combat the background noise by means of the schemes of time correlation detections. In principle, either autocorrelation or cross-correlation schemes can be adopted. However, the former have a better adaptivity of channels than those of latter, which are more suitable to be employed in violently random spatial-temporal variable channels, while the latter have a better anti-noise performance, which is equivalent to the well-known matched filter.

We know that adaptive noise cancellation schemes, in particular ALE (adaptive line enhance), have efficiently been employed in communication sonars, which are also based on the differences of the correlativity between signal and noise (refer to Chapter 3).

Finally, the prediction of the background noise in the sea would be paid attention to, but that is a difficult subjection. In fact, it is quite complicated and difficult to determine *NL*, which is one of the sonar equation parameters, for a specific communication sonar because it has large dynamic ranges.

References

[1] W. Dezhao, S. Erchang, Underwater Acoustic, Science Press, Beijing, 1981.
[2] Y. Shie, Principles of Underwater Acoustic Propagation, Harbin Engineering University Press, 1994.
[3] L.M. Brekhovskikh, Y. Lysanov, Fundamentals of Ocean Acoustics, China Ocean Press, Beijing, 1985.
[4] L.M. Brekhovskikh, Waves in Layered Media, Academic Press Inc., New York, 1960.
[5] R.J. Urick, Principles of Underwater Sound, McGraw-Hill Book Company, 1975.
[6] L. Bosheng, L. Jiayu, Principles of Underwater Acoustics, Harbin Engineering University Press, 1993.
[7] L.M. Brekhovskikh, Ocean Acoustic, Science Press, Beijing, 1983.
[8] P. Hirsch, A.H. Curter, Mathematical models for the prediction of SOFAR propagation effects, JASA 37 (1965) 91.
[9] F.E. Hale, Long-rang sound propagation in the deep sea, JASA 33 (1961) 456.
[10] M.B. Porter, H.P. Bucker, Gaussian beam tracing for computing ocean fields, JASA 82 (4) (1987) 1349–1359.
[11] J. Northrop, Long-range Pacific acoustic multipath identification, JASA 75 (1984) 1760.
[12] Z. Renhe, et al., The multipath structures of signal waveforms in shallow water with thermocline, Acta Oceanologica Sinica 3 (1) (1981) 57–69.
[13] Z. Ye, Z. Renhe, Pulse propagation in shallow water with thermocline, Science in China, Series A 26 (3) (1996) 271–279.
[14] M. Chitre, et al., Recent advances in underwater acoustic communications and networking, Marine Technology Science Journal 42 (1) (2008).
[15] H. Yi, Z. Renhe, et al., WKBZ mode approach to pulsed propagation in ocean channels, Acta Acustica 19 (6) (1994) 418–424.
[16] Z. Ye, Z. Renhe, et al., Theoretical analysis on pulsed waveforms in shallow water with an ideal thermocline, Acta Acustica 20 (4) (1995) 289–297.
[17] Л, А.оцернов, РасЦростриение валн в среде со случайными неоднороднстями , Изд, АН СССР (1958).
[18] C. Eckart, The scattering of sound from sea surface, JASA 36 (1964) 8.
[19] H.W. Marsh, Reflection and scattering of sound by the sea bottom, JASA 36 (2003) 1964.
[20] C. Geng, X. Junhua, A method of measuring correlation between signal pulses for study of time-variant characteris-ties of underwater sound channel, Acta Acustica 6 (3) (1981) 158.
[21] Rose, Principle of Underwater Noise, China Ocean Press, 1983.
[22] V.O. Kundson, R.S. Alford, J.W. Emling, Underwater ambient noise, Journal of Marine Research 7 (1948) 410.
[23] G. Wenz, Acoustic ambient noise in ocean specters and sources, JASA 34 (1936) 1962.
[24] Z. Minqiang, et al., The marine biological evening choruses in Xiamen Harbor, Acta Oceanologica Sinica 6 (1) (1984) 10–15.
[25] Z. Minqiang, et al., Marine animal noise observed in Zhoushan islands waters, Journal of Oceanography in Taiwan Strait 6 (2) (1987) 127–131.
[26] Z. Minqiang, et al., A new record on marine animal noise, Acta Oceanologica Sinica 2 (2) (1983) 19–21.
[27] T.E. Heindsman, Effect of rain upon underwater noise levels, JASA 27 (1955) 378.
[28] Z. Zhaoning, X. Dawei, The Theory of Passive Detection and Parameter Estimation for Underwater Acoustic Signals, China Science Press, 1983.
[29] X. Tianzeng, A digital time correlative accumulation (I) auto-correlative accumulation, Chinese Journal of Acoustics 4 (1990) 365–371.

CHAPTER 3

Digital Underwater Acoustic Communication Signal Processing

Digital underwater acoustic communication signal processing connects underwater acoustic communication channels with digital underwater acoustic communication equipment.

In the previous chapter, we have detailed sound transmission laws in underwater acoustic channels and their effects on digital underwater acoustic communications; how to adapt to the peculiarities encountered in the channels has also been briefly discussed. By connecting the special requirements for underwater acoustic communication equipment, the signal-processing systems suitable to the peculiar channels will be explored in this chapter to provide the technical fundamentals for designing them for robust performance.

Most civil communication sonars, the communication ranges are nearer (within 10 km) and are operated at fixed points or low-speed mobile communications. Moreover, we can develop series productions having varied specifications, such as different communication ranges, data rates, fixed points or mobile communications, required communication medium and possible multimedia combinations to adapt to a variety of applied communication fields. The design for special civil communication sonar is relatively simple, because the factors that must be considered in it are few.

However, in some communication sonars employed in civil underwater acoustic networks, and in the communications between surface ships and autonomous underwater vehicles (AUVs), higher performance is also necessary. For example, communications may require longer ranges and higher data rates, or have the ability to adapt to random spatiotemporal–frequency variable parameter channels even if they are operated in a mobile communication state. Moreover, it is better having a multimedia-compatible communication model.

The concept of communication is the process of transmission and exchange of information. The properties of digital communication passing through radio or underwater acoustic channels have an aspect of generality, but they possess another aspect of many essential differences, as mentioned in the previous chapter. Therefore, we introduce some advanced and effective digital signal-processing systems from the former to the latter. At the same time, it must be emphasized that innovative developments are needed for digital underwater acoustic communication signal-processing systems based on information theories and peculiar underwater acoustic transmission laws. It is also a basic thread that will be discussed in this chapter.

Digital Underwater Acoustic Communications
http://dx.doi.org/10.1016/B978-0-12-803009-7.00003-9

3.1 SOME SIGNAL-PROCESSING TECHNIQUES IN RADIO COMMUNICATIONS POSSIBLE TO EXTENSIBLE TO UNDERWATER ACOUSTIC COMMUNICATIONS

Principally, underwater acoustic communications are extensions and developments from radio communications; many advanced signal-processing systems in the latter may be introduced to the former, which will first be discussed in the following section. The multiple-frequency shift keying (MFSK) system has been known, which will not be specifically described here.

3.1.1 Optimum Linear Filter [1]

3.1.1.1 Linear System

The systems in signal processing may be divided into two kinds: linear and nonlinear. Because of the effective applications of Fourier analysis, the theories for linear systems have nearly been perfected, the relationships between input and output spectra and correlation function have definitely been established. However, in the case of the nonlinear system, the corresponding theoretical research is still ongoing.

As viewed from theoretical analyses, when we refer to a system, that means a certain transform from input $x(t)$ to output $y(t)$, which can be shown as $x(t) \rightarrow y(t)$.

A system is called linear if it satisfies additive and homogeneous conditions, which are expressed by Eqs. (1.6) and (1.7).

When a unite pulse $\delta(t)$ inputs to a linear time-invariant system, its output is referred to as an impulse response function $h(t)$. $h(t)$ plays an important role as a "bridge" to connect input with the output ends of the system. In fact, if $x(t)$ is an arbitrary input signal that can be expressed by the sum of a series of $\delta(t)$, that is

$$x(t) = \lim_{\Delta \to 0} \sum_{n=-\infty}^{\infty} x(n\Delta)\delta(t - n\Delta)\Delta \tag{3.1}$$

According to the time-invariant property, we have $\delta(t - n\Delta) \rightarrow h(t - n\Delta)$; moreover, based on the properties of linear systems, we get $x(n\Delta)\delta(t - n\Delta) \rightarrow x(n\Delta)h(t - n\Delta)$. Making linear operations to both sides of Eq. (3.1), we obtain

$$\begin{aligned} y(t) &= \lim_{\Delta \to 0} \sum_{n=-\infty}^{\infty} x(n\Delta)h(t - n\Delta)\Delta \\ &= \int_{-\infty}^{\infty} x(\tau)h(t - \tau)\mathrm{d}\tau \end{aligned} \tag{3.2}$$

The relationship between input and output for the linear time-invariant system is indicated by Eq. (3.2), this integral operation is referred to as convolution. It means that the output for a linear time-invariant system equals the convolution between input and impulse response function. Because the order of the convolution permits to be transformed, that is

$$y(t) = x(t) * h(t) = h(t) * x(t) \tag{3.3}$$

so that, provided $h(t)$ is known, the system may also be determined as shown in Fig. 3.1A.

The Fourier transformation of $h(t)$ is called frequency response function.

$$H(f) = \int_{-\infty}^{\infty} h(t)e^{-i2\pi ft}\mathrm{d}t \tag{3.4}$$

$$h(t) = \int_{-\infty}^{\infty} H(f)e^{i2\pi ft}\mathrm{d}f \tag{3.5}$$

(A) $x(t) \rightarrow$ [$h(t)$] $\rightarrow y(t)$ (B) $X(f) \rightarrow$ [$H(f)$] $\rightarrow Y(f)$

FIGURE 3.1 Relationship between input and output for a linear time-invariant system: (A) time domain, (B) frequency domain.

The relationship between the input and output for a linear time-invariant system can clearly be expressed by means of $h(t)$ and $H(f)$.

In the case of a regular input signal, the Fourier transformation exists, by considering Eq. (3.2) and using convolution theorem in time domain, we obtain

$$Y(f) = X(f)H(f) \tag{3.6}$$

in which $X(f)$ and $Y(f)$ are the spectra of input and output signals, respectively, as shown in Fig. 3.1B. Eq. (3.6) can be rewritten as $H(f) = Y(f)/X(f)$, it shows that $H(f)$ equals a ratio of output and input spectra; therefore, $H(f)$ is also called transfer function.

3.1.1.2 *Optimum Linear Filter*

Considering a kind of optimum linear filter, provided the signals, which are covered by background noise, pass through that, the maximum output SNR can be obtained, this is the conception of a matched filter.

Letting an input signal be $s(t)$ with a spectrum $S(f)$, input additive white noise be $n(t)$ with a power spectrum density $N_0/2$, now we attempt to find a linear system $h(t)$ with which the maximum output SNR at a certain time (observing time) will obtain provided input is $s(t) + n(t)$ as shown in Fig. 3.2.

Assuming that $x(t) = s(t) + n(t)$, $y(t)$ is the output of the filter, $H(f)$ is the transfer function of the system, according to the theory of the linear system, we have

$$\begin{aligned} y(t) &= x(t) * h(t) \\ &= \int_{-\infty}^{\infty} h(\tau)[s(t-\tau) + n(t-\tau)]\mathrm{d}\tau \end{aligned} \tag{3.7}$$

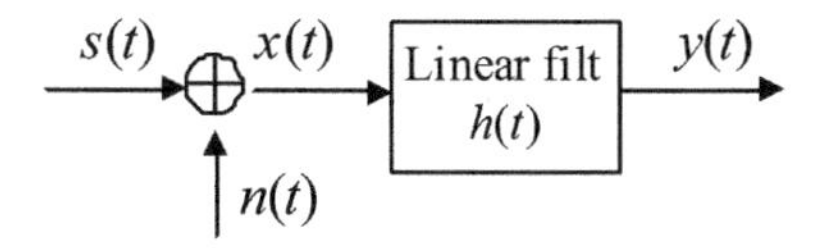

FIGURE 3.2 Matched filter.

the output signal at time t_0 is given by:

$$\begin{aligned} y_s &= \int_{-\infty}^{\infty} h(\tau)s(t_0-\tau)\mathrm{d}\tau \\ &= \int_{-\infty}^{\infty} H(f)S(f)e^{i2\pi f t_0}\mathrm{d}f \end{aligned} \tag{3.8}$$

the output noise

$$y_n' = \frac{N_0}{2}\int_{-\infty}^{\infty} |H(f)|^2\mathrm{d}f \tag{3.9}$$

so that the output SNR may be expressed as follows:

$$L_0 = \frac{y_s^2}{y_n'} = \frac{\left[\int_{-\infty}^{\infty} H(f)S(f)e^{i2\pi f t_0}\mathrm{d}f\right]^2}{\frac{N_0}{2}\int_{-\infty}^{\infty} |H(f)|^2\mathrm{d}f} \tag{3.10}$$

by means of Schwarrz's inequality, we obtain

$$\begin{aligned} &\left[\int_{-\infty}^{\infty} H(f)S(f)e^{i2\pi f t_0}\mathrm{d}f\right]^2 \\ &\leq \left[\int_{-\infty}^{\infty} |H(f)|^2\mathrm{d}f\right] \times \left[\int_{-\infty}^{\infty} \left|S(f)e^{i2\pi f t_0}\right|^2\mathrm{d}f\right] \end{aligned} \tag{3.11}$$

therefore,

$$L_0 = \frac{\int_{-\infty}^{\infty} |S(f)|^2\mathrm{d}f}{\frac{N_0}{2}} = \frac{2E}{N_0} \tag{3.12}$$

in which $E_0 = \int_{-\infty}^{\infty} |S(f)|^2 df$ is signal energy. This equality in the expression will hold if, and only if, the following condition is satisfied

$$H(f) = cS^*(f)e^{-i2\pi f t_0} \tag{3.13}$$

in which c is a constant. $H(f)$ expressed by Eq. (3.13) is the transfer function. Eq. (3.13) may be expressed in time domain

$$h(t) = cs(t_0 - t) \tag{3.14}$$

it means the impulse response of an optimum filter is the time reverse of the input signal. For this reason, the filter is set to be matched to the input signal, and it is referred to as a matched filter.

Now, let us calculate the gain of the matched filter G_t. Assuming that signal duration is T, bandwidth is B, therefore the average power of signal is E/T and that of noise is N_0B, we obtain the input SNR

$$L_i = \left(\frac{E}{T}\right) \Big/ (N_0 B)$$

the gain of the system

$$G_i = \frac{L_0}{L_1} = \frac{2E/N_0}{E/TN_0B} = 2TB \tag{3.15}$$

we see that G_i is proportional to TB. To acquire larger G_i, the signals having a larger TB must be provided.

Provided that $n(t)$ is nonwhite noise and its mean power spectrum is $K_n(f)$, in this case, the output signal is still as follows:

$$y_s = \int_{-\infty}^{\infty} H(f)S(f)e^{i2\pi ft_0}\mathrm{d}f$$

the noise of output

$$y_n' = \int_{-\infty}^{\infty} |H(f)|^2 K_n(f)\mathrm{d}f$$

the output SNR of the system

$$L_0 = \frac{y_s^2}{y_n'} = \frac{\left[\int_{-\infty}^{\infty} H(f)S(f)e^{i2\pi ft_0}df\right]^2}{\int_{-\infty}^{\infty} |H(f)|^2 K_n(f)df}$$

utilizing Schwartz's inequality again, we get

$$L_0 \le y_s = \int_{-\infty}^{\infty} \frac{|S(f)e^{i2\pi ft_0}|^2}{K_n(f)}\mathrm{d}f$$

The equality in this expression will hold if the following condition to be satisfied

$$H(f) = \frac{c}{K_n(f)} S^*(f)e^{-i2\pi ft_0} \tag{3.16}$$

This is the transfer function of a matched filter in a nonwhite noise background. According to the convolution theorem for a linear system, it may be thought that the system expressed by Eq. (3.16) is connected by two linear systems: one has a transfer function $1/\sqrt{K_n(f)}$ that is called prewhitening filter, provided a nonwhite noise passing through this filter, it will be transformed to white noise. The other has a transfer function of $cS^*(f)e^{-i2\pi ft_0}/\sqrt{K_n(f)}$, that will match to the output of prewhitening filter.

The realization of the optimum detection by means of a matched filter is commonly made use of an equivalent correlator in communication engineering.

According to Eqs. (3.8) and (3.14), we obtain

$$y(t) = x(t) * h(t) = \int_{-\infty}^{\infty} x(\tau)h(\tau - t)\mathrm{d}\tau$$
$$= c\int_{-\infty}^{\infty} x(\tau)s(t - t_0 + \tau)\mathrm{d}\tau \tag{3.17}$$

At a certain observing time $t = t_0$

$$y(t) = c\int_{-\infty}^{\infty} x(\tau)s(\tau)\mathrm{d}\tau \tag{3.18}$$

It is a cross-correlation operation between input signal and local reference signal $s(t)$ as shown in Fig. 3.3. The correlator has a better adaptability than that of the matched filter, while a difficult problem, that is, accurate synchronization must be required for correlation operations.

3.1.2 Adaptive Filter and Its Applications [1,2]

3.1.2.1 *Adaptive Filter*

Adaptive filter has been employed in wide fields and now has developed to an important branch of the digital signal processing.

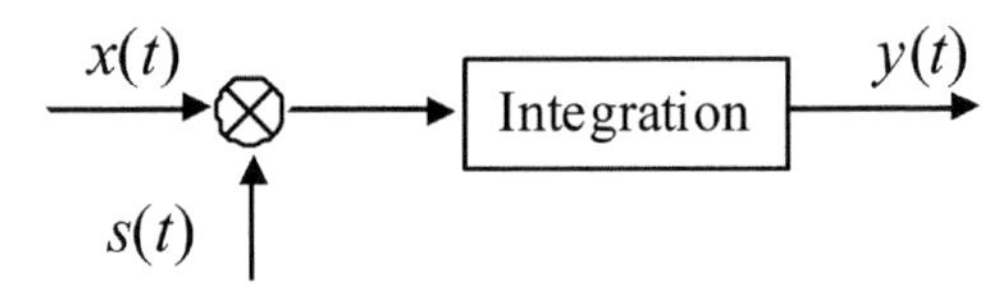

FIGURE 3.3 Correlator equivalent to a matched filter.

An adaptive filter is an optimum linear filter the parameters for which are not preset, while they are continuously adjusted by studying its communication environment to achieve stationary state according to a certain optimum criterion. The simplest adaptive filter with single input–output state will be here introduced below, which can be extended to general situations with multiple input–output states.

The model for an adaptive filter is given by Fig. 3.4. The input

$$x(t) = s(t) + n(t)$$

in which $s(t)$ is signal and $n(t)$ is noise.

Assuming that bath $s(t)$ and $n(t)$ are stationary random processes and independent each other, and the output of linear filter is $y(t)$. Let $d(t)$ be expected signal, the requirement for an adaptive filter is to adjust $H(f)$ to make

$$I = E\left[\varepsilon^2(t)\right] = E\left[(y(t) - d(t))^2\right] \tag{3.19}$$

$$\begin{aligned} I &= E\left[d^2(t)\right] - 2\int H(f)K_{xd}(f)\mathrm{d}f + \int |H(f)|^2 K_{xx}(f)\mathrm{d}f \\ &= E\left[d^2(t)\right] - \int H_{\mathrm{opt}}(f)K_{xd}(f)\mathrm{d}f + \int |H_{\mathrm{opt}}(f) - H(f)|^2 K_{xx}(f)\mathrm{d}f \\ &= I_{\min} + \int |H_{\mathrm{opt}}(f) - H(f)|^2 K_{xx}(f)\mathrm{d}f \end{aligned} \tag{3.23}$$

to be minimum. Expanding Eq. (3.19), we obtain

$$\begin{aligned} I &= E\left[(y(t) - d(t))^2\right] \\ &= E\left[d^2(t)\right] - 2E[y(t)d(t)] + E\left[y^2(t)\right] \\ &= E\left[d^2(t)\right] - 2\int K_{yd}(f)\mathrm{d}f + \int K_{yy}(f)\mathrm{d}f \\ &= E\left[d^2(t)\right] - 2\int H(f)K_{xd}(f)\mathrm{d}f \\ &\quad + \int |H(f)|^2 K_{xx}(f)\mathrm{d}f \end{aligned} \tag{3.20}$$

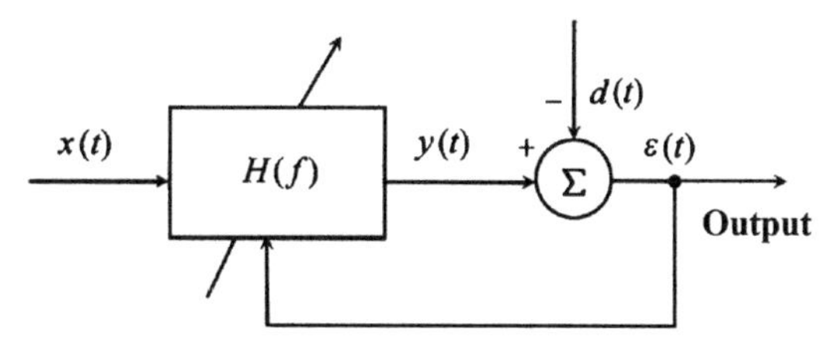

FIGURE 3.4 Block diagram of an adaptive filter.

in which K_{ij} are cross power spectral density functions. The pole value solution of Eq. (3.19) is as follows:

$$H_{\mathrm{opt}}(f) = \frac{K_{xd}^*(f)}{K_{xx}^*(f)} \tag{3.21}$$

thus

$$I_{\min} = E\left[d^2(t)\right] - \int H_{\mathrm{opt}}(f)K_{xd}(f)\mathrm{d}f \tag{3.22}$$

We can use a crosswise filter to approach $H_{\mathrm{opt}}(f)$. So that a formula to conveniently estimate the deviation of mean square error between practical filter $H(f)$ and optimum filter $H_{\mathrm{opt}}(f)$ may be introduced. It can be proved that

The block diagram of the crosswise filter is shown in Fig. 3.5, the impulse response function for which

$$h(t) = \sum_{k=0}^{L-1} w_k \delta(t - k\Delta) \tag{3.24}$$

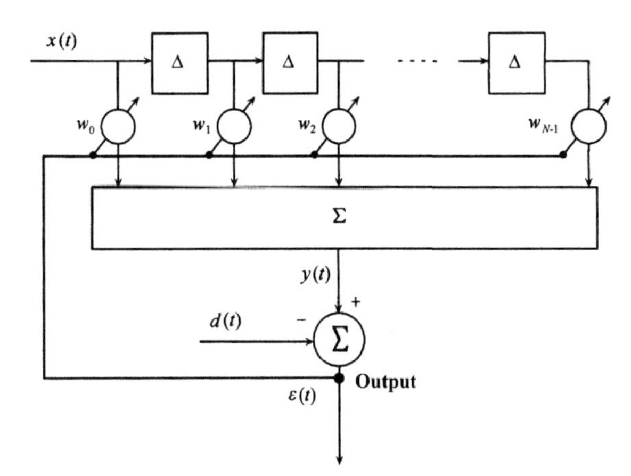

FIGURE 3.5 Adaptive filter realized by means of tap delay line.

transfer function

$$H(t) = \sum_{k=0}^{L-1} w_k e^{-i2\pi fk\Delta} \tag{3.25}$$

It should be noted that Eq. (3.20) can be used to find $H_{opt}(f)$ for general linear filters and to make I to be minimum. When a tap delay line is used to approximately express a linear system, the solutions for it must be limited in all $(N-1)$-order crosswise filters. To distinguish them, writing

$$J = E\left[(y(t) - d(t))^2\right] \tag{3.26}$$

Obviously, if the solution exists and equals J_{min}, the following relationship must hold:

$$J_{min} \geq I_{min} \tag{3.27}$$

How to find the solution based on an approach method will be discussed as follows.

Supposing weighted and observed vectors are expressed, respectively, as follows:

$$\mathbf{w} = \begin{pmatrix} w_0 \\ \vdots \\ w_{N-1} \end{pmatrix}, \quad \mathbf{x} = \begin{pmatrix} x_0 \\ \vdots \\ x_{N-1} \end{pmatrix}$$

in which

$$\begin{aligned} x_1 &= x(t - n\Delta) \quad n = 0, 1, 2, \cdots, N-1 \\ J &= E\left(\varepsilon^2(t)\right) = E\left(\mathbf{w}^T\mathbf{x} - d(t)\right)^2 \\ &= E\left(d^2(t)\right) - 2\mathbf{R}_{xd}^T\mathbf{w} + \mathbf{w}^T\mathbf{R}_{xx}\mathbf{w} \end{aligned} \tag{3.28}$$

in which $\mathbf{R}_{xx}$ are autocorrelation matrixes, $\mathbf{R}_{xd}$ are cross-correlation vectors of x and expected signal. We see that J is 2nd models of $\mathbf{w}$. To find J_{min}, the gradient of J with respect to $\mathbf{w}$ will first be found, and then letting that to zero, the solution for $\mathbf{w}$ will be obtained.

To find the gradient of J, the following formula will be used:

$$\nabla_{\mathbf{w}}\left(\mathbf{w}^T\mathbf{A}\mathbf{w}\right) = \mathbf{A}\mathbf{w} + \mathbf{A}^T\mathbf{w}$$

in which $\mathbf{A}$ is a matrix, $\mathbf{w}$ is a vector. We have that

$$\begin{aligned} \nabla_{\mathbf{w}}(J) &= -2\mathbf{R}_{xd} + \mathbf{R}_{xx}\mathbf{w} + \mathbf{R}_{xx}^T\mathbf{w} \\ &= -2\mathbf{R}_{xd} + 2\mathbf{R}_{xx}\mathbf{w} \end{aligned}$$

therefore,

$$\mathbf{w}_{opt} = \mathbf{R}_{xx}^{-1}\mathbf{R}_{xd} \tag{3.29}$$

Substituting Eq. (3.29) into (3.28), we obtain

$$J_{min} = E\left(d^2(t)\right) - \mathbf{w}_{opt}^T\mathbf{R}_{xd} \tag{3.30}$$

If the transfer function of the crosswise filter, which is obtained by optimum weight coefficient vector, is denoted by $H_0(f)$, Eq. (3.23) becomes

$$J_{min} = I_{min} + \int \left|H_{opt}(f) - H_0(f)\right|^2 K_{xx}(f)df \tag{3.31}$$

To calculate $\mathbf{w}_{opt}$ in Eq. (3.29), an arbitrary $\mathbf{w}$ will first be selected, and then starting from that the optimum solution will be found by using iterative method in the negative gradient direction of error function $\varepsilon(t)$.

Letting lth order $\mathbf{w}$ be $\mathbf{w}(l)$, thus

$$\mathbf{w}(l+1) = \mathbf{w}(l) + \mu\nabla(l) \tag{3.32}$$

in which ∇l expresses the gradient of $E(\varepsilon^2(l))$. We can get

$$\nabla(J) = -2\mathbf{R}_{xd} + 2\mathbf{R}_{xx}\mathbf{w}(l) \quad (3.33)$$

$\mathbf{R}_{xd}$, $\mathbf{R}_{xx}$ in Eq. (3.33) can be replaced by moment quantities, that is

$$\nabla(l) = \nabla\left(\varepsilon^2(l)\right) = 2\varepsilon(l)\mathbf{x}(l)$$

therefore,

$$\mathbf{w}(l+1) = \mathbf{w}(l) + 2\mu\varepsilon(l)\mathbf{x}(l) \quad (3.34)$$

in which $\varepsilon(t) = y(l) - d(l)$, μ is a factor controlling $\mathbf{w}(l)$ to converge. Provided that μ is small enough, we can obtain

$$\lim_{l\to\infty} E(\mathbf{w}(l)) = \mathbf{w}_{\text{opt}}$$

it means that if l is large enough, $\mathbf{w}(l)$ will finally converged to $\mathbf{w}_{\text{opt}}$ and is independent of selecting initial $\mathbf{w}$.

The relationship between J and $J_{\min}$ can be obtained as follows:

$$\begin{aligned} J = E\left(\varepsilon^2(l)\right) &= E\left(d^2(l)\right) - 2\mathbf{R}_{xd}^T\mathbf{w}(l) + \mathbf{w}^T\mathbf{R}_{xx}\mathbf{w}(l) \\ &= J_{\min} + (\mathbf{w}(l) - \mathbf{w}_{\text{opt}})^T\mathbf{R}_{xx}(\mathbf{w}(l) - \mathbf{w}_{\text{opt}}) \end{aligned} \quad (3.35)$$

this is an important expression to analyze an adaptive process. $J - J_{\min}$ is generally called excess mean squared deviation. J' as a function of l, is called the learning curve of the adaptive process.

Adaptive filter has widely been employed in sonars, radars, and so on.

The block diagram of an adaptive noise canceller is shown in Fig. 3.6A. Assuming that $m(t)$ is correlative to $n(t)$, but independent of $s(t)$, adjusting $m(t)$ to make

$$J = E\left(\varepsilon^2(t)\right) = E\left[(s(t) + n(t) - x(t))^2\right]$$

to be minimum, in which $x(t)$ is the output of $m(t)$ by passing through the filter $H(f)$. We see that

$$J = E\left(s^2(t)\right) + E\left[(n(t) - x(t))^2\right]$$

when J achieves minimal value, the output SNR

$$(\text{SNR})_{\text{out}} = \frac{E\left(s^2(t)\right)}{E\left[(n(t) - x(t))^2\right]}$$

to be maximum. In this case, the minimum mean squared deviation criterion and maximum output SNR criterion are equivalent.

The block diagram of an adaptive channel equalizing system is given in Fig. 3.6B. It has widely been employed in radio communications. The premise condition by using this system is that input signals are known. In this case, it is easy to prove that provided

$$H(f) = \frac{1}{X(f)}$$

$E(\varepsilon^2(t))$ has a minimal value.

We see that if a transfer function $H(f)$ has zero points, it has corresponding pole points, the order numbers of an adaptive crosswise filter would be much larger, and it is difficult to find its solutions.

The third application of the adaptive filter is adaptive line-spectra enhancer, which has excellent antinoise performance, and which will be discussed in detail in the next section.

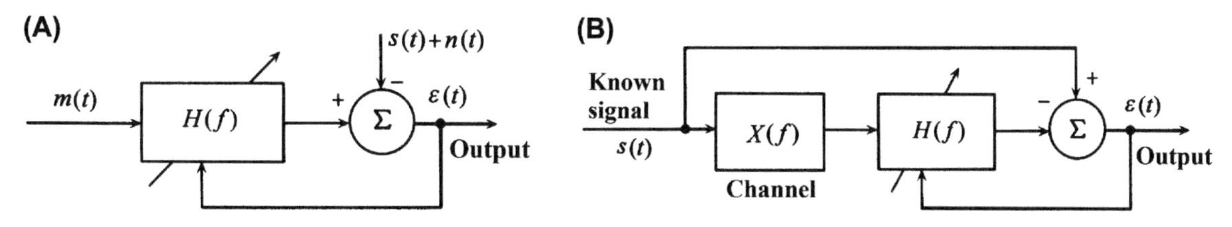

FIGURE 3.6 Applications of adaptive filter: (A) adaptive noise canceller; (B) adaptive channel equalization.

3.1.2.2 *Adaptive Line Spectra Enhancer*

The principal block diagram of an adaptive line-spectra enhancer (ALE) is given in Fig. 3.7. The original signals consist of broadband noise and single-frequency signal $A\cos(2\pi f_0 t+\varphi)$. Because the single-frequency signal has a period of 2π, any delay time Δ does not change its correlativity. On the contrary, the correlativity is dependent on Δ for a broadband noise. When Δ is large enough, their correlativity will be remarkably reduced until perfectly independent of each other. Based on the difference of correlation characteristics between single-frequency signal and broadband noise, we can enhance the former and suppress the latter by means of adaptive-filtering processing.

Let the original signal be delayed by Δ (refer to Fig. 3.7), and then take the difference between that and the output of the adaptive filter to make error signal

$$\varepsilon(t) = n(t) + A\cos(2\pi f_0 t+\varphi) - [n'(t) + B\cos(2\pi f_0 t+\theta)]$$

at minimum by adjusting the adaptive filter.

The quantity $n(t)$ is independent of $s(t)$; moreover, taking a suitable Δ to make $n(t)$ is independent of $n(t\text{-}\Delta)$, thus $n(t)$ and $n'(t)$ are also independent of each other. Therefore,

$$E(\varepsilon^2(t)) = E(n^2(t)) + E(n'^2(t)) + E[A\cos(2\pi f_0 t+\varphi) - B\cos(2\pi f_0 t+\theta)]^2 \tag{3.36}$$

$E(n^2(t))$ is a constant, which is independent of the adaptive filter. Provided that

$$E(n'^2(t)) + E[A\cos(2\pi f_0 t+\varphi) - B\cos(2\pi f_0 t+\theta)]^2 \tag{3.37}$$

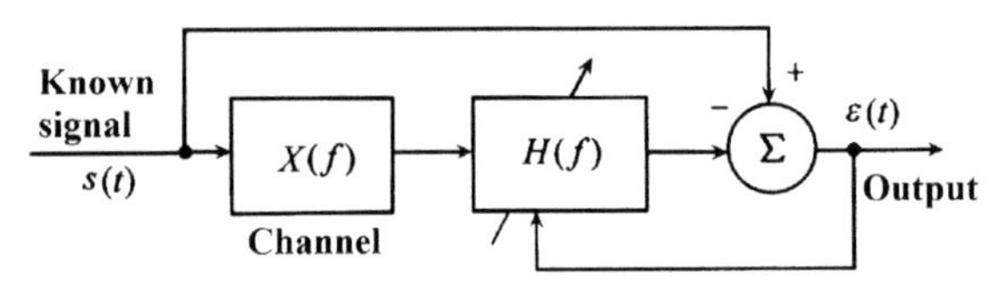

FIGURE 3.7 Principal block diagram of ALE.

is minimum, $E(\varepsilon^2(t))$ also tends to the minimal value. By adjusting the adaptive filter to make $\theta=\varphi$ and adjusting ALE to achieve the signal output

$$y(t) = B\cos(2\pi f_0 t+\theta) + n'(t)$$

with a small enough $n'(t)$ and signal component $B\cos(2\pi f_0 t+\theta)$ as tending to $A\cos(2\pi f_0 t+\varphi)$ as much as possible, so that the maximum output SNR can be obtained.

The remarkable advantage by using ALE is that the parameters including frequency, amplitude, and phase in the ordinal signal are unnecessary to be known in advance, rather they will be determined through the adaptive-adjusting process.

The qualitative analyses for an ALE have been given earlier. Now let us analyze its stable state characteristics.

The adaptive filter shown in Fig. 3.7 is generally formed by means of tapped delay lines (TDL) given in Fig. 3.5. Assuming that sampling period is T_s, the decorrelation delay time

$$\Delta = d_0 T_s$$

in which d_0 is a positive integer. Let

$$x(t) = n(t) + A\cos(2\pi f_0 t+\varphi), \quad y(t) = x(t-\Delta)$$

$$\mathbf{w} = (w(0), w(1), \cdots, w(L-1))^T$$

the optimum solution (refer to Eq. 3.19)

$$\mathbf{w}_{\text{opt}} = \mathbf{R}_{yy}^{-1}\mathbf{R}_{yx} \tag{3.38}$$

the (k, l)th elements of $\mathbf{R}_{yy}$ are as follows:

$$R_y(k-l) = E[y(k)y(l)] = R_s(k-l) + R_n(k-l) \quad k,l = 0,1,\ldots,L-1$$

the lth element of correlation vector $\mathbf{R}_{yx}$ is as follows:

$$E(y(k-l)x(k)) = R_s(l+d_0) \quad l = 0,1,\cdots,L-1$$

Eq. (3.38) can be written as

$$\sum_{k=0}^{L-1}[R_s(k-l)+R_n(k-l)]w(k) = R_s(d_0+l)$$

$$l = 0,1,\cdots,L-1 \qquad (3.39)$$

writing the solutions of Eq. (3.39) are as follows:

$$\mathbf{w}_{\text{opt}} = (\mu_0,\mu_1,\ldots,\mu_{L-1})^T$$

the output mean squared deviation $I=\mathrm{E}(\varepsilon^2(k))$ for ALE. When $\mathbf{w}=\mathbf{w}_{\text{opt}}$, I approaches a minimal value that is given by

$$\begin{aligned} I_{\min} &= \mathrm{E}(x^2(k)) - \mathbf{w}_{\text{opt}}^T\mathbf{R}_{yy}\mathbf{w}_{\text{opt}} \\ &= \mathrm{E}(x^2(k)) - \sum_{k=0}^{L-1}\sum_{l=0}^{L-1}\mu_k\mu_l R_x(k-l) \end{aligned}$$

Taking note of

$$x(k) = s(k)+n(k), \quad s(k) = A\cos(2\pi f_0 kT_s+\varphi)$$

we get

$$R_s(k) = \frac{A^2}{2}\cos(2\pi f_0 kT_s)$$

$$R_n(k) = \sigma_n^2\delta(k)$$

therefore, Eq. (3.39) becomes

$$\sigma_n^2 w(l)+\sum_{k=0}^{L-1}w(k)\frac{A^2}{2}\cos[(k-l)\alpha]$$

$$= \frac{A^2}{2}\cos[(l+d_0)\alpha] \quad l = 0,1,\ldots,L-1 \quad (3.40)$$

in which $\alpha=2\pi f_0T_s$. It may be proved that the approximate solutions of Eq. (3.40) is given by

$$\mu_k = w_{\text{opt}}(k)\approx Q\cos[(k+d_0)\alpha]$$

$$k = 0,1,\ldots,L-1$$

in which

$$Q = \frac{2(\text{SNR})_{\text{ALE,in}}}{2+L(\text{SNR})_{\text{ALE,in}}}$$

$(\text{SNR})_{\text{ALE,in}}$ expresses the input SNR of ALE. Obviously,

$$(\text{SNR})_{\text{ALE,in}} = \frac{A^2}{2\sigma_n^2}$$

the mean squared value of output for TDL is given by

$$D = \sum_{k=0}^{L-1}\sum_{l=0}^{L-1}\mu_k\mu_l R_{xx}(k-l) = D_s+D_n$$

in which D_s, D_n are signal and noise parts, respectively; moreover,

$$D_s = \frac{A^2}{2}\sum_{k=0}^{L-1}\sum_{l=0}^{L-1}\mu_k\mu_l\cos[(k+l)\alpha],$$

$$D_n = \sigma_n^2\sum_{k=0}^{L-1}\mu_k^2$$

We obtain

$$(\text{SNR})_{\text{ALE,out}} = \frac{D_s}{D_n} = \frac{2Lp+L^2+q}{2(L+p)}(\text{SNR})_{\text{ALE,in}}$$

in which

$$p = \frac{\cos[25+(L-1)\alpha]\sin L\alpha}{\sin\alpha}$$

$$q = \left(\frac{\sin L\alpha}{\sin\alpha}\right)^2$$

the system gain of ALE

$$G_{\text{ALE}} = \frac{(\text{SNR})_{\text{ALE,out}}}{(\text{SNR})_{\text{ALE,in}}} = \frac{2Lp+L^2+q}{2(L+p)} \qquad (3.41)$$

We see that, the system gain of ALE is approximately proportional to the node numbers L of TDL.

It should be noted that the larger the L, the higher the iterative noise of adaptive arithmetic, this is a factor to limit the system gain to be raised. Obviously, if $(\text{SNR})_{\text{ALE,in}}$ is too low, the system gain will be considerably reduced. This problem must be considered for designing ALE.

Because the gain of ALE is a function of input SNR, in particular when the input SNR is reduced to a certain value, the gain would be much lower. To improve the efficiency of ALE, the input SNR must possibly be raised. Therefore, an adaptive noise canceller adopted before using the ALE is a suitable selection.

3.1.3 Diversity Techniques for Fading Channels [3]

When a channel appears deep fading, received signals will encounter severe attenuation, thus bit error rate (BER) will rise. If we can supply several independent fading channels through which an identical signal is transmitted, the signals that are received from these channels combined and then decided, that is called diversity technique. Because the probability of several independent fading channels all too deeply fading simultaneously is small enough, the robustness will greatly be improved provided a suitable diversity technique is used.

3.1.3.1 Diversity Methods

Several methods are commonly employed in forming the diversity techniques for fading channels.

3.1.3.1.1 FREQUENCY DIVERSITY

That is, an identical information-bearing signal is transmitted on L carriers in which the frequency gap between successive carries equals or exceeds channel coherence bandwidths $(\Delta f)_c$. In this case, the L carries are equivalent to L independently fading channels.

3.1.3.1.2 TIME DIVERSITY

That is, the same information-bearing signal is transmitted on L different time slots, in which the space between successive time slots equals or exceeds channel coherence times $(\Delta t)_c$. The L time slots are equivalent to L independently fading channels.

3.1.3.1.3 MULTIPLE ANTENNA SPACE DIVERSITY

This method adopts multiple antennas the arrangement for which remains the space of at least 10 wavelengths to satisfy that the signals received are from independently fading channels. If multiple receiving antennas are used, that is called space-receiving diversity. If multiple transmitting antennas are used, that is called space-transmitting diversity. If a system in which multiple antennas are employed as both receiver and transmitter, that is referred to as a multiple input—output system.

Evidently, in the case of frequency diversity, if many more frequency channels must be used, frequency superfluity is thus introduced. Similarly, time superfluity is introduced by using many more time slots in the time diversity. A power superfluity also appears in the space-transmitting diversity. Time, frequency, and power are the communication resources that must be utilized effectively. These superfluities are unwanted in the multiple-antennas receiving diversity, but that requires increasing corresponding installations and occupying much larger spaces.

3.1.3.1.4 RAKE RECEIVE

A more ingenious method for obtaining good diversity results is that using a signal having a bandwidth B much larger than the channel coherence bandwidth $(\Delta f)_c$, so that it can be resolved into multipath components, and thus provide several independently fading signal paths. The time resolution is $1/B$. Consequently, there are $L = T_M/(1/B) = B/(\Delta f)_c$ independently fading paths that supply L orders of diversity for a channel with a multipath spread of T_M, which is called rake receiving technique or paths diversity.

3.1.3.2 Models of Combination in Diversity Techniques

There are two meanings for the diversity techniques, one is that an identical information-bearing signal is transmitted in several independently fading channels that means the signal is "dispersed"; the other is that the several replicas of the identical signal transmitted in different independently fading channels are combined and processed, it means that received signals from these channels are "centralized."

Some models of combination are briefly introduced as follows:

3.1.3.2.1 SELECTIVITY COMBINATION

Assuming a signal is transmitted in L independently fading channels. Let the equivalent low-pass outputs for kth channel be as follows:

$$r_{lk}(t) = \alpha_k e^{j\varphi_k} s_l(t) + z_k(t), \quad k = 1, 2, \ldots, L$$

in which α_k, φ_k are amplitude attenuation and phase shift, respectively, in the kth channel, $z_k(t)$ is equivalent low-pass complex Gaussian noise and assuming that it has the same power spectrum N_b in both real and complex parts, so that the output SNR of a receiver (correlator or matched filter) in the fading channel may be expressed by

$$\gamma_k = \frac{\alpha_k^2 E_s}{N_k} \tag{3.42}$$

in which E_s is the energy of complex envelope signal $s_l(t)$. In the case of the selectivity combination, we measure SNR in every fading channel and then select a maximal one to carry out the corresponding signal decision.

3.1.3.2.2 MAXIMAL RATIO COMBINATION

In the case of the maximal ratio combination, the received signals from every independently fading channel are linearly combined, that is

$$r_l(t) = \sum_{k=1}^{L} \beta_k r_{lk}(t) \tag{3.43}$$

the combination coefficient β_k is selected to make the SNR of $r_l(t)$ maximal. Therefore, a maximal ratio combiner, in fact, is a model of optimal combiner. The signal component in Eq. (3.43)

$$v(t) = s_l(t) \sum_{k=1}^{L} \beta_k \alpha_k e^{j\varphi_k}$$

The noise component

$$z(t) = \sum_{k=1}^{L} \beta_k z_k(t)$$

Because $z_k(t)$ $k = 1,2,\ldots, L$ are independent of each other, the total power of noise is given by

$$N = \sum_{k=1}^{L} \beta_k^2 N_k$$

thus, output SNR

$$\gamma = \frac{\left| \sum_{k=1}^{L} \beta_k \alpha_k e^{j\varphi_k} \right|^2 E_s}{\sum_{k=1}^{L} \beta_k^2 N_k}$$

by using Schwartz's inequality, we obtain

$$\left| \sum_{k=1}^{L} a_k^* b_k \right|^2 \le \left(\sum_{k=1}^{L} |a_k|^2 \right) \left(\sum_{k=1}^{L} |b_k|^2 \right) \tag{3.44}$$

the abundant conditions with which Eq. (3.44) is to be holed are

$$b_k = \eta a_k, \quad k = 1, 2, \ldots, L$$

Now let $a_k = \frac{\alpha_k e^{-j\varphi_k}}{\sqrt{N_k}}$, $b_k = \beta_k \sqrt{N_k}$, substituting them into Eq. (3.44), we get

$$\left| \sum_{k=1}^{L} \beta_k \alpha_k e^{j\varphi_k} \right|^2 \le \left(\sum_{k=1}^{L} \frac{|\alpha_k|^2}{N_k} \right) \left(\sum_{k=1}^{L} |\beta_k|^2 N_k \right)$$

therefore,

$$\gamma \le \sum_{k=1}^{L} \gamma_k \tag{3.45}$$

when

$$\beta_k = \eta \frac{\alpha_k}{N_k} e^{-j\varphi_k}, \quad k = 1, 2, \ldots, L \tag{3.46}$$

to be selected, the equality in Eq. (3.45) holds; that is to say, now the output SNR of the combiner is maximal and equals an optimal linear one, the condition achieving that is the phase of weighted coefficient β_k to be "$-\varphi_k$" that just counteracts the fading phase shift, thus the signals received from every branch overlap with the same phase. Moreover, the amplitude of β_k is proportional to α_k/N_k, therefore β_k is raised with increasing signal intensity, the noise appearing in every branch is relatively reduced.

3.1.3.2.3 EQUAL-GAIN COMBINATION

A maximal ratio combination can be regarded as an optimal one, which is based on the assumption that α has been accurately estimated; therefore the complexity to form that is much increased. Although the performance of an equal-gain combination is lower than the maximal ratio combination, they are approximately agreeable with each other. In the case of the equal-gain combination, the phases of the signals through every fading branch have been calibrated to achieve the same phase superposition, but weight gain is the same; therefore the combination coefficient $\beta_k = e^{-j\varphi_k}$. The output of the combinator is given by

$$r_l(t) = \sum_{k=1}^{L} \beta_k r_{lk} = s_l(t) \sum_{k=1}^{L} \alpha_k + \sum_{k=1}^{L} e^{-j\varphi_k} z_k(t)$$

SNR after combining can be

$$\gamma = \frac{E_s \left(\sum_{k=1}^{L} \alpha_k \right)^2}{\sum_{k=1}^{L} N_k} \tag{3.47}$$

3.1.3.2.4 SQUARE-LAW COMBINATION

In the case of a incoherent demodulation system, generally the parameters α and φ in fading channels are not perfectly estimated, but the equivalent low-pass signals received from every branch are squared and directly superposed, and then perform envelope detection. The input signals of a square-law detector are as follows:

$$r_l = \sum_{k=1}^{L} |r_{lk}(t)|^2 \tag{3.48}$$

3.1.3.3 *Rake Receive Technique for Frequency-Selective Slowly Fading Channels*

When the spread factor of a channel satisfies the condition: $T_m B_d \ll 1$, it is possible to select signals with a bandwidth $B \ll (\Delta f)_c$ and a signaling duration $T \ll (\Delta t)_c$. Thus, the channel belongs to frequency-nonselective and a slowly fading one. In such a channel, diversity techniques mentioned previously can be used to overcome the effect of the fading.

When a bandwidth $B \gg (\Delta f)_c$, the channel belongs to the frequency-selective one. In such a case, the channel can be subdivided into a number of frequency division multiplexed (FDM) subchannels having a mutual separation of at least $(\Delta f)_c$ in center frequencies. Then the same signal can be transmitted in the FDM subchannels, and thus the frequency diversity is obtained.

A more ingenious diversity method, which is called rake-receiving technique, will be introduced. This diversity technique utilizes multiple-path transmission signals, thus it is also called multiple-path diversity.

3.1.3.3.1 TAPPED DELAY LINE CHANNEL MODEL

A channel is still assumed to be slowly fading one, that is, the signaling duration $T \ll (\Delta t)_c$. Now suppose that the bandwidth occupied by the transmission signal $s(t)$ is B, then the bandwidth occupied by its equivalent low-pass signal $s_l(t)$ is $|f| \leq B/2$. According to sampling theorem, the low-pass signal

$$s_l(t) = \sum_{n=-\infty}^{\infty} s_l\left(\frac{n}{B}\right) \frac{\sin[\pi B(t - n/B)]}{\pi B(t - n/B)}$$

The Fourier transform of $s_l(t)$ is as follows:

$$S_l(f) = \begin{cases} \frac{1}{B}\sum_{n=-\infty}^{\infty} s_l\left(\frac{n}{B}\right)e^{-j2\pi fn/B}, & |f| \le B/2 \\ 0, & |f| > B/2 \end{cases} \quad (3.49)$$

If additive noise is disregarded, the received signals from frequency-selective channels are expressed in the form

$$r_l(t) = \int_{-\infty}^{\infty} C(f,t)S_l(f)e^{j2\pi ft}df \quad (3.50)$$

in which $C(f, t)$ is the time-variant transfer function. Inserting Eq. (3.49) into (3.50)

$$r_l(t) = \frac{1}{B}\sum_{n=-\infty}^{\infty} s_l\left(\frac{n}{B}\right)\int_{-\infty}^{\infty} C(f,t)e^{j2\pi f(t-n/B)}df$$
$$= \frac{1}{B}\sum_{n=-\infty}^{\infty} s_l\left(\frac{n}{B}\right)c(t-n/B,t) \quad (3.51)$$

in which $c(\tau, t)$ is the time-variant impulse response. We observe that Eq. (3.51) has the form of a convolution. Hence, it can also be expressed in an alternative form

$$r_l(t) = \frac{1}{B}\sum_{n=-\infty}^{\infty} s_l\left(t-\frac{n}{B}\right)c(n/B,t)$$

It is convenient to define a set of time-variant channel coefficients as

$$c_n(t) = \frac{1}{B}c(n/B,t)$$

therefore,

$$r_l(t) = \sum_{n=-\infty}^{\infty} c_n(t)s_l\left(t-\frac{n}{B}\right) \quad (3.52)$$

The form for the received signal in Eq. (3.52) implies that the time-variant frequency-selective channel can be modeled as a tapped delay line with tap spacing $1/B$ and tap weight coefficient $\{c_n,t\}$. In fact, we deduce from Eq. (3.52) that the low-pass impulse response for the channel is

$$c(\tau,t) = \sum_{n=-\infty}^{\infty} c_n(t)\delta\left(\tau-\frac{n}{B}\right)$$

and the corresponding time-variant transfer function is

$$C(f,t) = \sum_{n=-\infty}^{\infty} c_n(t)e^{-j2\pi fn/B}$$

thus, with an equivalent low-pass signal having a bandwidth $B/2$, in which $B \gg (\Delta f)_c$, we achieve a resolution of $1/B$ in the multipath delay profile. Because the total multipath spread is T_M, for all practical purposes the tapped delay line model for the channel can be truncated at $L = [T_M B] + 1$ taps. When the additive noise is neglected, the received signal can be expressed in the form

$$r_l(t) = \sum_{n=1}^{L} c_n(t)s_l(t-n/B) \quad (3.53)$$

The tapped delay line model is shown in Fig. 3.8. The time-variant tap weights $\{c_n(t)\}$ are zero that means they are complex-valued random processes. The magnitudes $|c_n(t)| = \alpha_n(t)$ are Rayleigh distributed and the phases of $c_n(t)$ are uniformly distributed.

3.1.3.3.2 RAKE-RECEIVING TECHNIQUE

We now consider the problem of digital signaling over a frequency-selective channel that is modeled by a tapped delay line with statistically independent time-varying tap weighs $\{c_n(t)\}$. That provides us with L replicas of the same transmitted signal at the receiver. Hence, a receiver that processes the received signal in an optimum manner will achieve the performance of an equivalent Lth-order diversity.

Let us consider binary signaling over the channel. We have two equal-energy signals S_{l1} and S_{l2} that are either antipodal or orthogonal. Their time duration T is selected to satisfy the condition of $T \gg T_M$. Thus, we may neglect

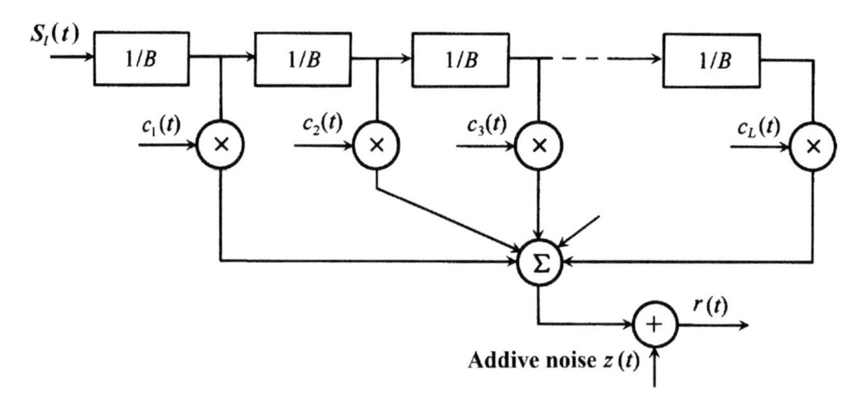

FIGURE 3.8 Tapped delay line model of frequency-selective channel.

any intersymbol interference due to multipath effect. Because the bandwidth of the signal exceeds the coherent bandwidth of the channel, the received signal is expressed as

$$\begin{aligned} r_l(t) &= \sum_{k=1}^{L} c_k(t) s_{lm}(t - k/B) + z(t) \\ &= v_m(t) + z(t) \quad (0 \le t \le T;\ m = 0,1) \end{aligned}$$

in which $z(t)$ is a complex-valued white Gaussian noise. Assume for the moment that the channel tap weights are known. Then the optimal receiver consists of two filters matched to $v_0(t)$ and $v_1(t)$, followed by samplers, and a decision circuit that selects the signal corresponding to the largest output. An equivalent optimal receiver employs cross correlation, instead of matched filtering. In either case, the decision variables for coherent detection of the binary signals can be expressed as

$$\begin{aligned} U_m &= \mathrm{Re} \int_0^T r_l(t) \mathrm{d}t \\ &= \mathrm{Re}\left[\sum_{k=1}^{L} \int_0^T r_l(t) c_k^*(t) \right. \\ &\qquad \left. \times s_{lm}^*(t - k/B) \mathrm{d}t \right], \quad m = 0,1 \end{aligned} \tag{3.54}$$

Fig. 3.9 illustrates the operations involved in the computation of the decision variables. In this realization of the optimal receiver, the two reference signals are delayed and correlated with the received signal $r_l(t)$.

An alternative realization of the optimal receiver employs a single-delay line through which is passed the received signal $r(t)$. The signal at each tap is correlated with $c_k^*(t)\ s_{lm}^*(t)$, in which $k = 1,2,\cdots, L$, and $m = 1,2$. The structure of this receiver is shown in Fig. 3.10. In fact, the tapped delay line receiver attempts to collect the signal energy from all the received-signal paths that fall within the span of the delay line and carry the same information.

In the case of slow-fading channel, the rake receiver with perfect estimates of the channel tap weights is equivalent to a maximal ratio combiner, which is optimal linear diversity in a system with Lth-order diversity.

3.1.4 Spread-Spectrum Technique [4]

3.1.4.1 Theoretical Fundamental of the Spread-Spectrum Technique

3.1.4.1.1 SHANNON FORMULA

Shannon theorem had indicated that the critical transmitting rate or channel tolerance c against a white Gaussian noise background is given by

$$C = B \log_2\left(1 + \frac{S}{N}\right) \quad \text{b/s} \tag{3.55}$$

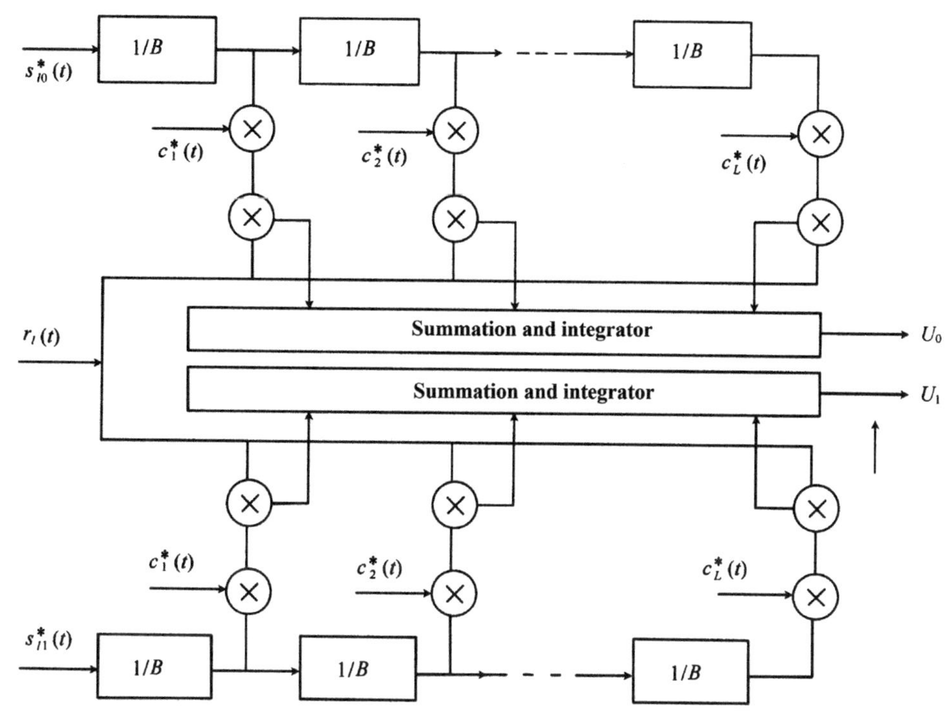

FIGURE 3.9 Optimal demodulator for wideband binary signals (delayed reference configuration).

in which B is signal bandwidth, S is signal average power, and N is noise power.

Assuming the power spectral density of the white Gaussian noise is N_0, thus its power $N = N_0B$. The channel tolerance can therefore be expressed as

$$C = B\log_2\left(1 + \frac{S}{N_0B}\right) \quad \text{b/s} \tag{3.56}$$

We see that if B, N_0, and S are determined, C is also determined. According to the second Shannon theorem, we have known that provided the data rate for an information source is equal to or less than C, the information can be transmitted over channels at an arbitrarily small error probability.

The Shannon formula shows that:

1. To increase the information rates, raising the channel tolerance C is required; the approaches for that make use of either increasing the signal bandwidth or the SNR. Evidently, it is more effective to increase B than to raise the SNR.
2. When C is a constant, we may adopt the approaches increasing the bandwidth B to lower the requirement for SNR, or by increasing the signal power S to reduce B.
3. Once B increases to a much larger value, the increase of C will slow down; because the noise power N will increase with increasing B, SNR will therefore decrease which will affect the increase of C. Let $B \to \infty$, the critical value of C is as follows:

$$\lim_{B\to\infty} C = 1.44\frac{S}{N_0} \tag{3.57}$$

We see that the channel tolerance C is limited provided the signal power S and noise power spectrum density N_0 are defined.

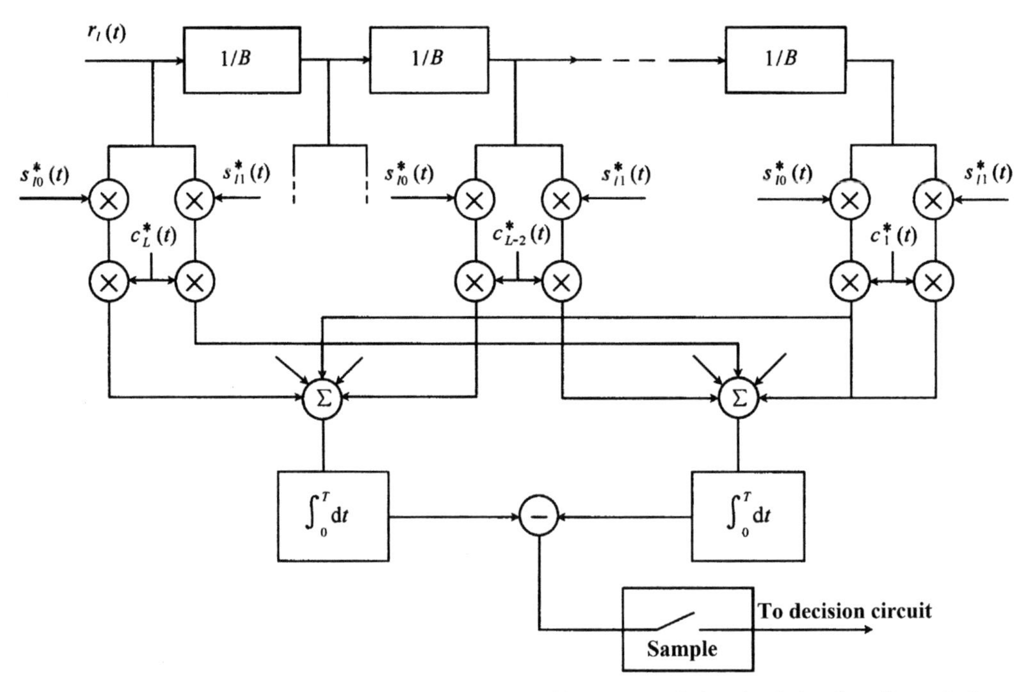

FIGURE 3.10 Optimum demodulator for wideband binary signals (received signal configuration).

3.1.4.1.2 PHYSICAL MODEL OF SPREAD-SPECTRUM SYSTEM

The physical model of spread-spectrum system is shown in Fig. 3.11. By passing through the first modulation, that is the information modulation, the signals generated by information source will be changed into digital signals, and then followed by the second modulation, that is the spread-spectrum modulation; the digital signals are spread to a very wide bandwidth by means of spread-spectrum codes. Finally, the third modulation will shift the spread-spectrum modulation signals to radio frequency and transmit them over the channels. Received signals are changed into medium frequency (IF) signals by a mixer in the receiver, and then the IF signals will perform correlation despreading by using local-spread frequency codes and recovered to narrowband signals. Finally, the digital signals will be restored after demodulation. It is necessary that the local S–S codes are perfectly synchronized with transmitting S–S codes in the receiving process.

3.1.4.2 Direct-Sequence Spread Spectrum

3.1.4.2.1 FORM OF A DIRECT-SEQUENCE SPREAD SPECTRUM (DS-SS) SYSTEM

The principal block diagram of a DS-SS system is given in Fig. 3.12. The signal $a(t)$ generated by an information source is an information stream that consists of the code elements with a duration T_a. The pseudorandom code $c(t)$ generated by a pseudorandom code generator has a duration T_c, sometimes that is referred to as a chip duration. By processing through modulo-2 addition between the signal code $a(t)$ and pseudorandom code $c(t)$, a spread-spectrum sequence is formed that has a rate the same with pseudorandom code. Then the carrier is modulated by using

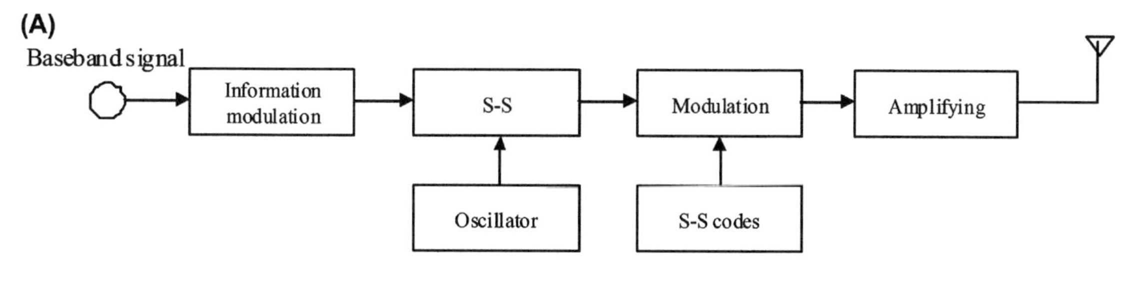

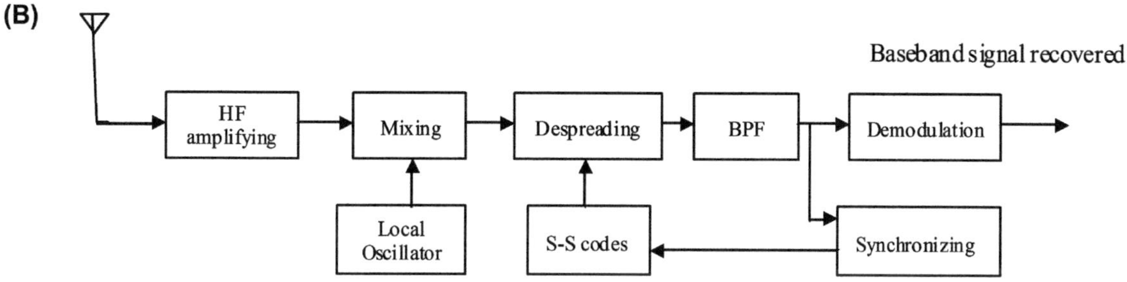

FIGURE 3.11 Physical model of S–S system: (A) transmitting, (B) receiving.

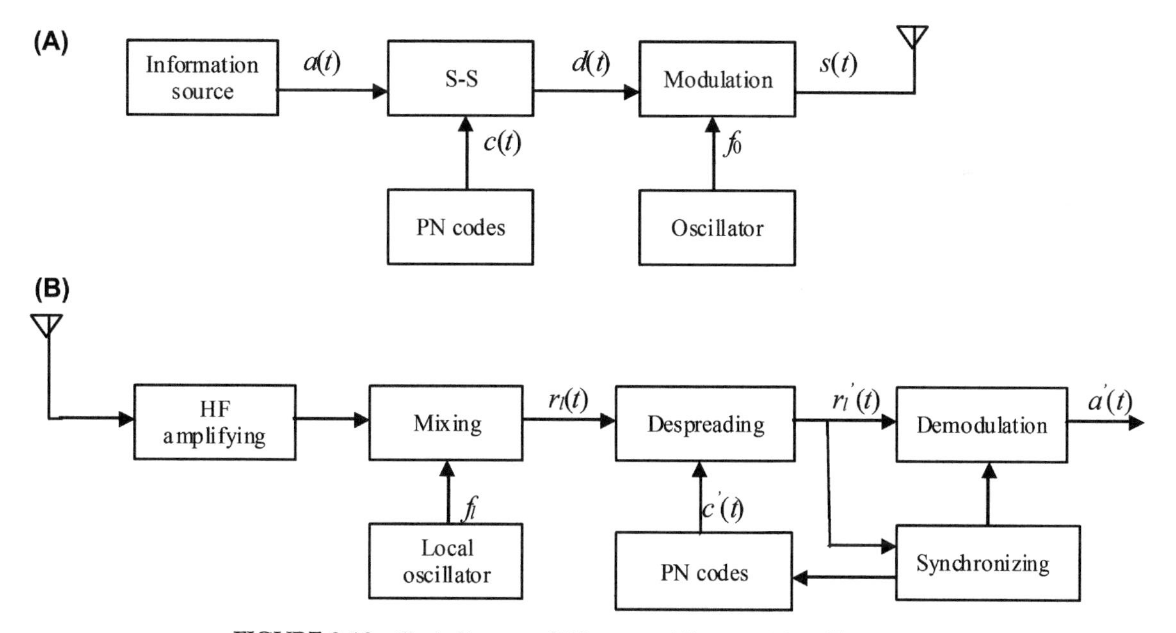

FIGURE 3.12 Block diagram of DS system: (A) transmitting, (B) receiving.

the spread-spectrum sequence, and we thus obtain an SS-modulated radio frequency (RF) signal that is transmitted in the channels.

The received spread spectrum signals by passing through a preamplifier and mixer, and then the IF S–S modulation signal is correlatively despread using the pseudorandom sequence, which is properly synchronized with that at the transmitting end. The signal bandwidth will be recovered as that of information sequence $a(t)$, that is the IF modulation signal, and then, followed by demodulator, the transmitted signal $a(t)$ will be recovered; therefore, the information transmissions have been performed. Jamming interference and noises, independent of the pseudorandom sequence are equivalent to making a spread spectrum by the effect of correlation despreading. the spectrum density for that will lower, so that the power of both the jamming interference and noise will greatly be reduced by passing through the signal pass band. As a result, input SNR and signal-to-interference ratio of the demodulator are improved.

3.1.4.2.2 SIGNALING ANALYSES OF DS-SS SYSTEM

Letting the information stream generated by information source be $a(t)$, the rate of code elements be R_a, the corresponding duration $T_a = 1/R_a$, thus

$$a(t) = \sum_{n=0}^{\infty} a_n g_a(t - nT_a) \tag{3.58}$$

in which a_n is information code and letting those probabilities P be 1 and $(1 - P)$ be -1, that is

$$a_n = \begin{cases} +1 & \text{for probability } P \\ -1 & \text{for probability } (1-P) \end{cases}$$
$$g_a(t) = \begin{cases} 1 & 0 \le t \le T_a \\ 0 & \text{others} \end{cases} \tag{3.59}$$

in which $g_a(t)$ is a gate function.

Let the pseudorandom sequence $c(t)$ generated by the pseudorandom sequence generator have a rate R_c and chip duration T_c, and $T_c = 1/R_c$, we get

$$c(t) = \sum_{n=0}^{N-1} c_n g_c(t - nT_c) \tag{3.60}$$

in which c_n is the code element of the pseudorandom code taking the values of $+1$ or -1; $g_c(t)$ is a gate function defined as Eq. (3.59).

The spread-spectrum process is substantially the modelo-2 addition or the product between the information stream $a(t)$ and the pseudorandom sequence $c(t)$. Because R_c is much larger than R_a, therefore the rate of the sequence after spread spectrum is still R_c. The spread-spectrum sequence is given by

$$d(t) = a(t)c(t) = \sum_{n=0}^{\infty} d_n g_c(t - nT_c) \tag{3.61}$$

in which

$$d_n = \begin{cases} +1 & a_n = c_n \\ -1 & a_n \ne c_n \end{cases} \qquad (n-1)T_c \le t \le nT_c$$

by using the spread-spectrum sequence to modulate the carrier, the signal will be removed to carrier frequency. Generally speaking, most digital modulation schemes, such as binary-phase shift keying (BPSK), minimum shift keying (MSK), quadrature-phase shift keying (QPSK), tamed frequency modulation (TFM), etc. can be employed in spread-spectrum system. However, we must select a suitable one to satisfy the specific requirement. The phase-shift keying (PSK) modulation will be adopted in our analysis in the next section, which can be performed by means of a balancing modulator. We get the signal being modulated as follows:

$$s(t) = d(t)\cos\omega_0 t = a(t)c(t)\cos\omega_0 t \tag{3.62}$$

in which ω_0 is the carrier angular frequency.

The inductive signals at the receiving antenna by passing through the amplifilter and mixer, we obtain the signals that include useful signals $s_I(t)$, channel noise $n_I(t)$, jamming signals $J_I(t)$, and spread-spectrum signals $s_J(t)$ from other networks, that is

$$r_I(t) = s_I(t) + n_I(t) + J_I(t) + s_J(t) \tag{3.63}$$

The pseudorandom sequence $c'(t)$ at the receiving end is the same as that at the transmitting end, but the original time or initial phase may possibly be different. The despreading process is identical with the spreading process, that is, using the local pseudorandom sequence $c'(t)$ to make the product with received signal, we obtain

$$\begin{aligned} r'_I &= r_I(t)c'(t) \\ &= s_I(t)c'(t) + n_I(t)c'(t) + J_I(t)c'(t) + s_J(t)c'(t) \\ &= s'_I(t) + n'_I(t) + J'_I(t) + s'_J(t) \end{aligned}$$

These four components will, respectively, be analyzed subsequently. First, we examine the signal component:

$$s'_I(t) = s_I(t)c'(t) = a(t)c(t)c'(t)\cos\omega_I t$$

If $c'(t)$ synchronizes with $c(t)$, that is $c(t) = c'(t)$, and thus $c(t)\cdot c'(t) = 1$, signal component will be given by

$$s'_I(t) = a(t)\cos\omega_I t \tag{3.64}$$

the following filter will just allow the signal to pass through, thus the signal will be demodulated by following a modulator.

The noise component $n_I(t)$, jamming signal component $J_I(t)$, and interferences due to other networks $s_J(t)$ will be largely degraded by the despreading processing. $n_I(t)$ is generally a white Gaussian noise; the spectrum density is not changed or is slightly reduced by despreading, but its relative bandwidth is changed; thus the noise power will be reduced. Jamming signal $J_I(t)$ is caused by intentional interference, because the pseudorandom codes are independent of $J_I(t)$ and $c(t)$, their product is equivalent to a spreading-spectrum process, the power of $J_I(t)$ is spread to a very broad bandwidth, the spectrum density is correspondingly reduced. The power of $J_I(t)$ is only allowed to pass through a bandwidth that is the same with that of $s_I(t)$; therefore, the power of $J_I(t)$ is greatly reduced after despreading processing, the input signal-to-interference at the demodulator will rise. Similarly, different spread-spectrum sequences exist for different networks that are equivalent to make another spread spectrum; the interference caused by them is thus reduced.

The waveforms and spectra for a direct-spread spectrum system are roughly plotted in Figs. 3.13 and 3.14, respectively.

Now let us analyze the power spectra of a DS-SS signal. The transmitted signal

$$s(t) = d(t)\cos\omega_0 t = a(t)c(t)\cos\omega_0 t \tag{3.65}$$

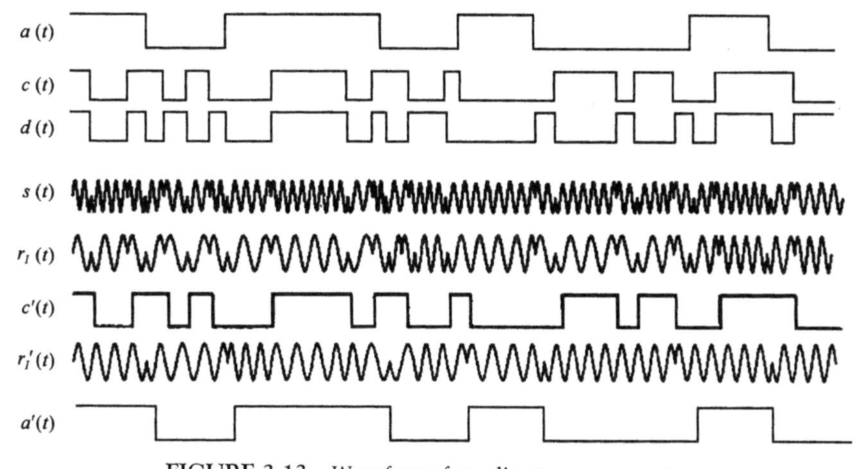

FIGURE 3.13 Waveforms for a direct sequence system.

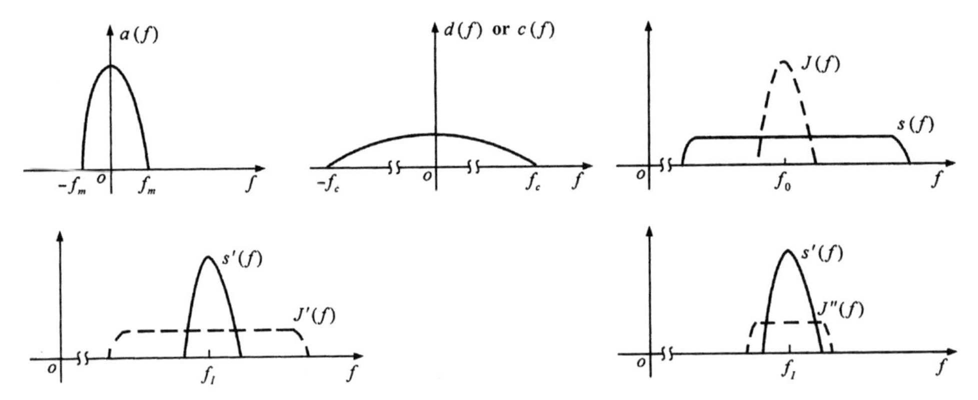

FIGURE 3.14 Spectra for a direct sequence system.

in which $a(t)$, $c(t)$, and $d(t)$ are determined by Eqs. (3.58), (3.60), and (3.61), respectively. The autocorrelation function $B_s(\tau)$ for $s(t)$ will first be found, and then following by fast fourier transformation (FFT), we can obtain its power spectra density $G_s(t)$. The autocorrelation function of $s(t)$ is given by

$$B_s(\tau) = \frac{1}{T}\int_{-T/2}^{T/2} s(t)s(t-\tau)dt = \frac{1}{2}B_d(\tau)\cos\omega_0\tau \tag{3.66}$$

Because the $a(t)$ and $c(t)$ are generated by two different information sources, they are independent of each other. We obtain

$$B_d(\tau) = B_a(\tau)B_c(\tau)$$

in which $B_a(\tau)$ and $B_c(\tau)$ are autocorrelation functions of $a(t)$ and $c(t)$, respectively, $c(t)$ is a periodic pseudorandom sequence with the length N, thus its autocorrelation function $B_c(\tau)$ is also a periodic function, the waveform for which is plotted in Fig. 3.15. To make FFT for $B_c(\tau)$, we get the power spectrum density of $c(t)$:

$$G_c(\omega) = \frac{1}{N^2}\delta(\omega) + \frac{N+1}{N^2}\mathrm{Sa}^2\left[\frac{\omega T_c}{2}\right] \times \sum_{\substack{k=-\infty \\ k\neq 0}}^{\infty} \delta\left(\omega - \frac{2\pi k}{NT_c}\right) \tag{3.67}$$

We see that the power spectra of pseudorandom sequence belongs to dispersed spectrum with a space of $\omega_1 = 2\pi/(NT_c)$, the amplitudes for which are determined by $\mathrm{Sa}^2(\omega T_c/2)$ as shown in Fig. 3.16A. The spectrum density for DS signal $s(t)$ may be found according to the properties of FFT, that is

$$\begin{aligned} G_s(\omega) &= \frac{1}{8\pi^2}G_a(\omega) \times G_c(\omega) \\ &\quad \times \pi[\delta(\omega-\omega_0) + \delta(\omega+\omega_0)] \\ &= \frac{1}{8\pi}G_a(\omega) \times [G_c(\omega-\omega_0) + G_c(\omega+\omega_0)] \end{aligned}$$

substituting Eq. (3.67) into this expression and considering single-side spectrum, we obtain

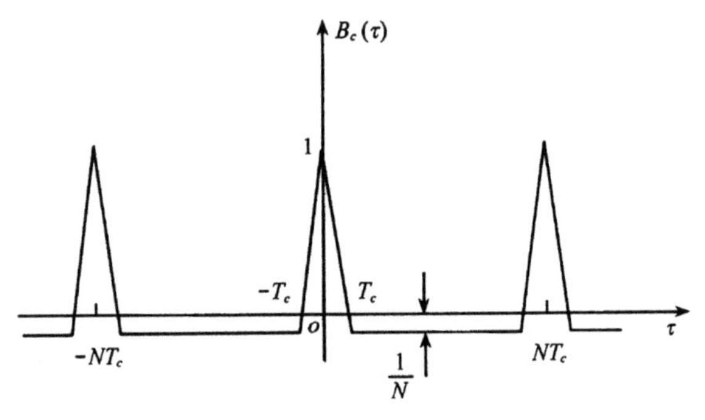

FIGURE 3.15 Waveform of $B_c(\tau)$.

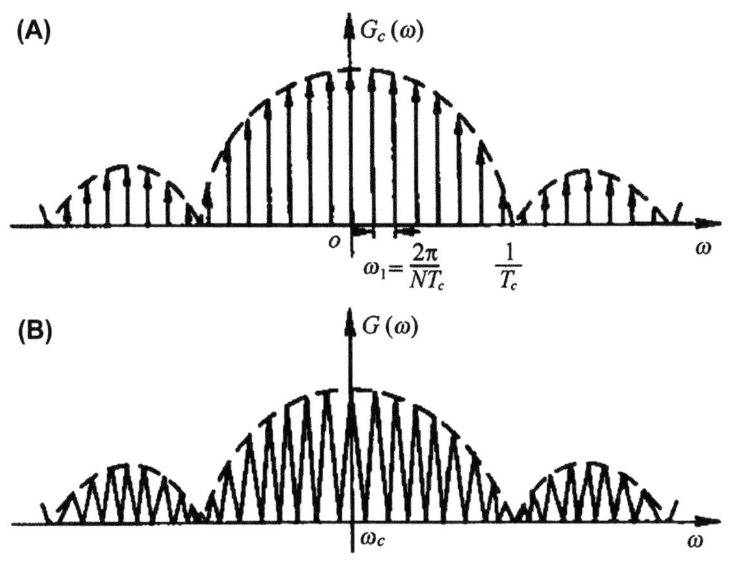

FIGURE 3.16 Power spectra of direct sequence signal: (A) power spectra of $c(t)$, (B) power spectra of $s(t)$.

$$G_s = \frac{1}{4\pi N^2}G_a(\omega - \omega_0) + \frac{N+1}{4\pi N^2} \times \sum_{\substack{k=-\infty \\ k\neq 0}}^{\infty} \text{Sinc}^2\left(\frac{\pi k}{N}\right) G_a\left(\omega - \omega_0 - \frac{2\pi k}{NT_c}\right) \tag{3.68}$$

it is plotted in Fig. 3.16B. Referring to Fig. 3.16, we can see that the larger the N, the denser the spectrum line; the smaller T_c, the wider the bandwidth of power spectra and thus the lower the spectrum density, and $c(t)$ is more approximately white noise.

3.1.4.2.3 PROCESSING GAIN

In a spread–spectrum (SS) system, the performance of antiinterference will largely be improved by the processes of the spread and despread spectra. The amount of improvement that is achieved by making use of the spread spectrum is defined as the processing gain of the spread-spectrum system. The processing gain for a DS-SS system is given by

$$G_P = \frac{R_{ch}}{R} \tag{3.69}$$

in which R_{ch} is chip rate, R is data rate. G_p is generally expressed by dB, that is

$$G_P = 10\lg\frac{R_{ch}}{R}\ \text{dB} \tag{3.70}$$

The power of the output signal for despreading does not change, whereas the power of the interference is spread into a significantly large bandwidth through despreading processing, thus the input power of the interference at the demodulator is greatly reduced in comparison with that of despreading, that is to say, interference power is greatly reduced by the despreading process. The processing gain for a DS system is therefore equal to the decreased times of the interference power.

It is necessary to point out that the processing gain mentioned earlier is not a measuring standard for different communication systems. For a communication system without using the spread-spectrum technique, the processing gain is therefore absent, but its SNR at intermediate frequency is the same as that of a spread-spectrum system, that is $G_s = 1$. In the case of white noise background, we cannot obtain a processing gain provided a narrow bandwidth system is spread to a broad bandwidth one. That is to say, an SS system lacks the superiority to combat white noise. In fact, it is impossible to perfectly avoid the synchronical error of pseudocodes in an SS system; the performance combating white noise for an SS system would be slightly lower than that of a narrow bandwidth system.

3.1.4.2.4 ADVANTAGES OF DIRECT-SEQUENCE SPREAD SPECTRUM SYSTEM

As mentioned previously, in the case of white noise background, the ability of antinoise for a DS-SS system practically is slightly lower than that of a narrow bandwidth system. However, the DS-SS system communication possesses some remarkable advantages, which will briefly be introduced as follows:

1. The DS-SS system has a much better ability to suppress jamming signals, which can be exhibited by processing gain G_s. The excellent performance to suppress then is the basic superiority by making use of a DS-SS system.
2. The spectrum density of received signal is quite low for a DS-SS system, even if the signal has been hidden by background noise; therefore, there exists low probability of interception. For example, a DS-SS system receiver can normally detect a transmission signal at an input SNR range from -15 to -10 dB. By the aid of the processing gain obtained from a matched filter or cross-correlator in the receiver, the weak signals can be recovered against a high-noise background.
3. The DS-SS system processes an ability to select addresses, and can thus be employed in a code division multiple access (CDMA) network that allows multiple users to simultaneously use a common channel for information transmission.
4. It has an ability to combat the fading of signals, especially frequency-selective fading, because the DS-SS system operates at a very broad bandwidth, a part of that being fading would not much impact the whole performance.
5. Based on the correlation characteristics of pseudorandom codes, it has an ability to combat multipath interference; that is, provided the multipath temporal delays exceed chip duration, the multipath interference will be suppressed by passing through correlation processing.

Furthermore, in applications other than communications, the DS-SS system may be used to obtain accurate ranging and positioning in radar and navigating systems according to the excellent correlation characteristic of pseudorandom codes.

3.1.4.3 Frequency-Hopped Spread-Spectrum System

3.1.4.3.1 FORM FOR A FREQUENCY-HOPPED SPREAD SPECTRUM (FH-SS) SYSTEM

In FH-SS system communications, the available channel bandwidth is subdivided into a large number of continuous frequency slots. In any signaling interval, the transmitted signal occupies one or more of the available frequency slots. The selection of the frequency slot in each signaling interval is made pseudorandomly according to the output from a pseudonoise (PN) code generator.

A block diagram of the transmitter and receiver for an FH-SS system is shown in Fig. 3.17. The carrier frequencies generated by the frequency synthesizer are modulated by the information stream generated by the information source; the radio frequency signals are obtained. The carrier frequencies are controlled by pseudorandom codes and let the frequencies hop according to a certain law. The noncoherent demodulation schemes, as frequency shift keying (FSK), amplitude shift keying (ASK), and so on are usually adopted in the FH system.

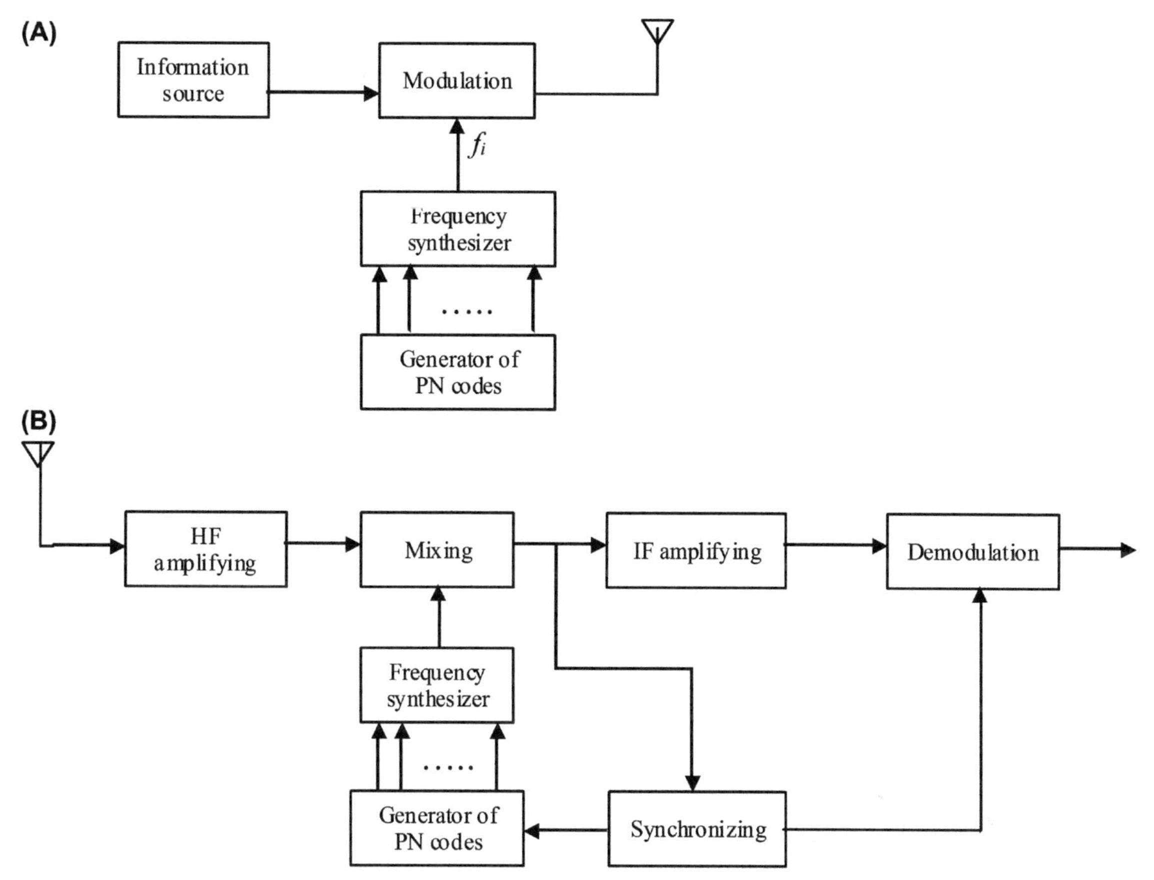

FIGURE 3.17 Block diagram of frequency-hopping system: (A) transmitting, (B) receiving.

At the receiving end, the received signal and interference pass through a high-frequency amplifier and filter, and then are input to a mixer. The local oscillating signal at the receiving end is also an FH signal with the hopping law that is the same with the FH signal in the transmitting end, but a frequency difference f_i exists between them that is just right to equal IF at the receiving end. Provided the pseudorandom codes between the transmitting and receiving ends are synchronized to each other, the hopping frequencies generated by the frequency synthesizer at both ends will also be synchronized, thus the output from the mixer will be IF signal. Finally, the transmission signal will be recovered by passing through the demodulation for the IF signal. In the case of interference, the FH law would be unknown, thus it is not relative with the frequencies generated by local frequency synthesizer and cannot pass through IF passage followed by the mixer, thus the interference cannot affect the high frequency (HF) system. In fact, the mixer here plays a frequency-dehopped role; once the synchronization between the transmitting and receiving ends is established, the FH signal will be changed to a definite IF signal.

3.1.4.3.2 SIGNALING ANALYSES FOR FREQUENCY-HOPPING SYSTEM

Let us assume that the signal $a(t)$ generated by information source is a double-polarity digital signal as

$$a(t) = \sum_{n=0}^{\infty} a_n g_a(t - nT_a) \tag{3.71}$$

in which a_n is information code with values of $+1$ or -1; while

$$g_a(t) = \begin{cases} 1 & 0 \leq t \leq T_a \\ 0 & \text{others} \end{cases}$$

in which T_a is the duration of information code element.

The modulation scheme of FSK is adopted here. The frequency f_i generated by the frequency synthesizer is as follows:

$$f_i \in \{f_1, f_2, f_3, \ldots, f_N\}$$

that is, f_i is one frequency selecting from frequency set $\{f_1, f_2, \ldots, f_N\}$ in the time range of $(i-1)T_h \leq t \leq iT_h$ and is determined by pseudorandom code, T_h is chip duration. Thus, by using $a(t)$ to modulate f_i, we obtain RF signal

$$s(t) = a(t) \cos \omega_i t \tag{3.72}$$

The received signal

$$r(t) = s(t) + n(t) + J(t) + s_J(t) \tag{3.73}$$

in which $s(t)$ is signal component, $n(t)$ is noise component (white Gaussian noise), $J(t)$ is jamming component, $s_J(t)$ are FH signals due to other networks.

The frequency f_i' is one of the frequency set generated by the frequency synthesizer at the receiving end, which is controlled by the pseudorandom code generator that is the same as that at the transmitting end, that is

$$f_j' \in \{f_1 + f_I, f_2 + f_I, f_3 + f_I, \ldots, f_N + f_I\}$$

in which f_I is IF. The received signal is multiplied by the signal generated by local oscillator in the mixer, we obtain

$$\begin{aligned} r(t) \cos \omega_j'(t) &= s(t) \cos \omega_j' t + n(t) \cos \omega_j' t \\ &\quad + J(t) \cos \omega_j' t + s_J \cos \omega_j' t \\ &= s'(t) + n'(t) + J'(t) + s_J'(t) \end{aligned} \tag{3.74}$$

The four components in Eq. (3.74) will be analyzed subsequently. We first examine the signal component

$$s'(t) = s(t) \cos \omega_j' t = a(t) \cos \omega_i t \cos \omega_j' t$$

We have known that the frequencies generated by frequency synthesizer at both transmitting and receiving ends correspond to each other and are controlled by the same pseudorandom code; moreover, the controlled model is also the same. Of course, the initial phase for both the pseudorandom codes may differ. Provided making the initial phase for both frequencies is also the same, that is, to be synchronical, therefore $i = j$, the frequency generated by the frequency synthesizer at the receiving end is thus just high enough to be an IF than that at the transmitting end; by passing through the following mixer and taking below sideband, we obtain the signal component as follows:

$$\begin{aligned} s'(t) &= a(t) \cos \omega_i t \cdot \cos \omega_j' t \\ &= \frac{1}{2} a(t) \left[\cos\left(\omega_j' - \omega_i\right) t + \cos\left(\omega_j' + \omega_i\right) t \right] \end{aligned} \tag{3.75}$$

By passing through the filter, we get

$$s''(t) = \frac{1}{2} a(t) \cos\left(\omega_j' - \omega_i\right) t = \frac{1}{2} a(t) \cos \omega_I t \tag{3.76}$$

This is a definite IF signal. Following by a demodulator, the transmission information $a(t)$

will be recovered and the purpose of the transmission information has been achieved.

When $n(t)$ is a white Gaussian noise in the component $n'(t)$, it is not changed by passing through the mixer, as well as a system without using FH; that is to say, in the case of a white noise background, the processing gain is absent for an FH system, as well as a DS system.

In the case of jamming component $J'(t)$, it lacks prior knowledge of the frequency-hopping law and thus will be removed from the IF bandwidth by passing through the mixer, and cannot get into the demodulator; the purpose suppressing that is achieved.

The component $s_J'(t)$ are generated by the FH signals from other networks, although different networks have different frequency-hopping laws and therefore other networks cannot cause interference.

The spectra of frequency generated by a frequency synthesizer and that of the RF signal for an FH system are plotted in Fig. 3.18. The frequency spectrum generated by an ideal frequency synthesizer is a discrete line spectrum with identical space and amplitude distributions, the space each other equals ΔF, the whole occupying bandwidth $B = f_N - f_1$. The frequency at a certain moment is one of N frequencies that is determined by pseudorandom code as shown in Fig. 3.18A. The frequency spectrum for an FH signal is plotted in Fig. 3.18B. It is narrow bandwidth at a certain moment for an FH system, but it is broad bandwidth for the whole processing time because the signals hop over the whole spreading bandwidth.

The FH system can generally be divided into two types of frequency hopping according to the frequency-hopping rates R_h. Provided R_h are faster than the information rate R_a, that is $R_h > R_a$, it is called a fast FH-SS system; on the contrary, provided $R_h < R_a$, it is called a slow FH system. There exist different characteristics combating with the interference for the different FH rates; the complexity and cost to form them are also different.

In an FH system, the frequency hopping is controlled by the pseudorandom code. There exist the pseudorandom codes with different phases at different moments; the frequencies generated by the corresponding frequency synthesizer are also different. The frequency hopping law for an FH system is called a frequency-hopped pattern; an example is shown in Fig. 3.19. The properties of an FH pattern will directly affect the whole performance of an FH system.

3.1.4.3.3 ADVANTAGES OF FREQUENCY-HOPPING SPREAD SPECTRUM (FH-SS) SYSTEM

By examining the basic principle for an FH-SS system, we can see that there is no processing gain against a white noise background, as well as for a DS-SS system. However, the FH system has some remarkable advantages, which will briefly be introduced subsequently.

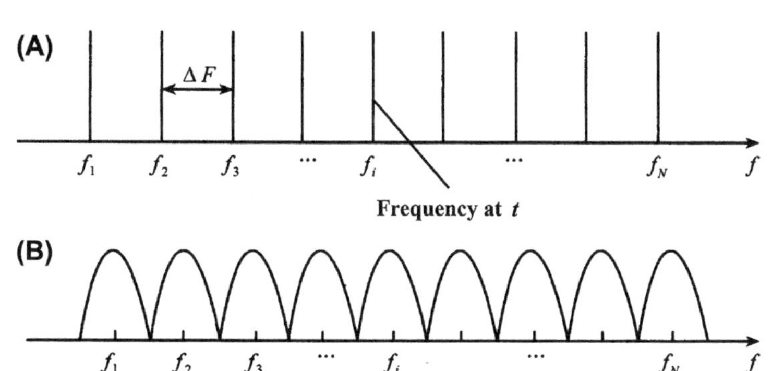

FIGURE 3.18 Frequency spectra for frequency-hopping (FH) system: (A) frequency synthesizer, (B) FH signal.

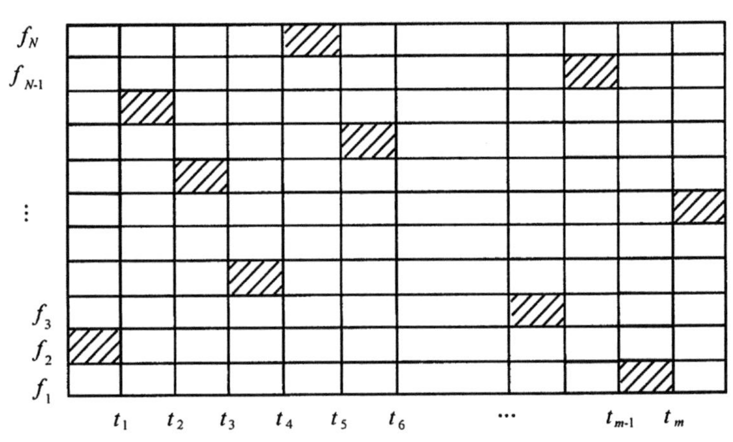

FIGURE 3.19 Frequency-hopping pattern in time–frequency plane.

1. The FH system is remarkably able to overcome violent interference, because interference only arises if the frequencies are the same between the interference and the FH signals; thus, it is widely employed in military communications. The larger the numbers of hopping frequencies and the higher the rates, the better the antiinterference ability of an FH-SS system.
2. It is easy to form a communication network and to realize CDMA for an FH-SS system. Moreover, the frequency spectrum efficiency is slightly higher than that for a DS-SS system.
3. The system has good compatibility. At present, all FH radio stations have better compatibility, such as fixable frequency and FH, digital and analog, and voice and data are compatible.
4. The near–far problem has been solved for an FH system. This problem is relatively difficult to solve for a DS system.
5. The pseudorandom code rates are much lower for an FH system than those for a DS system; therefore, to achieve synchronizing operations in an FH system is also easier than that in a DS system.

The correlation between despreading and demodulation in an SS system discussed previously are based on the supposition that the synchronizations, including carrier, code element, and code word are perfectly achieved. Achieving accurate synchronization, including capture and accurate tracking, is a key technique for an SS communication system. Because we have put forward that by using total-time sampling processing instead of conventional synchronizing schemes, this important topic will not be discussed here; the latter may be found in a number of books.

3.2 SOME DIGITAL UNDERWATER ACOUSTIC COMMUNICATION SYSTEMS

The noncoherent detection systems, as MFSK, and coherent detection systems as PSK, had been adopted as basic systems for digital underwater acoustic communications. At present, the advanced communication systems in radio, such as spread spectrum and Orthogonal Frequency Division Multiplexing (OFDM) systems, are also introduced in digital underwater acoustic communications.

Now by combining our research on digital underwater acoustic communications, both MFSK and FH-SS systems employed in them will be introduced here. Some problems in employing optimal linear filter in the fields will also be discussed.

3.2.1 Multiple-Frequency Shift Keying (MFSK) System

An MFSK system digital underwater acoustic communication system belongs to a noncoherent detection system. There are some potential advantages in this system; for example, it can eliminate the difficult problem of carrier phase tracking, the higher communication data rates would be realized in some operating conditions. Moreover, if the gaps between the frequency codes are availably chosen, this system can adapt to frequency excursions at a certain range, especially if it has the potential ability to combat intersymbol interference (ISI) caused by multipath effects.

There are two major shortcomings for an MFSK system employed in digital underwater acoustic communication. First, its communication rates in united bandwidth are low; to achieve higher data-rate underwater acoustic communication, broad bandwidths must be provided; but generally it is difficult to satisfy this condition because the underwater acoustic communication channels have the strict band-limited characteristic as mentioned in Chapter 2. Then the MFSK system belongs to a noncoherence detection one, the higher input SNR is necessary to satisfy an expected error bit rate.

It would be noted that to adapt to multipath effects, an available condition is that the amplitudes of multipath interference are much less than those of the direct signal. In this case, ISI caused by multipath propagations would therefore be eliminated by directly using an amplitude decision. This approach of antimultipath may principally be realized for the communication state at high operating frequencies and correspondingly short communication ranges. However, in the case of long-range communications, the direct signals are absent due to sound refraction effect; moreover, the maximum amplitude for a multipath structure generally does not appear at the first pulse; thus, if some effective approaches are not adopted, to adapt to this communication circumstance, the data rates would correspondingly be reduced.

Two typical examples of MFSK system underwater acoustic communications will briefly be introduced as follows:

Example 1: A prototype of high data-rate underwater acoustic image communication at shorter distances [5].

The block diagrams of transmitting and receiving principles for the image communication prototype are shown in Figs. 3.20 and 3.21, respectively.

The pseudorandom codes that have a better autocorrelativity are adopted as synchronical signals to improve the performances of the image communication.

The different gray scales of image elements are expressed by different frequencies. There

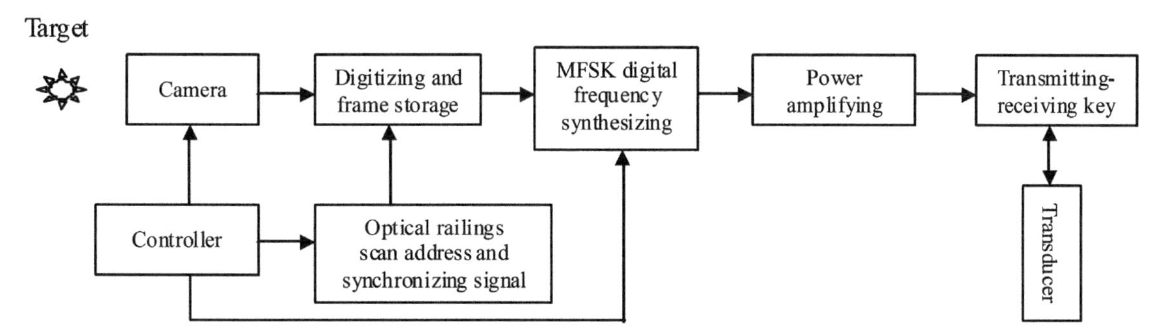

FIGURE 3.20 Block diagram of transmitting principle.

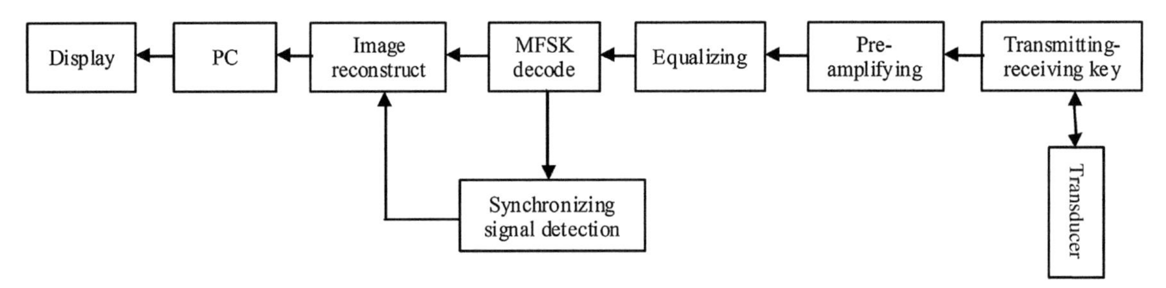

FIGURE 3.21 Block diagram of receiving principle.

are 16 frequencies to express 16 quantized gray scales in this prototype.

The image element densities are 320 × 200 points/picture, the lower densities of 160 × 100 points/picture are also formed.

In the case of a shorter-range communication, the higher operating frequencies and relative broad bandwidths can be adopted to reduce the multipath effects and obtain high data rates, as up to 8 kbps.

The field tests are performed in a shallow-water acoustic channel in Xiamen Harbor where a typical multipath structure is shown in Fig. 3.22, which belongs to a weaker multipath propagation at higher operating frequencies.

The experimental records having higher image-element densities at different distances in Xiamen Harbor are shown in Fig. 3.23, in which 1 is original image; 2, 3, and 4 are at 4, 7, and 10 km, respectively. Experimental results had demonstrated that the distinct images can be obtained even if at the distance of 7 km.

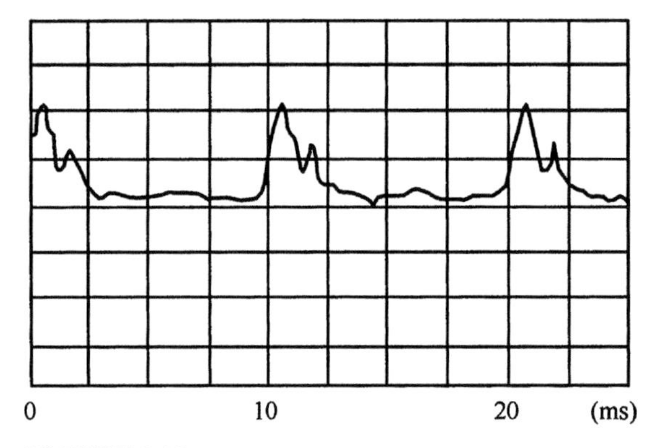

FIGURE 3.22 Record for a weaker multipath structure.

Fig. 3.24 presents the image records with lower image-element density at different distances. Of course, the images are rougher than those of higher-density ones, but the frame rates are raised by four times, corresponding frame frequency is 8 s.

It should be noted that the transmitting images mentioned previously are not adopted image data compression processing, therefore the error bit rates (EBR) are permitted to be higher; for example, we can obtain the clear image records in situ when EBR is about 5×10^{-2}. Provided the image data compression is used, the frame rates will greatly be raised. In such a case, EBR must correspondingly be lowered (refer to Chapter 4).

Example 2: An MFSK system underwater acoustic communication at long distances [6].

The communication distances are expected as far as 80 km in this communication prototype; the lower operating frequencies must be selected, therefore the multipath total delay spread T_M would greatly be increased. To adapt to this condition, the code duration τ_s would correspondingly be increased, that is, the data rates are decreased in the prototype. To improve the ability of antimultipath, the eight frequencies with equal gap in effective bandwidth are chosen as four-frequency shift keying (4FSK) modulation frequency points and divided them into

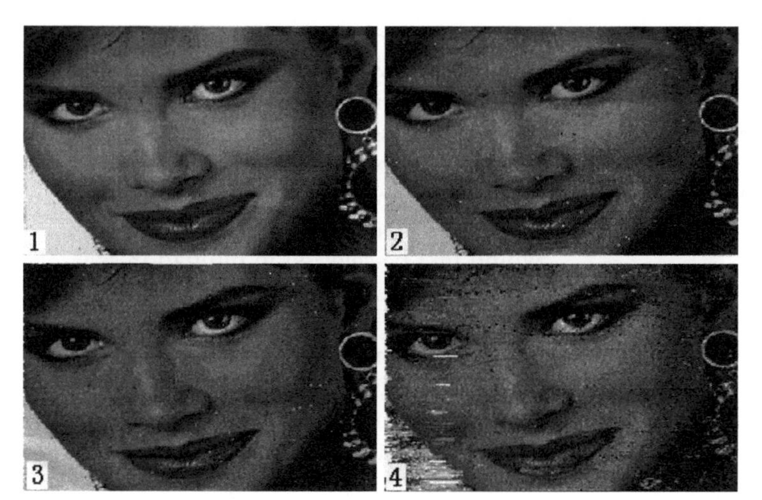

FIGURE 3.23 Image communication records at different distances.

two transmitting sequences to satisfy the condition of $2\tau_s > T_M$. The information will be reconstructed by using spectrum analysis scheme to demodulate the multi frequency signaling (MFS) signal.

Field tests were performed in the sea area with the depths range from 800 to 1000 m and sea state was 4. Transmitting and receiving transducers are placed at depths of 100 and 300 m, respectively. Multipath total delay spread T_M is up to 300 ms in the test circumstance.

The experimental results had demonstrated that EBR is zero at the communication distance $r = 80$ km and data rate $R = 4$ bps. The tests were also performed at different r and R, for example $r = 50$ km, $R = 6.7$ bps; and $r = 30$ km, $R = 20$ bps, the same results, that is EBR $\rightarrow 0$, are also obtained.

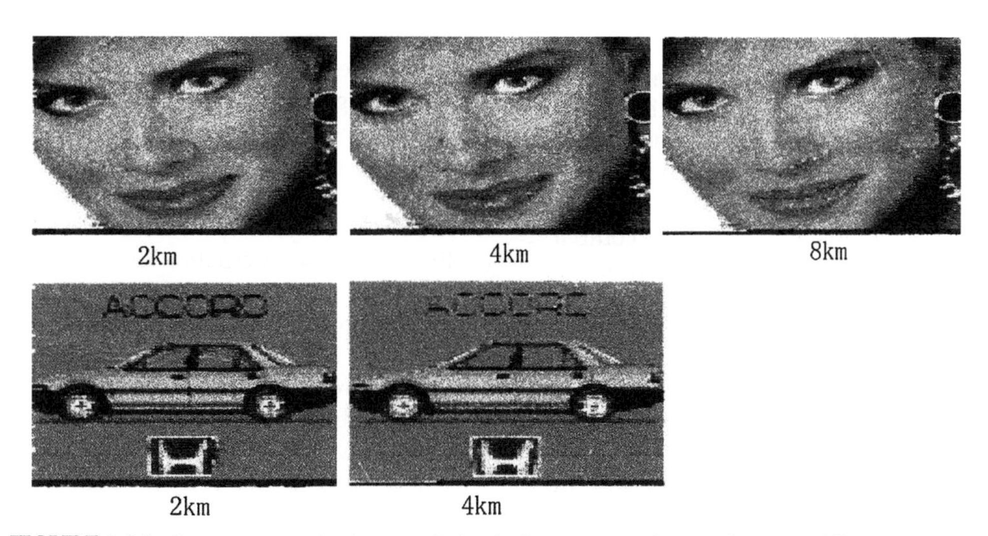

FIGURE 3.24 Image communication records for the lower image-element density at different distances.

The experimented results mentioned earlier state that if using the simple approaches by increasing τ_s and frequency point numbers for an MFSK system to combat ISI caused by multipath effects, increasing r means decreasing R, the contrary is true, too. It is identical with the transmission law shown in Eq. (1.1).

3.2.2 Applications of Spread Spectrum Systems on Digital Underwater Acoustic Communications

3.2.2.1 DS-SS System Underwater Acoustic Communications

3.2.2.1.1 FEASIBILITY ANALYSES

We have discussed the sound transmission laws in the underwater acoustic channels and their effect on digital underwater acoustic communications in a previous chapter. Moreover, the configurations and features for DS system radio communications have also been described in Section 3.1.4. We can make a probably prediction for the performance by employing a DS system in digital underwater acoustic communications.

1. A DS system may be operated at very low-input SNR, say −15 dB; therefore, it is able to accurately detect the signals that are covered by the noise background. This is a remarkable advantage, in particular for military underwater acoustic communications.
2. In comparison with a narrow bandwidth system, although a DS system has not improved the performances to combat white noise, it has an excellent feature to combat against the jamming signals and pulse interference that generally appear in underwater acoustic channels (refer to Chapter 4), thus the robustness of underwater acoustic communications will be improved.
3. We can realize CDMA underwater acoustic communications by means of a DS system. This is a remarkable advantage because the data rates are generally low in underwater acoustic communications; time division multiple access (TDMA) is thus not much fitted for use; moreover, frequency division multiple access (FDMA) is also not well adapted to them because the underwater acoustic channels have a strictly band-limited peculiarity. Therefore, a CDMA scheme would first be selected in multiaddress underwater acoustic communications, such as in network configurations, and so on.
4. Generally, there is a rake receiver in a DS system. It may be expected to improve the performances of communication sonar by employing path diversity technique, which is able to use multipath energy.

The main limits by employing a DS system in underwater acoustic communications are as follows:

1. The basic contradiction is that a DS system requires spreading frequency spectra to a much larger duration, but one of the basic characteristics for underwater acoustic channels is that it is strictly band-limited. To obtain a larger spreading frequency gain for long-distance underwater acoustic communications, the data rates would thus be lowered. If communication sonar expects to achieve a longer range, better antiinterference, and privacy operation, while permitting operation at low data rates, such as a text communication [8], employing a DS system for that is quite suitable.
2. A DS system based on the coherent detection has less channel adaptability. We have known that the phase fluctuation of sound signals will appear when they transmit in underwater acoustic channels in which severe reflection and scattering effects are encountered; particularly, it is able to appear as phase hopping up to 180 degrees in multipath propagations. If we expect to have longer-range underwater acoustic communication, the received signals, in fact, would be randomly spatiotemporal (phase)-

frequency variable multipath structures; whether to realize accurate phase tracking depends on specific communication conditions, in particular on the characteristics of channels. Therefore, careful considerations must be made provided that a coherent detection scheme will be employed in communication sonar.

3. In the case of DS system underwater acoustic communication, the usual rake receiver mentioned previously is difficult to adapt to complex and randomly variant multipath structures, particularly which have the features of large time delay spread, rapid fluctuation, and dense distributions. Generally, provided a rake receiver is employed in these conditions, an adaptive operating approach must be adopted, which is possible to adapt to such multipath propagation circumstances.

3.2.2.1.2 APPLIED EXAMPLES

Example 1: A DS-SS system underwater acoustic communication [9].

Gerard Loubet et al. had developed a DS-SS system underwater acoustic communication sonar. A rake receiver is employed in it to make use of multipath energy, and spatial diversity is utilized to combat with the signal fading caused by multipath effects.

Different receiver patters may be used according to channel hypothesis:

1. Known and stable amplitudes and phases: the receiver is the filter matched to the estimated medium.
2. Known amplitudes and random phases: the filter is matched to the modulus of the medium-impulse response.
3. Random amplitudes and phases: suboptimal filter is one matched to the square modulus of the medium-impulse response.

If the paths are not well known or fluctuating too fast, a simple integration may also be an interesting solution. This processing has been used for the Loracom experiment because the transmitted energy was essentially focused round the sound fixing and ranging axis and it was impossible to discriminate the different energetic propagation paths which have quite the same propagation delays.

The fading phenomenon is a critical problem for underwater communications, particularly at short or medium ranges.

When propagation occurs through the SOFAR channel, the multiple paths are rarely separated; that makes time diversity unavailable. Then, frequency diversity obtained by spreading the signal's bandwidth gave disappointing results. In this case, spatial diversity seems to be the most efficient solution to improve the receiver's performance. The basic idea is that amplitude fluctuations might be uncorrelated if the array elements are sufficiently separated. Fig. 3.25 shows the fluctuations of the energy received at 100 and 150 m depths, respectively, for the 50 km experiment. They are quite correlated but the fading is more important for the 150 m element.

The experimental results had demonstrated that a DS-SS system communication is more robust because only the sign of the symbol has to be detected. A 20 km communication was maintained for several minutes with an error rate inferior to 0.2% and a very low SNR, about -14 dB. But the data rate was weak: 7.2 bps.

Example 2: A DS-SS system underwater acoustic receiver for multiaccess networking [10].

An adaptive-array receiver architecture that utilizes direct sequence-code division multiple access (DS-CDMA) and spatial diversity combining has been proposed for reliable low data-rate multiuser communications in an asynchronous shallow-water network. The most outstanding feature of the receiver algorithm is the approach that integrates three fundamental communications functions into one structure: despreading, equalization, and multiaccess

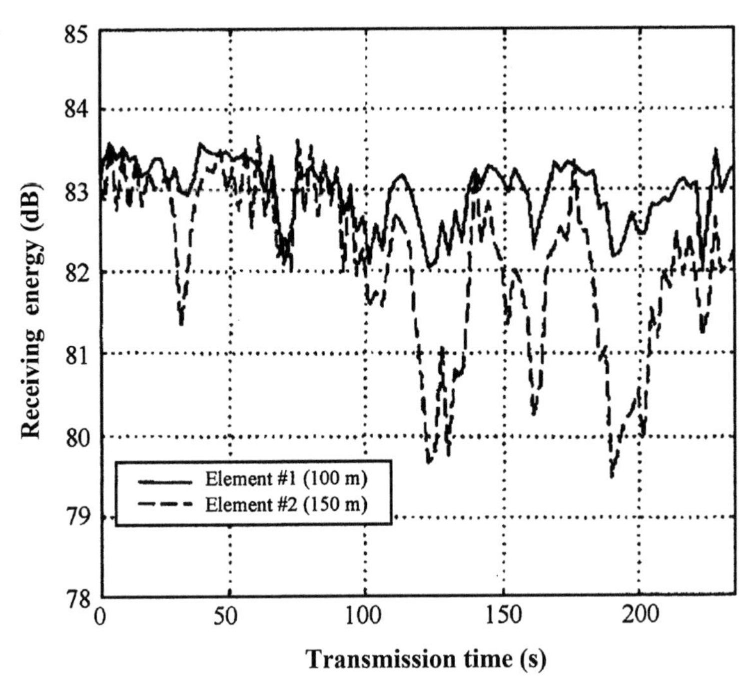

FIGURE 3.25 Fluctuation of sound energy at different depths.

interference rejection. Moreover, the only information required by the receiver is the knowledge of the distinct training sequences utilized to detect the presence of a user. These are required to train the adaptive equalizers at the beginning of the transmission.

To test the performance of the proposed receiver structure in multipath and multiaccess interference scenarios, recorded data were processed offline. The distinct receiver–transmit positions give an impression of the simulated multiuser shallow-water network. The network was preset to cover ranges between 1 and 10 km whereby different scenarios were set up to test both the near–far problem and the angular user separating capability of receiver algorithms.

The receiver array employed was positioned in approximately 40–50 m deep water. It consisted of eight omnidirectional sensors comprising a horizontal plane of four elements and a vertical plane to provide both beamforming and spatial diversity reception. The two identical transmitters were positioned at 5 m distance from the seabed. The transmitting power was fixed at 190 dB re 1 μPa at a carrier frequency of 9.6 kHz maximally bandlimited at 4 kHz. The data rate is of 267 or 534 bps. The modulation scheme employed was DS-CDMA QPSK.

The performance of the proposed receiver architecture had been verified by means of offline processing of data acquired during sea trials. Results show that this computationally efficient structure is near–far resistant and provides successful multiuser operation in the shallow-water channel.

3.2.2.2 *Applications of FH System on Underwater Acoustic Communications*

3.2.2.2.1 FEASIBILITY ANALYSES

It may be expected that there exist some remarkable advantages for an FH system communication sonar.

1. A potential feature to adapt to the multipath effects is the first reason to employ an FH system in communication sonars.
   Letting the duration for a signal-code elements be τ_s, the frequency numbers that are not repeatedly used be n, the total delay spread of the multipath effects be T_M, provide $n\tau_s > T_M$ to be satisfied, ISI due to multipath effects will disappear. For example, taking $\tau_s = 5$ ms, which is generally equivalent to a data rate $R = 200$ bps, and $n = 25$, thus it is able to combat against ISI with $T_M = 125$ ms, the whole bandwidth occupied is 5 kHz; there are more suitable specifications for some civil communication sonars. To adapt to longer-range underwater acoustic communications in which T_M may be spread to a larger value, as up to several 100 ms, an adaptive bandwidth compression scheme would be adopted (see Chapter 4).
2. An FH system has the latent effect of frequency diversity, one that combines with channel coding, the compacts of frequency-selective fading; particularly the deep fading caused by interference effect would much be reduced. Because of the complexity utilizing spatial diversity, it is more reasonable to use the FH system.
   Provided frequency gaps are suitably selected in an FH system, the correlativity of frequency-selective fading will remarkably weaken. Fig. 3.26 shows an experimental result in a shallow underwater acoustic channel, in which the fading of amplitude have a different feature between the pulses signed by Refs. [1] and [2] that have a frequency gap of 300 Hz and a relative time delay of 10 ms. Another experimental result of the frequency diversity is given in Fig. 3.27, although the frequency gap between pulses [1] and [2] is only 300 Hz, the amplitudes have a large difference. It states that employing frequency diversity is effective in an FH system.
3. The noncoherent detection schemes are generally employed in an FH system; therefore, it has better channel adaptability as a result to improve the robustness of the system. Of course, an FH system generally lacks a feature to realize privacy in underwater acoustic communications, because the input SNR is higher in the FH system than that in the DS one.
   In the cases of some underwater acoustic channels having better transmission circumstances, as weak multipath propagations, the coherent detection scheme can also be adopted and operated at low SNR conditions (refer to Chapter 4).
4. It is able to realize CDMA underwater acoustic communications, as well as the DS system.

3.2.2.2.2 THE LIMITATIONS

1. A basic contradiction is that an FH system requires spreading spectrum, but the underwater acoustic channels have a strictly band-limited characteristic. It is similar as well as by using a DS system.
2. Provided T_M is quite large, up to several 100 ms for a low-frequency, long-range communication [6], in this case, by adopting the scheme of $n\tau_s > T_M$ to combat such an ISI is difficult; for example, let T_M be 300 ms, R still be 200 bps, therefore $n = 60$, the whole bandwidth will thus be spread to 12 kHz. Obviously, it is too broad to achieve long-range underwater acoustic communications.
3. It has not a feature to perform privacy communications as mentioned earlier. Perhaps it may be realized by adopting the fast-hopping frequency model combined with some special signal-processing schemes for lower data-rate underwater acoustic communications.

To sum up, we can see that the FH system possesses a better channel adaptability; the communication distances and data rates may

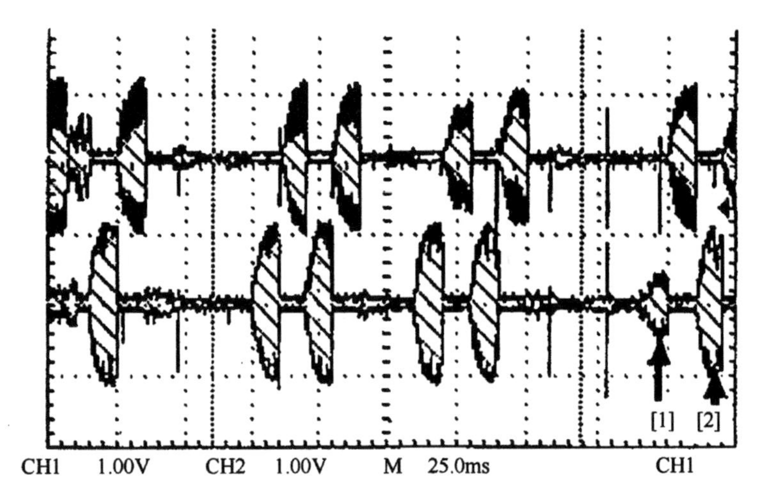

FIGURE 3.26 Frequency correlativity for amplitude fading.

be compatible at a certain range. Provided the limitations mentioned previously would better be resolved based on the guides of the physical fundamental of the sound transmission and practical requirements of underwater acoustic communication engineering, it would be expected to develop an improved FH system available to be employed in digital underwater acoustic communications.

3.2.2.2.3 APPLIED EXAMPLE

Example 1: FH system voice communication [5].

The principal block diagram of the transmitter for an FH system underwater acoustic voice communication sonar is shown in Fig. 3.28. Corresponding receiver is plotted in Fig. 3.29.

Input signals include voice and Chinese, the receiving terminal correspondingly includes both voice and Chinese display.

The operating bandwidth can be selected either 6–11 kHz or 12–17 kHz to adapt to different communication distances.

To improve input SNR, an instantaneous time spectrum analysis technique is adopted to detect the received signals. Fig. 3.30 gives the section of period sinusoidal pulses at a priori signal-time duration that is recorded at 9 km in situ. It shows that the signal pulse is covered by noise

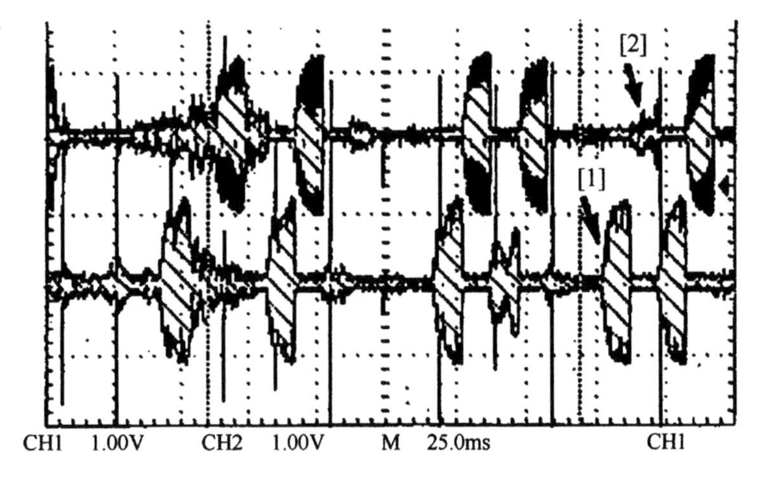

FIGURE 3.27 Experimental result for frequency diversity.

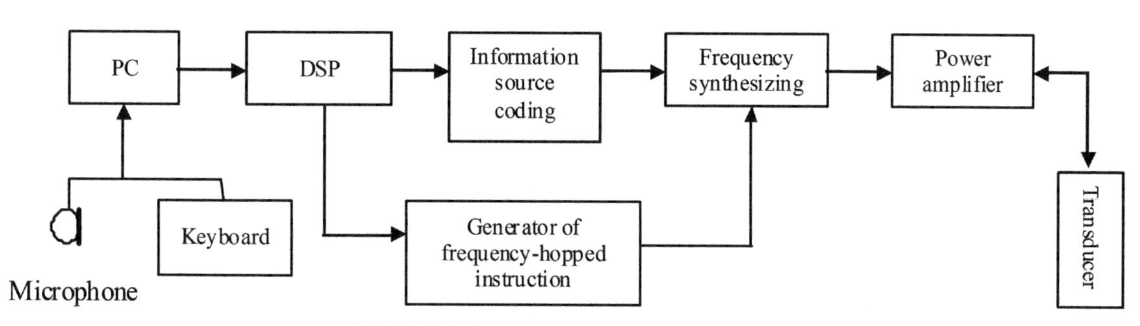

FIGURE 3.28 Block diagram of transmitter.

background in time domain. The instantaneous time–frequency spectrum for the section ("which area" in Fig. 3.30) is shown in Fig. 3.31. We see that the expected signal can correctly be detected by means of the instantaneous time spectrum analysis technique.

The experiments for the FH system voice communication were performed in Xiamen Harbor with the transmission conditions like those for the image communication mentioned previously. Experimental results had demonstrated that the voice information may reliably be received at the distance of 7.5 km.

Example 2: E-mail transmission in shallow-water acoustic channels [7].

The slow HF system is employed in an underwater acoustic E-mail communication prototype to improve its channel adaptability. Designing a suitable frequency-hopping pattern to combat ISI caused by multipath effects is first implemented. Moreover, a frequency diversity is utilized to adapt to severe signal fading in which identical information is transmitted by two frequencies, one is lower, the other is higher frequency in the FH pattern.

The principal block diagram of transmitter for the prototype is given in Fig. 3.32. By passing

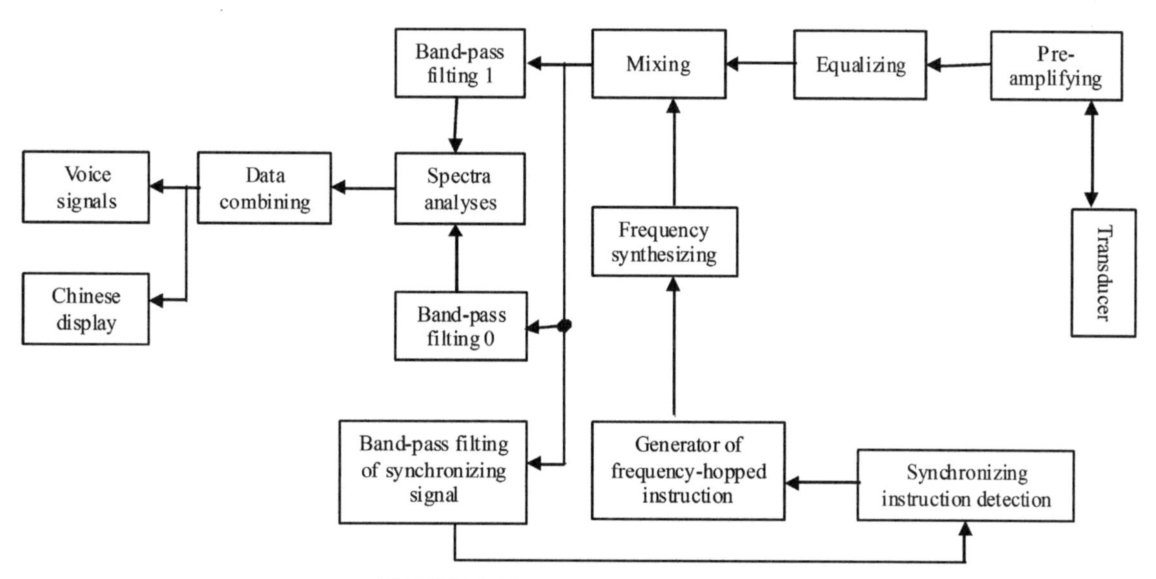

FIGURE 3.29 Block diagram of receiver.

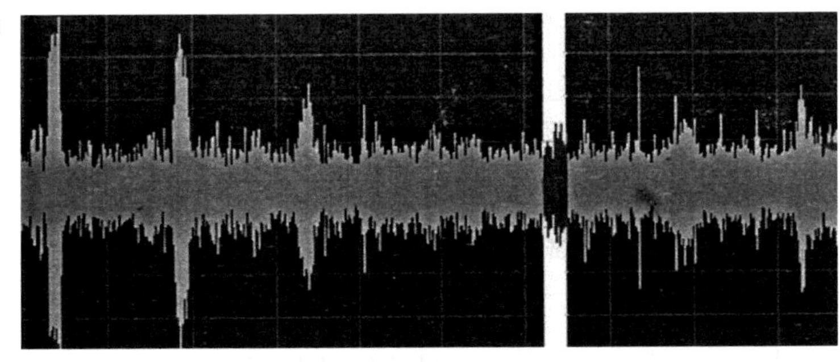
FIGURE 3.30 Section of period sine pulse at a priori time duration.

through E-mail transmitting–receiving software based on PC, the data are sent to microcontroller. When full data are received in a register, the data channel coding will be carried out; moreover, the microcontroller generates a control signal and lets frequency synthesizer operate according to the FH pattern. By passing through power amplifier and transducer, FH signals are transmitted in the shallow-water acoustic channels.

The principal block diagram of receiver for this prototype is shown in Fig. 3.33. The accurate detection of synchronizing signals is a key technique in the receiver. Provided it is realized, the microcontroller will generate control signals to start the frequency synthesizer, and to demodulate the data signals.

The design for the E-mail transmitting–receiving software is based on PC software scheme. The servicer transmitting E-mails obeys simple mail transfer protocol (SMTP). The servicer receiving E-mail adopts post office protocol - version 3 (POP3) protocol. Visual Basic 6.0 provides Winsock control parts to hold up the mutual communications based on transmission control protocol/internet protocol (TCP/IP) among computers in computer network.

The field test for E-mail communication prototype had been implemented in Xiamen Harbor. Experimental conditions are similar to that in the image communication. Transmitting acoustic power is about 70 W. The expected results are obtained at the distance of 10 km.

Besides carrying out hardware correlation detections, the software demodulation by means of digital signal processing (DSP) is also adopted in the prototype. The frequency spectra by using Chirp-Z transform (CZT) analyses are given in Fig. 3.34. The spectra at the left-hand side of the figure are "1" data that are acquired by frequency diversity using two

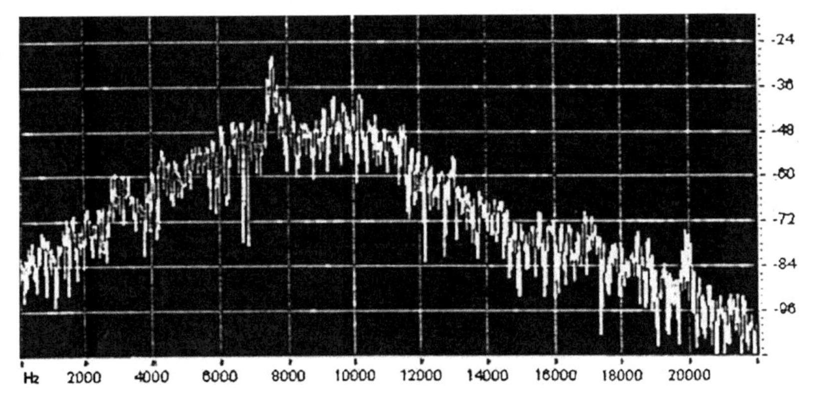

FIGURE 3.31 Spectrum corresponding to the section shown in Fig. 3.30.

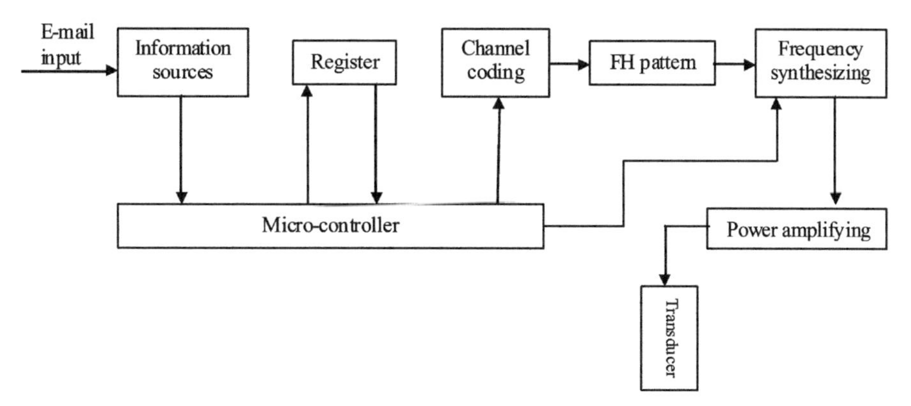

FIGURE 3.32 Principal block diagram of transmitter.

different frequencies that transform according to the FH pattern. The spectra of "0" are shown at its right-hand side that will be eliminated by preset threshold provided those levels are lower than this threshold.

3.2.3 Underwater Acoustic Communication by Using OFDM System

High data rate is desired in many data communications. However, as the symbol duration reduces with the increase of data rate, the systems using single-carrier modulation suffer from more severe ISI caused by the dispersive fading of wireless channels, thereby needing more complex equalization. Orthogonal Frequency Division Multiplexing (OFDM) modulation divides the entire frequency selective fading channel into many narrowband flat-fading subchannels (subcarriers) in which high bit-rate data are transmitted in parallel and do not undergo ISI due to the long symbol duration. By overcoming the impairments of radio channels to OFDM signals, including Doppler shift, dispersive fading, timing, and frequency offsets, sampling clock offset, and nonlinear distortion due to large peak-to-average-power ratio of the OFDM signal, OFDM modulation has been chosen for many standards, including Digital Audio Broadcasting (DAB) and terrestrial TV in Europe, and wireless local area network (WLAN). Moreover, it is also an important

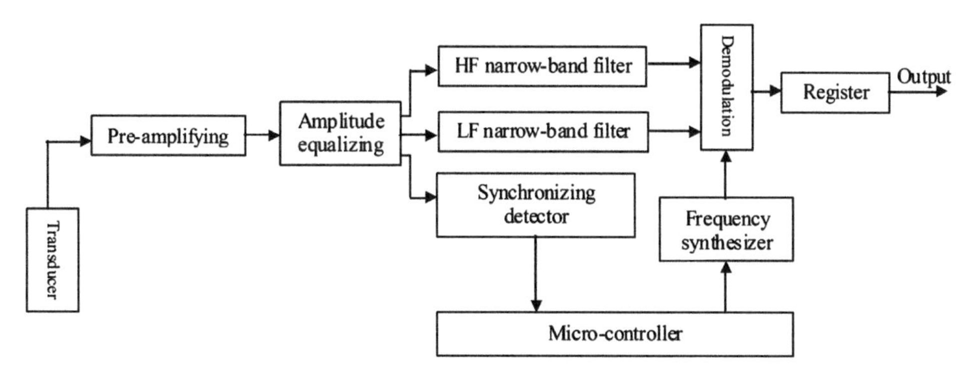

FIGURE 3.33 Principal block diagram of receiver.

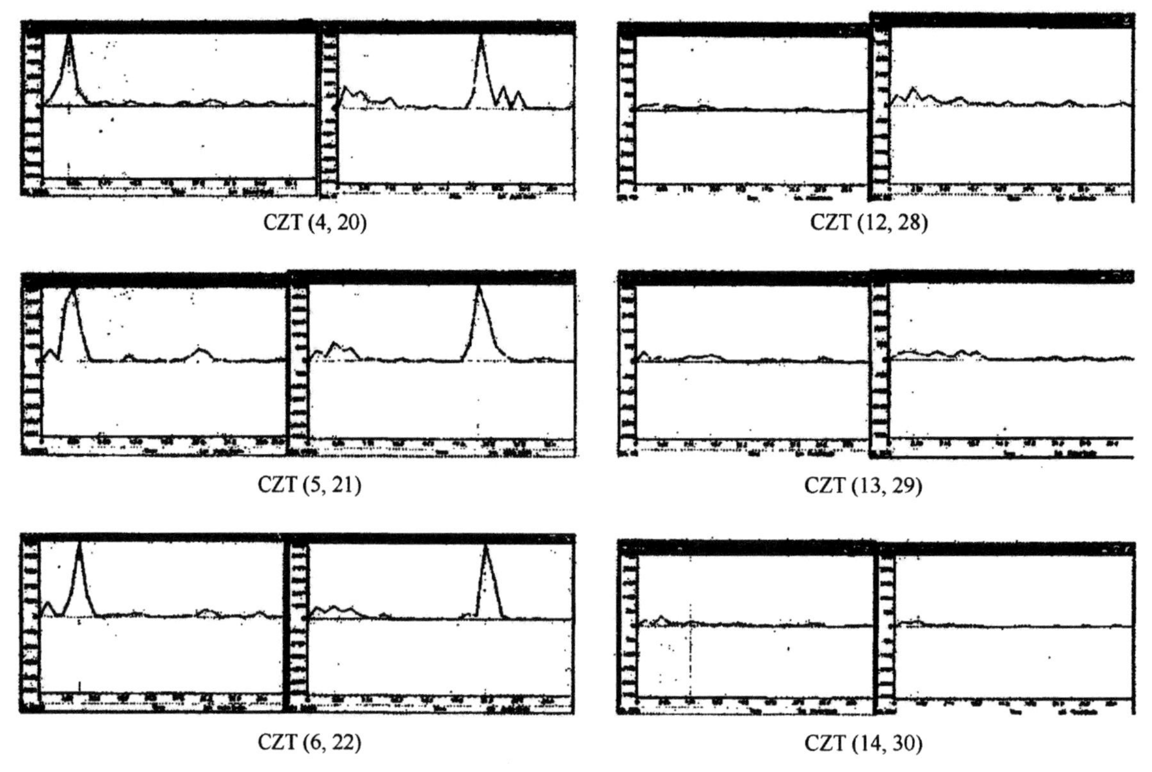

FIGURE 3.34 Typical Chirp-Z transform (CZT) spectra.

technique for high data-rate transmission over mobile wireless channels.

OFDM is an attractive technique beginning to be used in underwater acoustic communication, because it has three obvious advantages: good performance against multipath interference, ability to combat frequency-selective fading, and the high-frequency band efficiency.

We have known that, as mentioned in Chapter 2, there exist a lot of the peculiarities for underwater acoustic communication channels in comparison with radio communication ones, as random spatiotemporal—frequency variable parameters, severe multipath interference, rapid signal fluctuations, strictly limited bandwidth and low sound velocity. Therefore, many signal-processing systems, like OFDM in radio communications, cannot directly be employed in digital underwater acoustic communications. In fact, applying OFDM to underwater acoustic channels is a challenging task because of its sensitivity to frequency offset that arises due to motion. In particular, because of the low speed of sound and the fact that acoustic communication signals occupy a bandwidth that is not negligible with respect to the center frequency, motion-induced Doppler effects result in major problems such as nonuniform frequency shift across the signal bandwidth and intercarrier interference.

Time-varying multipath propagation and limited bandwidth place significant constraints on the achievable throughput of underwater acoustic communication systems. To support high spectral efficiencies, communication systems employing adaptive modulation schemes have to be considered. However, the performance of an adaptive system depends on the transmitter's knowledge of the channel which would be provided via feedback from the receiver. Because sound propagates at a very low speed, the design and implementation of an adaptive system essentially relies on the ability to predict the channel at least one travel time ahead. This is a very challenging task for communications in the longer ranges, as above several kilometers impose significant limitations on the use of feedback.

It seems that OFDM-based underwater acoustic communications are more suitable for shorter distance (as several km) data transmission. At this communication condition, we can use higher operating frequencies above 20 kHz, ISI caused by the multipath propagation will reduce, efficient bandwidths would be extended to about 10 kHz, and more subcarriers cab be arranged. Moreover, because signal fluctuations in both amplitude and phase are proportional to transmission distances at some certain conditions (refer to Section 2.4.1), especially the received signals would be come from directed paths, in water-depth acoustic channels, the impacts of sound scattering from random boundaries can be neglected, channel estimation is, hence, easier.

Recently, research with respect to underwater acoustic communication by using OFDM system focus on short-range condition. Jianguo Huang et al. pay attention to medium-range (10 km) high-speed underwater acoustic communication based on OFDM [11]. An OFDM system is designed and realized.

Fig. 3.35 shows simplified block diagrams of underwater acoustic OFDM system.

In the transmitter, the input data are a serial bit-stream that is to be mapped. After that, the data streams are converted from serial to N-parallel data streams and then a pilot signal is inserted. The modulation of OFDM is conducted by using inverse fast fourier transformation (IFFT). The cyclic prefix signal is used to reduce the effect of ISI and ensure the orthogonality of subcarriers. It is followed by parallel to serial conversion (P/S) and digital to analog (D/A) conversion.

In the receiver, according to Doppler shift, especially the Doppler shift index Δ is estimated by using cyclic prefix and is compensated effectively by resampling (using a sampling rate $[(1+\Delta)/T]$ in the receiver different from the rate $1/T$ in the transmitter) the received signal. It is followed by S/P and FFT. The channel is estimated by using the pilot signal, and the phase variety must be tracked in OFDM system. Equalization is then implemented in time domain.

The parameters of OFDM system used for performance simulation are shown in Table 3.1.

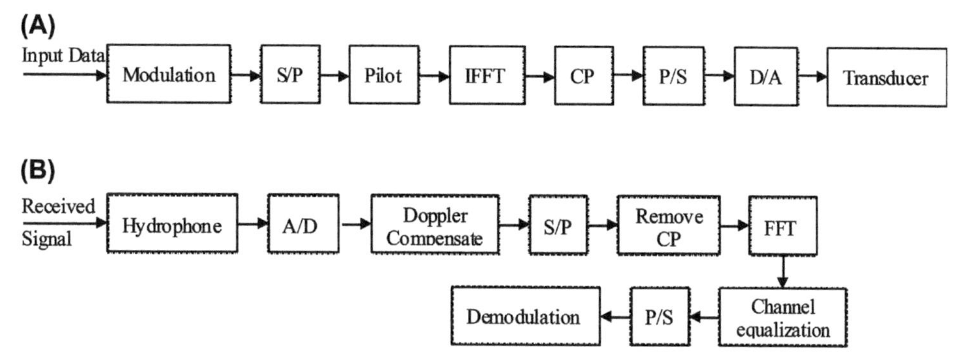

FIGURE 3.35 Block diagrams of underwater acoustic OFDM system: (A) transmitter, (B) receiver.

TABLE 3.1 OFDM Underwater Acoustic Parameters

Guard Interval (s)	FFT/IFFT (s)	Symbol Duration (s)	Bandwidth (kHz)	Bandwidth Efficiency	Number of Sub Carrier
0.05	0.2	0.25	3	1.6	600

A linear frequency modulation signal is used for synchronization. The underwater acoustic channel is based on the ray model, including eight propagation paths and the maximum time delay is below 50 ms.

M-ary digital modulation can increase the data rate; meanwhile, the distribution of points in the constellation will affect the system performance. Fig. 3.35 shows the comparison of different OFDM systems using binary-phase shift keying (BPSK), four-quadrature amplitude modulation (4QAM), and 8QAM modulation. From Fig. 3.36 it can be seen that this system is suitable for underwater acoustic communication due to the low BERs.

Compared to two-frequency shift keying (2FSK) and 4QAM, using 8QAM, the data rate rises at the price of increasing EBR. By using error control coding, the system performance can be improved.

Fig. 3.37 shows the improvement of system performance by using one-half rate turbo error control coding. The effect of Doppler shift is shown in Fig. 3.38. The Doppler shift index is assumed 0.001. For multicarrier systems, the shift of carrier frequency will lead to inter-channel interference (ICI), and will be more serious for an OFDM system. It is a disadvantage that OFDM system is very sensitive to frequency shift. If there is no compensation for frequency shift, correct decoding will not be more efficient. In addition, Fig. 3.38 shows the system performance after Doppler compensation. The effect of Doppler can be removed by this method.

To evaluate the performance of the OFDM system, an experiment of underwater acoustic

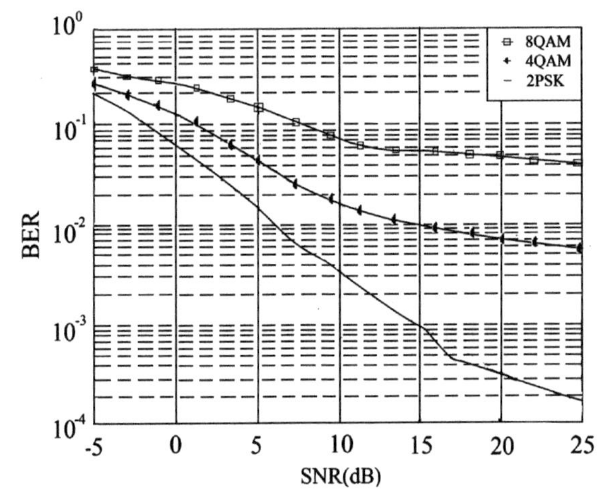

FIGURE 3.36 Bit error rate (BER) performance of three kinds of modulation.

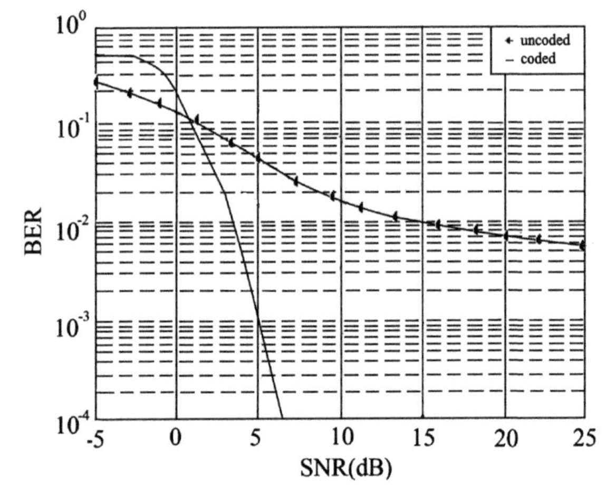

FIGURE 3.37 Improvement of performance after turbo coding.

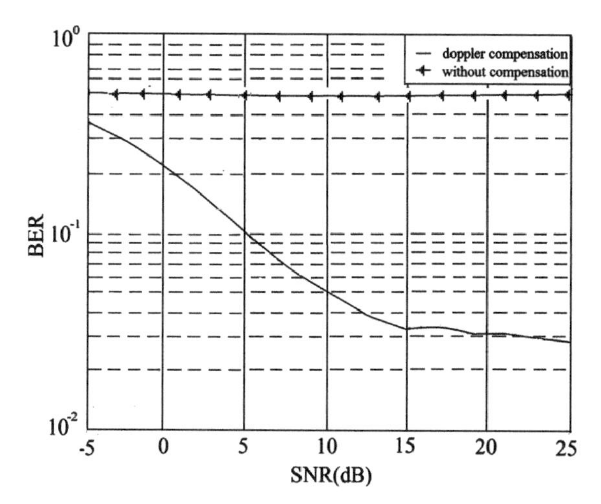

FIGURE 3.38 Effect of Doppler shift.

communication in a lake was conducted. The experimental results demonstrate that at the distance of 5 and 10 km, the data rate can reach 9 and 2.8 kps, respectively, with BER less than 10^{-4}. Of course, the signal-transmission conditions in lake are better than in the sea.

Our key laboratory of underwater acoustic communication has implemented some research on OFDM-based underwater acoustic communication for many years. Main researching topics are as follows.

1. Implementation of timing synchronization for OFDM underwater acoustic communication system on field programmable gate array (FPGA) [12].
   Based on characteristics of underwater acoustic channel and Linear Frequency Modulation (LFM) signal, LFM signal is adopted for timing synchronization signal of OFDM underwater communication system. The generation and detection principle of LFM signal are summed up. Then the scheme based on the earlier theory is implemented on FPGA. Finally, the result of pool test shows the feasibility of the scheme.
2. Simulation of OFDM systems based on LDPC [13].
   LDPC (low-density parity-check) is an effective error-correcting code. Its performance is very close to the Shannon limit and it has lower decoding complexity. This error-correcting code is used in OFDM. The performance of Low-density Parity-check–Coded Orthogonal Frequency Division Multiplexing (LDPC-COFDM) system with different modulations have been simulated and analyzed. The comparison between LDPC-COFDM and Turbo-COFDM in dissimilar channels has also been implemented. It shows that the LDPC-COFDM method is effective in high-speed underwater acoustic data transmission.
3. The simulation study of the cooperative diversity based on OFDM modulation [14].
   Cooperative diversity is widely used in wireless communication. On the other hand, OFDM is highly competitive due to its advantages in systematic spectral efficiency, power utilization, and systematic complexity. In this paper, two methods to realize cooperative diversity in OFDM systems and the frame of the OFDM are proposed; compared the cyclic redundancy check (CRC) coding OFDM system with the convolutional coding OFDM system, and analyzed the performance of the CRC coding OFDM system and convolutional coding OFDM system, convolutional codes to achieve the decode and forward (DAF) of the cooperative diversity are adopted. Two cooperative protocols in OFDM cooperative diversity system are simulated. The simulation results show that the performance of the OFDM system with the cooperative diversity accesses to a significant diversity gain.
4. The implementation of the encoding and decoding OFDM technique based on digital signal processing (DSP)6711 [15].
   Adopting the idea of the software radio, an OFDM communication system implemented by floating-dot digital signal-processing chip

DSP6711 has been designed. Based on the result of the simulation of this system with Matrix Laboratory (MATLAB), the OFDM communication system with DSP hardware had been implemented, and the C language of the professional software Command, Control, and Subordinate Systems had been used. In addition, the performance of the encoding and decoding of the OFDM communication system by utilizing the channel simulated with MATLAB was tested. The results indicate that the encoding and decoding scheme designed possesses the fine performance of antagonism to Gauss's Fading Channel.

5. Based on the fundamental reaches mentioned earlier, a prototype for underwater acoustic image communication by using OFDM system has been established; its performance has also been examined in shallow-water acoustic channels. Transmitting transducer was located on a sea beach (as shown in Fig. 4.41). A hydrophone was hung in the depth of 9 m, shown in Fig. 3.39. Fig. 3.39A shows the original image; Fig. 3.39B shows a received image in which BER $= 3.9 \times 10^{-3}$; Fig. 3.39C shows another received image in which BER $= 3.1 \times 10^{-4}$. The communication range is 5 km and the data rate is 1900 bps. The expected experimental results have been achieved.

3.2.4 Underwater Acoustic Communication Channels and Optimum Linear Filter

We have mentioned that if underwater acoustic channel belongs to a linear and time-invariant system, the optimum detector for a definite known signal against the additive white noise background is a matched filter, or an equivalent cross-correlator with which the maximum-output SNR can be obtained.

Some remarkable peculiarities of sound transmission law in underwater acoustic channels have be introduced briefly (refer to Chapter 1) and then pointed out the channels do not always satisfy the additive principle. Now we can further discuss the nonlinear characteristics after studying sound transmission laws in the channels in Chapter 2.

3.2.4.1 Finite Amplitude Sound Wave and Nonlinear Effects [16,17]

We have known that there are some assumptions, such as bulk modulus k_0 is independent of sound pressure, and medium density ρ_0 is constant in elementary acoustic theory. It means that the theory developed is strictly valid only for sound waves of very low amplitude, because for sound waves of finite amplitude, both k_0 and ρ vary (if only slightly) by the effect of sound pressure. In particular, a lot of small bubbles appear in seawater medium that will remarkably

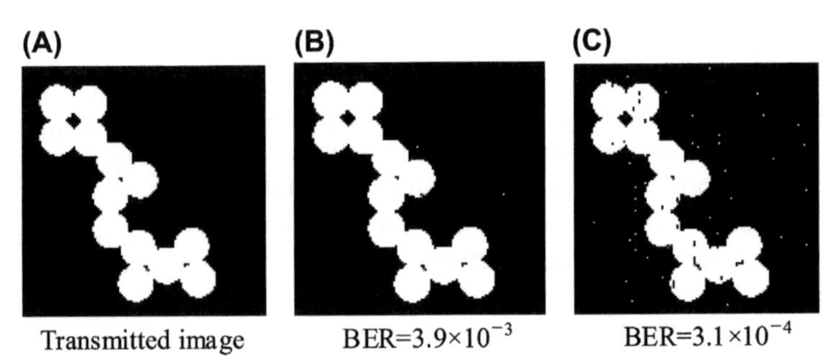

FIGURE 3.39 Typical records for image communication: (A) transmitted image, (B) BER $= 3.9 \times 10^{-3}$, (C) BER $= 3.1 \times 10^{-4}$.

change the compressibility of the medium. Thus, instead of the sound velocity c_0 being a constant for given conditions, it actually varies according to the waveform and amplitude of the sound wave. If at any point in the medium the sound pressure is p, then $c = \sqrt{k_0/\rho_0}$ would be replaced by

$$c^2 = (k_0/\rho_0)(1 + \alpha_1 p + \alpha_2 p^2 + \cdots)$$

in which α_1, α_2, etc., are constants determined by the parameters of the medium. The equation for a plane acoustic wave becomes

$$\frac{\partial^2 p}{\partial t^2} = \frac{k_0}{\rho_0}(1 + \alpha_1 p + \alpha_2 p^2 + \cdots)\nabla^2 p$$

Consider what happens as a finite sinusoidal plane wave propagates. In this case, the vibration velocity of medium u can not be neglect. Fig. 3.40A indicates that the crest of the wave is at higher pressure than the undisturbed medium; therefore, the sound velocity c is greater than c_A in which the particle velocity is zero. In a rarefaction, on the other hand, $c < c_A$ and u is negative. The net result of these different signal speeds at different parts of the wave is that the crests advance relative to the axial positions and the troughs lag behind. As the effect continues, the wave distorts to a form resembling Fig. 3.40B. Fig. 3.40C shows finite-amplitude wave having fully developed to shock wave, when the wave farther away from source. Fig. 3.40D shows aging

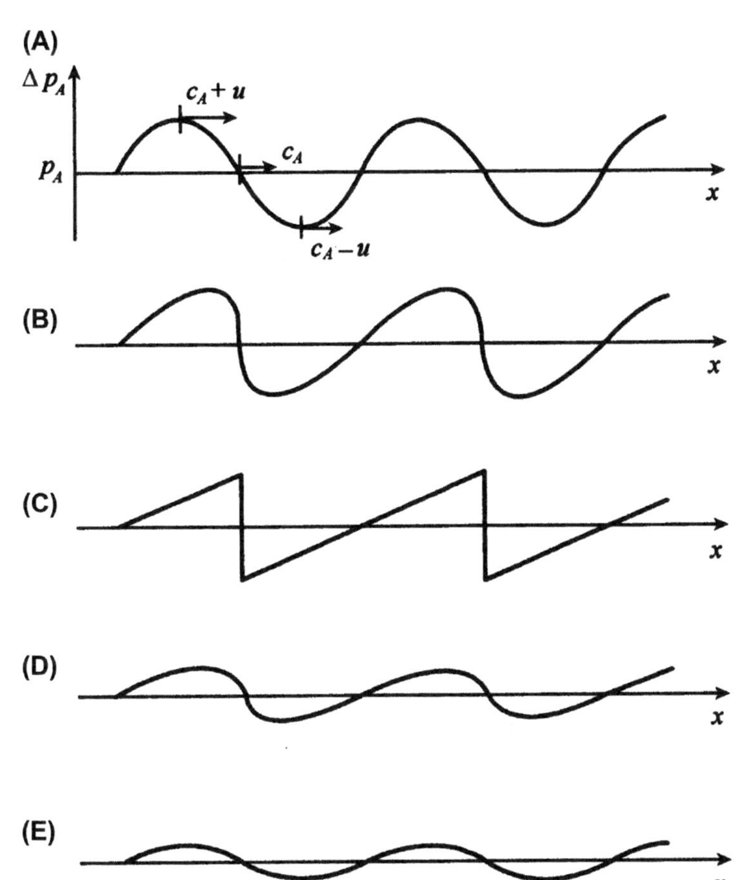

FIGURE 3.40 Finite amplitude sound wave at different stages.

shock wave due to high-frequency dissipation rate is greater than growth rate. Fig. 3.40E shows infinitesimal amplitude, degenerated shock wave.

Obviously, because high-frequency components appear for a finite-amplitude wave in a nonlinear medium, it means that it has a larger attenuation in amplitudes than that of the fundamental wave, therefore the communication distances are decreased.

When two sinusoidal sound waves with different frequencies are propagated through the nonlinear seawater in the same direction, they will interact each other and thus will form the components of sum and difference frequencies; the latter has some peculiar properties, such as narrow beam width, small-side maximum, and broad bandwidth. This component can consist of parametric sonars, but they have not been employed widely in applied engineering, because its acoustic–electric conversion efficiency is very low, therefore their transmission distances are generally nearer, say several 100 m, in seawater.

3.2.4.2 Nonlinear Effect due to Multipath Interference

We have known that the multipath propagations due to the sound reflections from both sea surface and sea bottom appear in shallow-water acoustic channels. Moreover, the sound refractions in deep-sea channeling would also generate the multipath effects. Therefore, complex and variable interference effects will appear by the coherent sound waves. In the case of a carrier frequency pulse signal, a multipath structure will be formed in which the sound interference effect is also included.

We see that whether the interference effect will appear that depends not only on the characteristics of sound channels and operating conditions, such as communication ranges, depths at which transducers are located, but also on pulse duration τ_s, that is efficient bandwidth B, for a certain communication sonar.

Assuming a carrier frequency pulse signal with finite pulse duration τ_s or finite bandwidth B corresponding to required data rate R is transmitted by communication sonar, obviously, the multipath interference will be reduced when τ_s is decreased or B is increased. Provided τ_s is less than the minimal time delay of adjacent multipaths, the multipath pulses (wave packets) will be separated from each other and form a comb multipath structure. In such case, the underwater acoustic channels would fit the linear superposition principle for an infinitesimal-amplitude sound wave. As mentioned previously, the nonlinear effect due to finite-amplitude sound wave would disappear at longer-range underwater acoustic communications in which the nonlinear effect is mostly caused by multipath effects.

The transmission loss of energy for band-limited sound signals may be determined [16] by

$$TL_B = 10 \lg \frac{E(1,z)}{E(r,z)}$$

in which $E(1,z)$ is signal energy at unit distance in the depth direction z.

TL_B for the signals having different bandwidths passing through a surface sound channeling in the Arctic Ocean at different communication distances were calculated; results were plotted in Fig. 3.36. The channel has a depth of 500 m, the sound velocity just below the surface is 1490 m/s; it is 1496 m/s near the sea bottom. Both transmitting and receiving transducers are played at the depth of 50 m. The center frequency of the signal is 150 Hz.

The real lines in Fig. 3.41A–D express TL_B for the signals having single frequency of 150 Hz, and bandwidths to be 10, 20, 50, and 100 Hz relative to the center frequency of 150 Hz, respectively. The virtual lines express distance average fields for the signal with frequency of 150 Hz at different ranges. We see that: (1) TL_B for pulse signals change to be smooth with increasing bandwidth. Remarkable convergence zones and attenuation zones appear due to multipath interference effects for a single-

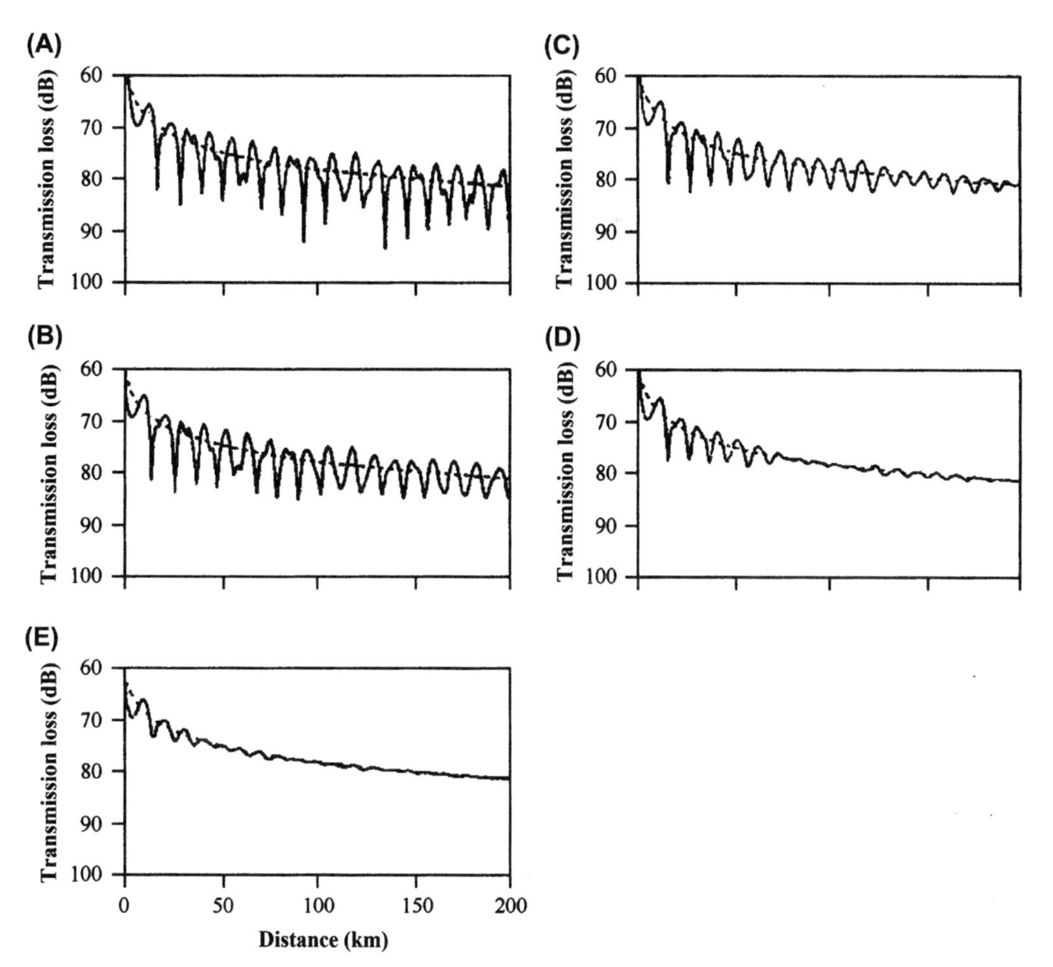

FIGURE 3.41 Comparison of TL_B between band-limited signal energy (*real lines*) and single-frequency signal average intensity (*virtual lines*) (A) single-frequency signal (B) 10 Hz bandwidth (C) 20 Hz bandwidth (D) 50 Hz bandwidth (E) 100 Hz bandwidth.

frequency signal; whereas in the case of pulse signals, both convergence and attenuation zones are gradually blurred with increasing the bandwidths and thus TL_B curves become smooth. (2) The smooth effect is more remarkable with increasing distances for the pulse signals with the same bandwidth. (3) Provided bandwidths are larger than a certain value (here to be 50 Hz), TL_B of pulse signals are consistent with the distance average field of the center frequency at longer ranges.

Similar results with respect to interference effect in shallow-water acoustic channels with a weaker multipath effect are obtained. The recorded waveforms for a pulse signal sequence are shown in Fig. 3.42. When the pulse duration τ_s is narrow (0.5 ms), the multipath pulses (wave packets) are separated from each other as shown in Fig. 3.42A. Once τ_s is larger (3 ms), direct signal pulses will be superposed by multipath pulses as shown in Fig. 3.42B. The characteristics of waveforms are similar with those shown in

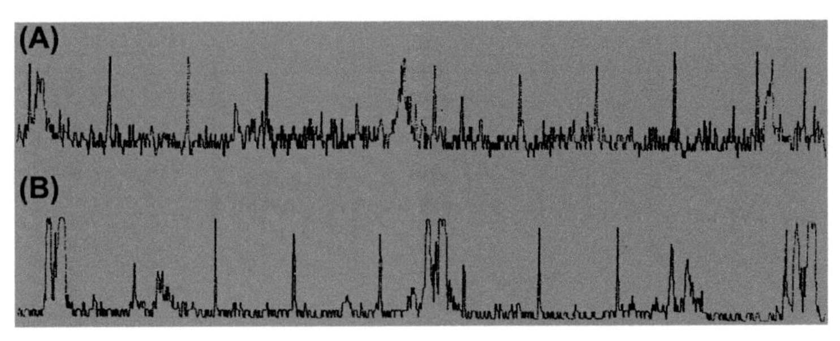

FIGURE 3.42 Multipath structures for different τ_s. (A) 0.5 ms; (B) 3 ms.

Fig. 1.9 but the durations of wave peaks are wider than those of the latter, which means that the time differences arriving at receiving point between the direct and multipath pulse reflected from sea surface are larger in the latter.

The interference effect of multipath propagations caused by sound refraction in a deep-sea sound channeling appears provided τ_s is larger than the time differences arriving at receiving point for two adjacent pulses as shown in Figs. 1.8 and 2.29, in which the waveforms with double peaks also appear.

The data rate R is a basic specification for an applied underwater acoustic communication sonar, that is corresponding to a certain τ_s or B. Therefore, for a specific communication channel and operation conditions, only τ_s is narrow enough, that is B is wide enough, the interference caused by multipath effects would disappear. In this case the underwater acoustic channels can be processed as a linear system, and described by means of the impulse response function $h(\tau,\mathbf{r},t)$ in time domain, or the transfer function $H(\omega,\mathbf{r},t)$ in frequency domain as mentioned previously.

Generally speaking, the underwater acoustic communication channels whether to satisfy the linear additive theorem depends not only on the transmission characteristics of the channels, but also on τ_s, that is bandwidth B for a specific communication sonar. Therefore, $h(\tau,\mathbf{r},t)$ and $H(\omega,\mathbf{r},t)$ corresponding to δ function having an infinite bandwidth as an input signal could not reflect the transmission characteristics of general communication conditions. Instead of them, a band-limited pulse response function $h_l(\tau,\mathbf{r},t,B)$ in time domain and a band-limited transfer function $H_l(\omega,\mathbf{r},t,B)$ in frequency domain for actual underwater acoustic communication channels are introduced. Moreover, the sound signals traveling over the channels would be changed into random processes; therefore, the reference signals in a correlation receiver are not suitable to use preknown transmitting signals as in a copy cross-correlator. In this case, the real-time output of a channel, that is field band-limited multipath structure, must be taken as the input of the communication receiver, and then by making use of some effective signal-processing schemes to adapt to that, as shown in Fig. 3.43, in which an additive white noise background is given.

The sign T in Fig. 3.43 represents a transform or operator through that input signal $s(t)$ is reflected to output signal of the channel obeying a certain law or formula. The theoretical analyses and preestimation models for multipath structures in Chapter 2 just attempt to reveal the relative law. In underwater acoustic communication engineering, the preestimated multipath structures can be used as primary references in designing communication sonar, and then based on the multipath structure acquired in situ, an adaptive rake receiver is used to adapt to that. So that matching the field multipath structure and a large processing gain would be obtained, transmitting signal $s(t)$ may be reconstructed at a preset BER P_e.

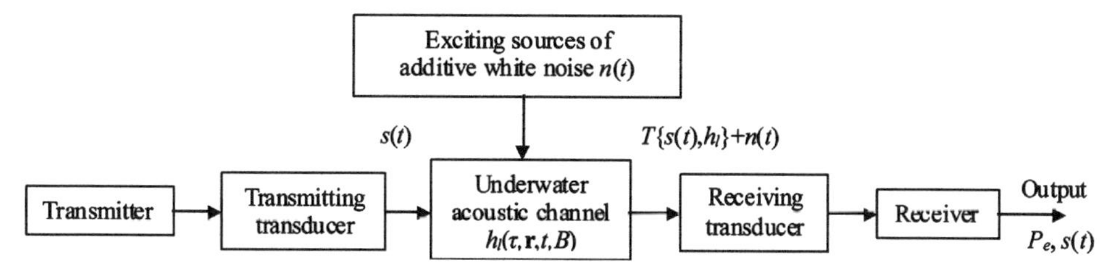

FIGURE 3.43 Block diagram of signal processing for band-limited and additive noise.

3.2.4.3 *Signal Detection Against Both Additive and Nonadditive Noise Backgrounds*

In such a case, the total noise

$$N(t) = n_a(t) + n(t)$$

The nonadditive noise $n_a(t)$ is mostly generated by the line spectra of the ships noise as mentioned earlier. Once the transmitting signals are modulated by both the nonlinear sound channels and the nonadditive noise, the multipath structures are more complex and variable, and theoretical analyses for them are very difficult. However, the exciting sources of $n_a(t)$ distribute around hydrophones, so that the essential features of multipath structures, such as its distribution in time domain, would not change; but a nonlinear modulation will act on them and thus attached distortions appear. In underwater acoustic communication engineering, the field multipath structure is taken as a reference to perform adaptive matched processing. Because the nonadditive noise of the ships appears only at some line spectra, by means of a rake receiver and potential feature of frequency diversity in the adaptive pseudo-random frequency modulation (APNFM) system would adapt to the effect of nonadditive noise. Of course, effective channel coding will also improve the channel adaptability of this system. Therefore, the transmitting information will also be reconstructed at a preset P_e, as shown in Fig. 3.44.

Of course, the effect of the nonadditive noise can generally be disregarded in civil underwater acoustic communication sonars, because it appears at low-frequency bands, as below 2 kHz; they are much lower than the efficient frequency bands employed in civil communication sonars in which the communication ranges are shorter.

3.3 EXPLORATIONS ESTABLISHING AN INNOVATIVE DIGITAL UNDERWATER ACOUSTIC COMMUNICATION SIGNAL-PROCESSING SYSTEM

3.3.1 Several Core Techniques Requiring Solution in Civil Underwater Acoustic Communications

Some basic signal-processing systems, including noncoherent systems as MFSK,

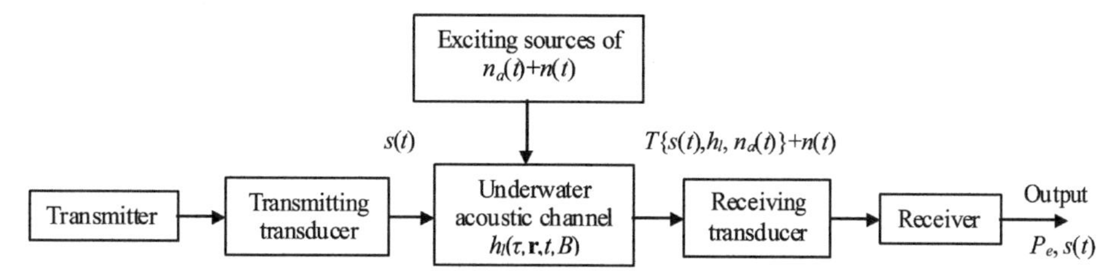

FIGURE 3.44 Block diagram of signal processing for band-limited and nonadditive noise.

coherent systems as multiple-phase shift keying (MPSK), optimum linear filter, DS-SS, and FH-SS, have been introduced previously. These systems cannot perfectly adapt to the peculiarities existing in underwater acoustic communication channels. So that present underwater acoustic communication equipment, which are established based on these systems, cannot perfectly adapt to practical requirements exploring and using ocean resources, and so on.

According to the peculiar transmission laws of underwater acoustic channels and the operating features of civil underwater acoustic communications, several core signal-processing techniques must efficiently be solved at present, which are the technical support to establish robust civil underwater acoustic communication equipment.

1. Adequately raising both the communication ranges r and data rates R
   Because the peculiar transmission laws, including great transmission loss, strict band-limited property, violent multipath effects and high-noise levels, and so on, are encountered in underwater acoustic communication channels, raising both r and R are quite difficult. How to develop a civil digital underwater acoustic communication signal-processing system that can adapt to these peculiarities and obtain both higher r and R would be difficult.
   It is noteworthy that underwater acoustic communication equipment is generally operated at a lower-input SNR, thus their bit error rates are higher. Therefore, error correction techniques would be adopted, for R to be further lowered.
2. Underwater acoustic communication channels have a violent time-variant characteristic. Therefore, the channels cannot be regarded as time-invariant ones in general communication periods of time. To adapt to the time-variant characteristic, one of the efficient signal-processing schemes is that by dividing the total communication period into several subperiods in which the channel can be regarded as a time-invariant ones. Once the changes for the channel parameters in a subperiod have impacted on whole communication quality, relative parameters of communication sonar must timely be adjusted and transformed to another subperiod to adapt to a new communication environment. However, the sound velocity in water is only 1.5 km/s, for longer-range communications, such as $r = 25$ km, the time exchanging information between transmitting and receiving ends needs 32 s, which is generally larger than the channel coherent time. How to adapt to the temporal variability of underwater acoustic communication channels, especially to fast time-variant multipath structures, for which usual channel equalizer is generally no longer valid, is a difficult problem.
3. In addition to the time-variant feature in underwater acoustic communication channels, there is a significant spatial-variant one. Therefore, how to adapt to the complicated and varied communication environments is also difficult.
4. In some cases, mobile communications are necessary, larger Doppler frequency shifts possibly appear. As noted earlier, usual signal-processing systems, such as matched filter and OFDM, have not frequency shift adaptability; moreover, in strictly band-limited underwater acoustic channels, leaving greater bandwidth room to accommodate the larger Doppler frequency shifts is generally not allowed. Therefore, complex adaptive Doppler frequency shift correction in situ is necessary.
   In mobile communications, noise levels (*NLs*) will rise, in which nonadditive noise components may also be contained. Moreover, the multipath structure will

remarkably be changed in the whole mobile communication process (refer to Figs. 1.3 and 1.5), how to adapt to the spatial-variant peculiarity is another difficult problem.

5. Different kinds of underwater acoustic communication equipment, which are associated with different applied fields and different communication media, have their own particular specificities; therefore, they are generally developed separately. It is feasible for civil ones. However, in some applied fields, such as the underwater acoustic communication between surface command ship and underwater AUV, still require multimedia. How to carry out compatible design to establish civil multimedia communication sonars is a problem worthy to be explored.

How to solve these core technical issues encountered in civil underwater acoustic communications will be explored subsequently. It would be expected to develop an innovative signal-processing system, and then based on that to establish compatible multimedia communication sonar that may adapt to complex and variable communication environments, so that robust communication results would be obtained by that.

3.3.2 Principles Establishing an Innovative Civil Digital Underwater Acoustic Communication Signal-Processing System

According to the transmission peculiarities existing in underwater acoustic communication channels and the several core techniques needing to be solved mentioned previously, based on the guidance of information theory and underwater acoustic physics, the establishment of the innovative digital underwater acoustic communication signal-processing system would obey the following principles:

3.3.2.1 Principle Selecting Signal-Modulation Modes

Because violent signal fluctuations in the amplitude and phase occur in underwater acoustic channels, amplitude modulation (AM) (such as amplitude shift keying (ASK), random amplitude modulation (RAM), etc.) should not be used. Phase modulation (PM) would be used with caution. It is better to select frequency modulation (FM), because it has a less sensibility for signal fluctuations than that for AM and PM.

The United States Navy's "Seaweb" program had carried out an experiment [18] to compare the robustness between frequency hopping—frequency shift keying (FH-FSK) and differential phase shift keying (DPSK) modes in shallow-water acoustic channels. Experimental results show that the changes of BERs are much greater by using DPSK mode. For example, BER is zero in some data packets; whereas others are up to 50%. In such a case, BER in FH-FSK mode will change in a way that is consistent with better predictability. Moreover, when appropriate channel error correction encoding schemes are used, BER will extend to zero by using this mode. The preliminary conclusion had been obtained: the robustness using FH-FSK mode is higher than that using DPSK in shallow-water acoustic channels.

3.3.2.2 Principle Suitably Matching With the Outputs of the Underwater Acoustic Communication Channels

Because underwater acoustic communication channels have the peculiarities of randomly spatiotemporal—frequency variant parameters, violent multipath effects, and signal fluctuations, there exist essential differences between the output and input signals of underwater acoustic communication channels. Therefore, detecting results by means of a signal-processing system based on linear and time-invariant channels, such as copy correlator, will remarkably be reduced. The actual outputs of the channels, which are randomly spatiotemporal–frequency variant multipath structures, must thus be taken as the input signals for an underwater acoustic communication receiver, and then using appropriate signal-processing schemes to adapt to them, it would be expected to obtain robust communication results (refer Figs. 3.43 and 3.44).

3.3.2.3 *Principle of Antimultipath and Using Multipath Energy*

Multipath propagations are the basic feature of the underwater acoustic communication channels.

As noted earlier, the multipath effects have duality. We must first reduce their negative impact. In the performance analyses of the FH-SS system underwater acoustic communication, we had pointed out that, provided the total delay spread of multipath effects T_M is shorter, we can let

$$n\tau_s \geq T_M$$

to combat ISI caused by multipath effects, in which τ_s is the duration of frequency codes with full duty ratio, and n are the numbers of nonrepetition used frequency codes in FH pattern. For nonfull duty ratio signal sequence, τ_s will be replaced by repetition period. This treatment usually means that the communication data rates will be reduced, and thus there is no universal adaptability. The experiments (refer to Section 3.2.1) had pointed out that T_M may be up to 300 ms in long-range underwater acoustic communications. Let communication rate $R = 200$ bps, that is, $\tau_s = 5$ ms, corresponding frequency gap is equal to 200 Hz, therefore required $n = 60$, whole bandwidth is up to 12 kHz, obviously, that is too wide to adapt to the long-range communications. However, the numbers of significant wave packets in a multipath structure generally are only 6–8. Therefore, we would fully use the sparse characteristics of multipath structures and then by means of adaptive multipath suitable matching schemes to achieve greater bandwidth compression, and then to adapt to strict band-limited underwater acoustic communication channels. It would be a basic premise to realize the long-range underwater acoustic communications.

It seems that the best scheme to use the multipath energy would be by means of rake receiver. The major wave packets in multipath structures will first be "rake taken," and then combined with an optimum mode in the rake receiver. It may be expected to obtain the optimum detection result at the criterion of the maximum output SNR against the multipath interference background.

3.3.2.4 *Principles by Means of Adaptive Signal-Processing Techniques*

For peculiar underwater acoustic channels with random spatiotemporal–frequency variant parameters and violent signal fluctuations, utilizing adaptive signal-processing techniques will naturally be selected, which would include the following five main aspects:

3.3.2.4.1 ADAPTIVE TOTAL TIME SAMPLING PROCESSING

It would realize the multifunctions as mentioned in Section 1.3.3. In particular, we may use that instead of conventional synchronous scheme to correctively detect PN sequences under the condition of violent signal fluctuations in time–amplitude domains, even if the fluctuations in time domain are up to one to two durations of code element. This adaptive processing scheme is a key part in the APNFM system to adapt to time-variant channels.

3.3.2.4.2 ADAPTIVE RAKE RECEIVER

The input of an underwater acoustic communication receiver, that is the multipath structure, is random spatiotemporal–frequency variable, its total delay spread T_M may also be quite long, such as up to several 100 ms. Therefore, by using the conventional rake receiver (refer to Fig. 3.9 or 3.10) to realize effective path diversity generally is not adaptive; but we can convert that into a "magic rake," in which the numbers of teeth, rake lengths (ie, tooth spaces), rake rates, etc. may be changed to follow field randomly variable multipath structures. It seemingly is complex, provided the receiver is properly designed, the excellent results of the path diversity without additional energy consumption would be obtained, because the numbers of multipath pulses (wave packets) that must be acquired are generally less than eight by passing through a preprocessing circuit; therefore, this core part would be established more easily.

3.3.2.4.3 ADAPTIVE DOPPLER FREQUENCY SHIFT CORRECTION

We can use a highly accurate discriminating frequency scheme to detect the frequency shifts for some typical frequency codes in situ, and then they will adaptively be corrected, respectively. This adaptive processing can still get the information of sail speeds.

3.3.2.4.4 CORRECTION OF ADAPTIVE SPATIOTEMPORAL VARIANT PARAMETERS

Another basic characteristic of underwater acoustic channels is their variability. From the point of view of signal processing, the variability is sometimes more difficult to adapt to that than the complexity. As mentioned previously, for common communication periods of time, underwater acoustic channels belong to time-variant ones. Therefore, it is necessary to divide them into several subperiods that are allowed to be considered as time-invariant ones, and then they will adaptively be transformed from one subperiod to another by studying relative field parameters, and then making relative signal processing.

3.3.2.4.5 REALIZING ADAPTIVE SUITABLE MATCH WITH THE CHANNELS WITHOUT PRIOR KNOWLEDGE

For complicated and variant underwater acoustic communication channels, if the communication equipment is required to supply relatively prior knowledge of the channels, it is not available for civil users. Therefore, we must make the communication signal-processing systems that can normally operate under the premise condition without the prior knowledge, and achieve the suitable match to the outputs of the channels, that is, the multipath structures.

3.3.2.5 *Principle of Seeking Common Ground While Reserving Differences*

Underwater acoustic communications have varied applied fields and by means of corresponding communication media, which seemingly are complex and variable, but they have a feature of "common" as main, and "difference" as deputy. Because no matter what kinds of communication media in digital communications, they will be converted into the bit streams without substantial differences by passing through a digital format. The "common" tasks of underwater acoustic communication equipment are how to transmit the bit streams as long communication ranges r as possible at an expected communication data rates R and a minimum BER. Of course, different applied fields and corresponding communication media have their own peculiarities, such as there are different requirements for r, R, and BER, etc., which can be solved by adjusting the relative parameters of underwater acoustic communication receivers, and using different information source and channel-coding schemes.

It is possible to compose more efficiently a high-performance and intensive multimedia communication sonar by using the principle of seeking common while reserving differences. In fact, the applications of an integrated communication sonar are an effective measure to improve robustness, because the different communication media can be used to adapt different marine communication environments, the probability of realizing reliable communications is therefore raised.

3.3.2.6 *Principle of Tolerance*

As noted earlier, there is a common disadvantage of digital communications: the "threshold effect"; that is, when the input SNR drops to a certain value, a sharp deterioration in communication quality will appear. For complex and varied underwater acoustic communication channels, we should pay attention to select relative parameters following the principle of tolerance, such as to leave adequate room for the input SNR to meet violent signal fluctuations and *NL* to unexpectedly be raised. Similarly, we would leave the room for one to two durations of the signal pulse to adapt to violent signal fluctuations in time domain in the total time sampling processing to ensure the signal can still be correctly detected in such a fluctuating state. Of course, such a tolerance design principle, required sound source levels *SLs* are thus higher, whole bandwidths are also widened, thus communication data rates and ranges would correspondingly be reduced; however, the probability of communication failure will also be lowered remarkably.

3.3.2.7 *Civil Communication Sonars*

Being low in cast, small in size, and light in weight are generally demanded for civil communication sonars, which restrict the applications of some advanced but complex signal-processing techniques. How to use the simpler signal-processing systems to compose the communication sonars having higher performances is also difficult.

By passing through the organic synthesis of the designing principles mentioned earlier, it is expected to form an innovative adaptive pseudorandom frequency modulation (APNFM) signal-processing system employed in digital underwater acoustic communication.

As noted previously, there are two specific approaches to implement the signal processing by employing the APNFM system in communication sonars. The first one is a perfectly adaptive processing scheme, which is more suitable to be used in military underwater acoustic communications, and thus it will not be discussed further in this book, which will focus on the civil ones. The second approach is that the field input signal, that is the multipath structure, is first acquired, and then the relative parameters of the system are adaptively adjusted to realize a suitable field channel match. The specific structures and performance analyses for the second processing approach will be summarized in the following section and Chapter 4.

3.3.3 Innovative APNFM Digital Underwater Acoustic Communication Signal-Processing System

3.3.3.1 *Basic Principle and Structure for the APNFM System*

The schematic block diagram for an innovative APNFM system digital underwater acoustic communication receiver is shown in Fig. 3.45.

Before transmitting formal communication information, a suitable calling signal is first transmitted by the side requiring communication; it is also used as a channel exciting source to obtain field channel parameters, including multipath structures, Doppler frequency shifts, and so on. This calling signal by passing through preset preamplifier and preset total time sampling processing parts (refer to Fig. 1.10), followed by the detection-processing part of channel parameters, a pseudo-random frequency modulation (PNFM) pattern suitably matching to the channel has thus been obtained; that is, corresponding local PNFM pattern. Moreover, these parameters acquired will also be used as the references to preset the relative adaptive processing parts (refer to Fig. 3.45).

Once the receiving side receives the calling signal, it transmits as soon as possible a replying signal that is agreeable to the calling one. When the transmitting side receives this signal, the detection and processing for channel

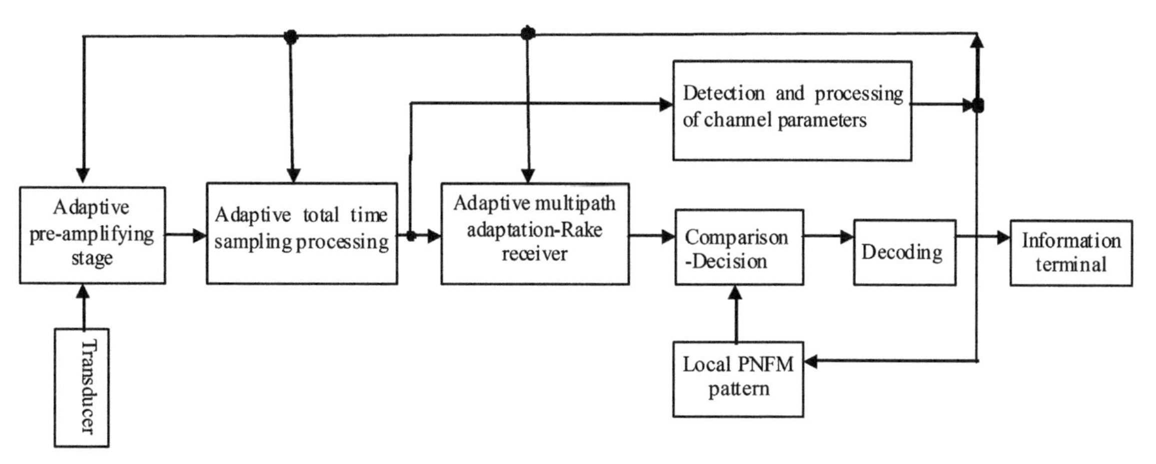

FIGURE 3.45 Schematic block diagram of APNFM receiver.

parameters, just as receiving side, will correspondingly be performed.

Subsequently, communication information that is modulated according to the PNFM pattern matching with the channel is transmitted, a formal underwater acoustic communication may be implemented.

The inputs of the communication receiver are multipath structures mixed with the noise in the sea. Preamplification part generally includes a fixed gain amplifier, adaptive gain control, channel, and transducer adaptive amplitude equalization, and ALE. For the latter, experiments had proved that it has a good antinoise performance, as shown in Fig. 3.46. Fig. 3.46A shows the received synchronization signal of an FH system underwater acoustic communication receiver with the input SNR = −12 dB; Fig. 3.46B shows the output of a usually filter; Fig. 3.46C is the output of ALE. Certainly, how to adapt to a quickly jumping and complicated PNFM pattern still needs careful design.

The total time sampling processing part, which is designed according to the principle of tolerance, would adapt to the multipath structures with violent fluctuations in time–amplitude domains, even if those fluctuations in time domain are up to one to two durations of frequency codes.

The core technique in the APNFM signal-processing system is the following adaptive rake receiver. It implements the multipath diversity under the condition of the violent random fluctuations of multipath structures. It would expect to obtain an optimal detection result against the background of multipath propagations. So that the sparse characteristics of multipath structures must first be identified, and then by means of the preprocessing of a multipath matching circuit, the adaptive multipath diversity would be realized in underwater acoustic communication engineering (refer to Chapter 4). The transmitted PNFM pattern may be correctly detected by using comparison-decision processing with the local preset PNFM pattern in comparison–decision part. It is necessary to make the calibration of Doppler frequency shift for mobile communications, as shown in Fig. 3.45. Finally, by passing through information source and channel decoding, it is expected to reset the messages of information source at a preset BER.

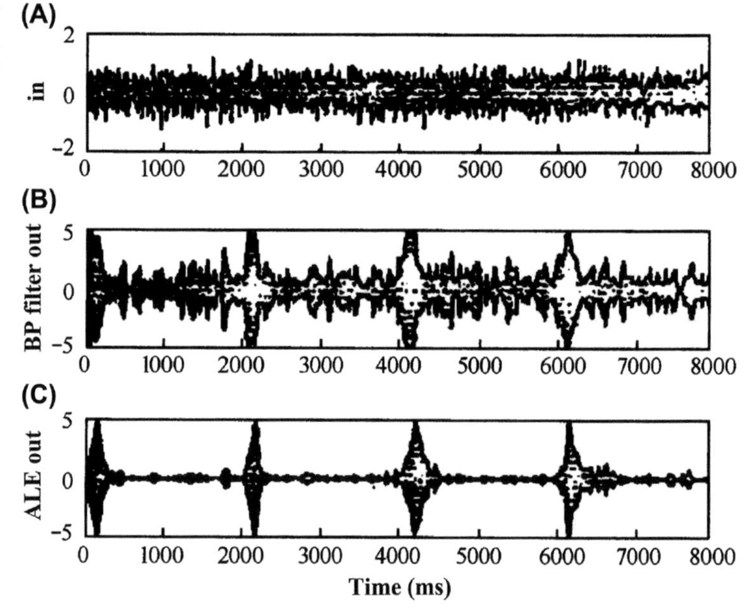

FIGURE 3.46 Experimental results for traditional filter and ALE. (A) The received signal; (B) the output of a usually filter; (C) the output of ALE.

3.3.3.2 Brief Analyses of Performances for the APNFM System

1. Multiorder multipaths to be sampled and then combined with an optimum mode in the adaptive rake receiver may be achieved in the APNFM system underwater acoustic communication. Therefore, it would be expected to obtain an optimal detection result against the background of multipath propagations. A typical multipath structure is shown in Fig. 2.39 in which the numbers of wave packets equals eight, which are unnecessary for additional energy consumption.
2. The APNFM system with an adaptive rake receiver would adapt to the nonlinear underwater acoustic channels, because its received signals are the main components of multipath structure. Perhaps one or two of them meet reverse phase interference condition and thus deep fading appears, whereas the probability would be small most multipath pulses all with reverse phase interference effect to exist, as shown in Fig. 1.3 Of course, for high-frequency and weak multipath communication conditions, multipath diversity results will be reduced. In this case, an appropriate error correction scheme is generally used to adapt to the violent signal fading caused by self-multipath interference effect.
3. As mentioned earlier, the nonadditive noise effect may generally be neglected in civil underwater acoustic communications. Even if in the case of low-frequency and long-range underwater acoustic communication, the APNFM system would also be able to adapt to the existence of some linear spectrum components in the noise of ships, because this system has a performance of implicit temporal–frequency diversity. Once by combining with a suitable channel coding, we can allow some frequency codes to be lost due to nonadditive noise parasitic modulation.
4. The system would adapt to underwater acoustic communication channels having the peculiarities of random

temporal–frequency–amplitude variability in a certain range, because the adaptive approaches have been used in the main parts of the system, in particular, that could adapt to time-variant multipath structures. Therefore, real-time civil underwater acoustic communications would be realized.

5. According to the basic principles mentioned previously, the system could achieve robust civil underwater acoustic communications without the prior knowledge of the channels, thus the usefulness of communication equipment will greatly be improved.

Some performances of the APNFM system underwater acoustic communications mentioned previously will be reflected in designing relative communication equipment; moreover, they will be proved by the laboratory simulations and preliminary experiments in shallow-water acoustic channels (see Chapter 4).

It is noteworthy that the APNFM signal-processing system may take into account different applied fields by using corresponding communication media. Based on that, an integrated digital communications sonar may be established (refer to Chapter 4). However, for much practical civil digital underwater acoustic communication equipment, especially for commercial ones, we can compose the products with different specifications according to different purposes and using corresponding communication media for different users, and thus it is unnecessary to establish such complex communication equipment. For high frequencies or operating in the conditions of deep-sea vertical links, the multipath interference would not be considered. Obviously, in these cases, using the rake receiver is meaningless, the signal processing will remarkably be simplified, or even coherent detection schemes can be used (refer to Chapter 4).

References

[1] L. Qihu, Introduction to Signal Processing of Sonar Written, China Ocean Press, Beijing, 2000.

[2] B. Sklar, Digital Communications, Publishing House of Electric Industry Press, Beijing, 2002.

[3] Q. Perliang, et al., Fundamentals of Digital Communications, Publishing House of Electric Industry Press, Beijing, 2007.

[4] X. Zeng, et al., Spread-Spectrum Communication and Multi-Addresses Technique, Xidian University Publishing House, 2004.

[5] X. Keping, X. Tianzeng, et al., Underwater acoustic wireless communications, Journal of Xiamen University (Natural Science) 40 (2) (2001) 311–319.

[6] M.A. Wen, Long-range underwater acoustic information transmission through MFSK modulation, Telecommunication Engineering 29 (1) (2007) 52–56.

[7] C. En, E-mail transmission in shallow water acoustic channels, The Collection of Thesis for First Oceanic Development Forum (2004) 234–238.

[8] A.I. Yuhui1, et al., A study of m sequence spread spectrum underwater acoustic communication, Journal of Harbin Engineering University 21 (2) (2000) 15–18.

[9] G. Loubet, V. Capellano, K. Filipiak, Underwater spread-spectrum communications, IEEE (1997) 574–579.

[10] C. Boulanger, G. Loubet, J.R. Lequepeys, Spreading sequences for underwater multiple-access communications, IEEE (1998) 1038–1042.

[11] J. Huang, et al., High-speed underwater acoustic communication based on OFDM, IEEE (2005) 1135–1140.

[12] X. Yongjun, et al., Implement of timing synchronization for OFDM underwater communication system on FPGA, Modern Electronics Technique 7 (2009) 1–3.

[13] Z. Jie, et al., Performance of OFDM systems based on LDPC, Wireless Communication Technology 15 (4) (2006) 1–4.

[14] L. Zhijuan, et al., The simulation study of the cooperative diversity based on OFDM modulation, Journal of Xiamen University (Natural Science) 50 (1) (2011) 24–27.

[15] W. Deqing, et al., The implementation of the encoding and decoding OFDM technique based on DSP6711, Ocean Technology 24 (2) (2005) 42–45.

[16] D.G. Tucker, B.K. Gazey, Applied Underwater Acoustics, Pergamon Press, 1977.

[17] R.J. Urick, Principles of Underwater Sound, McGraw-Hill Book Company, 1975.

[18] M.B. Porter, V.K. Mcdonald, P.A. Baxley, J.A. Rice, Signal ex: linking environmental acoustics with the signaling schemes, in: OCEAN' 2000 MTS/IEEE Conference and Exhibition, vol. 1, 2000, pp. 595–600.

CHAPTER

4

Digital Underwater Acoustic Communication Equipment

We have discussed the physical basis of sound transmission in underwater acoustic communication channels in Chapter 2. The signal-processing systems employed in digital underwater acoustic communications have also been described in Chapter 3. A basic purpose in researching them is to develop digital underwater acoustic communication equipment suitable for practical applications.

Many types of underwater acoustic communication equipment are available, including some simple commercial products to provide to different users. However, the research on this equipment is generally performed according to particular requirements, in particular in military applications; therefore, open reports in relation to them are few. Three types of civil digital underwater acoustic communication equipment being developed by authors will be discussed in this chapter: (1) Underwater acoustic telecontrol communications equipment in which a digital time correlation (including autocorrelation and cross-correlation) accumulation decision system has been employed. (2) Underwater acoustic multimedia communication equipment in which an improved FH-SS system has been employed. Moreover, (3) a digital underwater acoustic communication prototype in which an innovative adaptive pseudorandom frequency modulation (APNFM) system has been employed.

A transducer, as an antenna in radio communication devices, plays an important role in underwater acoustic communication equipment. Its structures and operating characteristics must be understood to help us design suitable and feasible underwater acoustic communication equipment to satisfy expected requirements. For example, a new model transducer used in a shallow-water target telemetry sonar and an array used in an ultrasonic sensor, which will be mentioned later, has been designed and developed by authors. They are key parts for these devices. Therefore, the basic requirements, operating principles, and characteristics of transducers employed in underwater acoustic communications will be described in this chapter, and then the developments with respect to the three typical communication systems previously mentioned will be described.

Digital Underwater Acoustic Communications
http://dx.doi.org/10.1016/B978-0-12-803009-7.00004-0

4.1 BRIEF INTRODUCTION TO UNDERWATER ACOUSTIC TRANSDUCERS EMPLOYED IN UNDERWATER ACOUSTIC COMMUNICATION EQUIPMENT

4.1.1 Some Specific Requirements for Underwater Acoustic Communication Transducers

4.1.1.1 Center Frequency

The center frequency, which generally is the fundamental frequency of a transducer, for underwater acoustic communication equipment, closely relates to its whole performance, such as communication distances *r*, data rates *Rs*, noise levels *NLs*, directivity, and so on. Making this frequency approach the optimum operating frequency is necssary, which may be roughly determined by the compromising operations according to the communication sonar equation (refer to Section 1.4).

4.1.1.2 Bandwidth

To design a bandwidth *B* available for deployment in digital underwater acoustic communication equipment is a complex problem, because it relates to not only *r* and *R*, but also the characteristics of underwater acoustic channels and signal-processing systems used. If communication equipment can be operated at a single frequency or a very narrow bandwidth, channel-amplitude equalization is unnecessary. Moreover, if the transducer can be operated at or near a resonance vibration state, its transmitting efficiency and receiving sensitivity would be optimal, making it easy to realize the impedance match with both the power amplifier in the transmitter and the preamplifier in the receiver, because such a transducer is approximately equivalent to pure resistance. In fact, communication sonar has an expected *R* or *B*, particularly once the sonar system is employed in that, *B* will be spread to a much broader band. Therefore, designing an available bandwidth for specific communication sonar is necessary to adapt to the band-limited peculiarity in underwater acoustic channels and avoid the excessive difficulty for developing a broadband transducer operating at large transmitting acoustic power.

4.1.1.3 Transmitting Acoustic Power

The transmitting acoustic power is the main component of the sound source level *SL*, and is mostly determined by the communication distance *r*. Of course, this is related to the preset input signal-to-noise ratio (SNR) of a specific receiver. The desired *SL* can generally be estimated by compromising operations according to the communication sonar equation. The cavitation effect must be considered for a transducer the size of which is strictly limited. When communication sonar is operated under the water for a long time, its energy is generally supplied by a battery set; particularly in the field of secret communications, the estimation of the acoustic power would leave enough margin. Obviously, these kinds of communication equipment will be operated under the condition of low-input SNR, the designs for which are more difficult.

4.1.1.4 Transmitting Efficiency and Receiving Sensitivity

The transducers used in underwater acoustic communication equipment generally are transmitting–receiving compatible. The operating properties of both transmitting and receiving states will be mutually considered. However, the high transmitting efficiency, say above 70%, must first be selected, in particular for a communication sonar operating under water for a long time, perhaps the receiving sensitivity needs to be lower, because raising that would not help to raise the input SNR in higher marine background-noise levels.

4.1.1.5 Impedance Characteristics

To easily realize the impedance match with both power amplifier and the input circuit of receiver to transducer, it is better that the transducer has more uniform impedance, thus the performance of communication equipment will improve and avoid the excessive difficulty of designing an amplitude equalizer.

4.1.1.6 Operating Depths

A transducer operated at great depths must specifically be designed to consider the possible reduction of corresponding operating performance.

4.1.1.7 Directivity

The directivity of a transducer closely relates to the performance of communication sonar. Because a suitable directivity (including suppressing side lobes) will raise *SL*, decrease *NL*, and delay total spread T_M due to multipath effects, so sharper directivity would first be considered. However, if the positions for underwater communication are unknown to each other, a transducer with omnidirectivity in the horizontal would generally be chosen. Moreover, possible rocking, the limitation in size for a transducer, and convenient operations (say for divers) would be considered.

According to the specific requirements for the directivity of communication sonar, a tubular transducer is generally utilized. Its specific structures and operating principles will be introduced in the following section. Moreover, a transducer array may sometimes be used, such as in a diverse space, to improve the spatial processing gains, and so on.

4.1.2 Tubular Piezoelectric Ceramic Transducer [1,2]

Some vibrating modes of piezoelectric ceramic transducers, such as thickness (shapes of plate, bar, disc, etc.), length (shapes of plate, bar, and tube), circumference (hollow sphere), and bent vibration models may be utilized in underwater acoustic communication equipment. As a typical example, we shall analyze the operating principles and characteristics of a tubular underwater acoustic communication transducer, which would help us design a power amplifier–transducer set suitable for a specific communication sonar.

Because the tubular piezoelectric ceramic transducers have a series of advantages, such as simple structure, steady electroacoustic characteristics, high receiving sensitivity, and uniform horizontal directivity, they are generally employed in underwater acoustic communication equipment.

The encapsulated mode for mounting the transducers is generally adopted as shown in Fig. 4.1; its vibrator consists of several identical short piezoelectric ceramic tubes polarized in the radial direction. A rubber washer is put between the two tubes for vibrating insulation and coupling elimination; moreover, some foam plastics fill the tubes for sound absorption. Castor oil is filled inside the metal external shell to help sound transmission. Of course, the

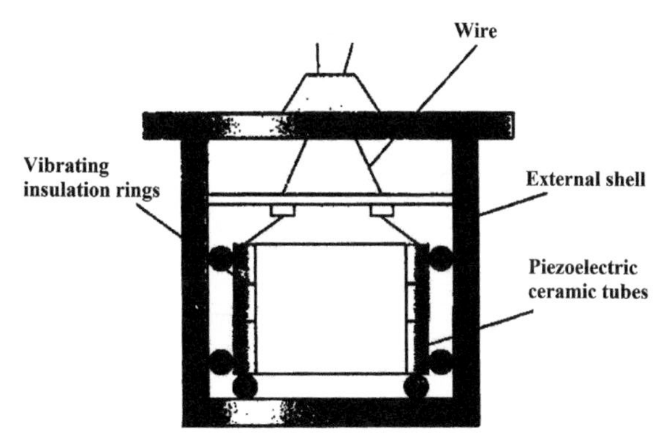

FIGURE 4.1 Tubular transducer encapsulated mode.

encapsulated mode is also adopted in communication sonars by using rubber with sound impedance approximately equal to that of a water medium.

4.1.2.1 *Vibration Equations and Piezoelectric Relation Equations*

Let us first analyze the characteristics of a single tubular piezoelectric ceramic transducer, the piezoelectric heat of an actual tubular transducer consisting of several superposed identical short tubes.

The relationships between deformations and displacements in cylinder coordinates are given by

$$\begin{aligned} S_1 &= \frac{1}{r}\frac{\partial \xi_1}{\partial \theta} + \frac{\xi_3}{r}, \quad S_2 = \frac{\partial \xi_2}{\partial z}, \quad S_3 = \frac{\partial \xi_3}{\partial r} \\ S_4 &= \frac{\partial \xi_1}{\partial r} + \frac{1}{r}\frac{\partial \xi_3}{\partial \theta} - \frac{\xi_1}{r}, \quad S_5 = \frac{1}{r}\frac{\partial \xi_2}{\partial \theta} + \frac{\partial \xi_1}{\partial z} \\ S_6 &= \frac{\partial \xi_3}{\partial z} + \frac{\partial \xi_2}{\partial r} \end{aligned} \tag{4.1}$$

in which S_i are components of strain, ξ_1, ξ_2, and ξ_3 are the components of displacements, and subscripts 1, 2, and 3 express the directions in θ, z, and r, respectively.

Assuming that a tubular transducer symmetrically vibrates relative to radial and axial directions, therefore, $\xi_1 = 0$, ξ_2 and ξ_3 are independent of θ, and we obtain

$$\begin{aligned} S_1 &= \frac{\xi_3}{r}, \quad S_2 = \frac{\partial \xi_2}{\partial z}, \quad S_3 = \frac{\partial \xi_3}{\partial r} \\ S_4 &= 0, \quad S_5 = 0 \quad S_6 = \frac{\partial \xi_3}{\partial z} + \frac{\partial \xi_2}{\partial r} \end{aligned} \tag{4.2}$$

To find its vibration equations, a small six-face volume element inside the transducer is selected, the exerting force states for which are shown in Fig. 4.2, in which T_i are the components of stress.

According to the analyses of force effect and Newton's second law, the radial and axial vibration equations for a piezoelectric ceramic tube are, respectively, shown as follows:

$$\rho \frac{\partial^2 \xi_3}{\partial t^2} = \frac{\partial T_3}{\partial r} + \frac{\partial T_6}{\partial z} + \frac{T_3 - T_1}{r} \tag{4.3}$$

$$\rho \frac{\partial^2 \xi_2}{\partial t^2} = \frac{\partial T_2}{\partial z} + \frac{\partial T_6}{\partial r} + \frac{T_6}{r} \tag{4.4}$$

in which ρ is the density of piezoelectric ceramic material. If piezoelectric ceramic tube is radially polarized and excited by an external electric field E_3 in this direction, the piezoelectric relation

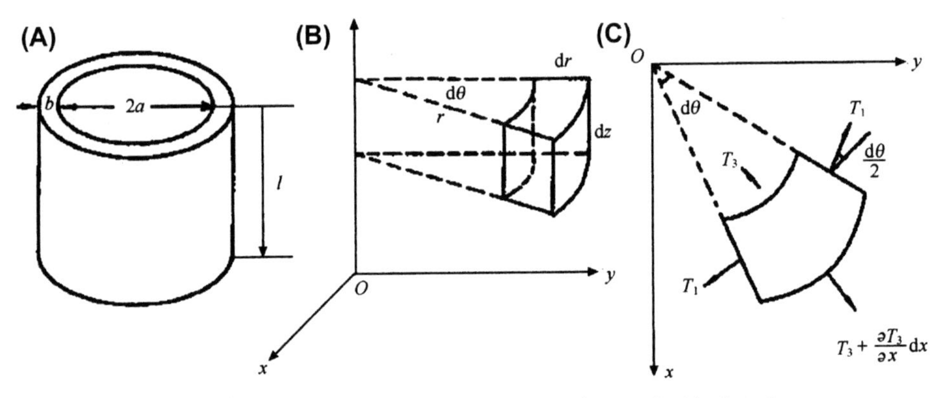

FIGURE 4.2 Force states for small-volume elements inside the tube.

equations for axial symmetric vibrations are given by

$$\begin{aligned} S_1 &= s_{11}^E T_1 + s_{12}^E T_2 + s_{13}^E T_3 + d_{31} E_3 \\ S_2 &= s_{12}^E T_1 + s_{11}^E T_2 + s_{13}^E T_3 + d_{31} E_3 \\ S_3 &= s_{13}^E T_1 + s_{13}^E T_2 + s_{13}^E T_3 + d_{33} E_3 \\ S_6 &= s_{44}^E T_6 \\ D_2 &= d_{15} T_6 \\ D_3 &= d_{31}(T_1 + T_2) + d_{33} T_3 + \varepsilon_{33}^T E_3 \end{aligned} \quad (4.5)$$

in which s_{ij}^E are cut-off compliance constants, D_i are electric displacements, d_{ij} are piezoelectric strain constants, and ε_{ij}^T are free dielectric constants for the piezoelectric ceramic tube.

4.1.2.2 Dynamic Theoretical Analyses for a Tubular Piezoelectric Ceramic Transducer Polarized in Radial Direction

An excited tubular piezoelectric ceramic element, in fact, will generate many complex three-dimensional coupling vibration models. To simplify the analyses for that, we assume: (1) the ceramic tube has a thin wall, the thickness for which $b \ll \lambda$ (wavelength in resonance state); therefore the radial stress waves do not exist, (2) because the rubber rings eliminating coupling vibration are used on both sides of each tube. Thus, $T_2 = 0$ at $z = \pm l/2$, in which l is the height of the tube, and (3) only radial vibrations symmetric to the z axis and longitudinal vibration along the z axis occur in the tube, provided that it is excited by E_3; ξ_2 and ξ_3 are thus only the functions of z. (4) The electric field would be uniform for a thin-wall tube, therefore $E_3 = V/b$, in which V is the externally exciting voltage.

According to the assumptions mentioned earlier, the vibration equations may be simplified as follows:

$$\rho \frac{\partial^2 \xi_3}{\partial t^2} = -\frac{T_1}{a} \quad (4.6)$$

$$\rho \frac{\partial^2 \xi_2}{\partial t^2} = \frac{\partial T_2}{\partial z} \quad (4.7)$$

in which a is the average radius of the tube. When the tube is at resonance vibration state, Eqs. (4.6) and (4.7) will be written as follows:

$$\rho \omega^2 \xi_3 = \frac{T_1}{a} \quad (4.8)$$

$$\rho \omega^2 \xi_2 = -\frac{\partial T_2}{\partial z} \quad (4.9)$$

The piezoelectric relation equations can also be simplified

$$S_1 = s_{11}^E T_1 + s_{12}^E T_2 + d_{31} \frac{V}{b} \quad (4.10)$$

$$S_2 = s_{12}^E T_1 + s_{11}^E T_2 + d_{31} \frac{V}{b} \quad (4.11)$$

$$S_3 = s_{13}^E T_1 + s_{13}^E T_2 + d_{33} \frac{V}{b} \quad (4.12)$$

$$D_3 = d_{31}(T_1 + T_2) + \varepsilon_{33}^T \frac{V}{b} \quad (4.13)$$

Let

$$Y_0^E = 1/s_{11}^E, \quad \sigma = -s_{12}^E / s_{11}^E$$

Referring to Eqs. (4.10) and (4.11), we can get

$$T_1 = \frac{Y_0^E}{1-\sigma^2}\left[S_1 + \sigma S_2 - d_{31}(1+\sigma)\frac{V}{b}\right] \quad (4.14)$$

$$T_2 = \frac{Y_0^E}{1-\sigma^2}\left[\sigma S_1 + S_2 - d_{31}(1+\sigma)\frac{V}{b}\right] \quad (4.15)$$

Substituting Eqs. (4.14) and (4.15) into Eq. (4.13), we get

$$D_3 = \varepsilon_{33}^* \frac{V}{b} + \frac{d_{31} Y_0^E}{1-\sigma}(S_1 + S_2) \quad (4.16)$$

$$\varepsilon_{33}^* = \varepsilon_{33}^T \left(1 - k_p^2\right)$$

in which ε_{33}^* are cut-off dielectric constants, and

$$k_p^2 = \frac{2 d_{31}^2 Y_0^E}{(1-\sigma)\varepsilon_{33}^T}$$

in which k_p is called plane electromechanical coupling coefficient. By finding the partial derivative of Eq. (4.10) with respect to z, according to Eq. (4.15), we obtain

$$\frac{\partial S_1}{\partial z} = \frac{\sigma v^2}{\omega^2 a^2 (1-\sigma^2) - v^2} \frac{\partial S_2}{\partial z} \tag{4.17}$$

in which $v^2 = Y_0^E/\rho$. Substituting Eq. (4.15) into Eq. (4.9), according to Eq. (4.17), we obtain

$$\frac{\partial^2 \xi_2}{\partial z^2} + k^2 \xi_2 = 0 \tag{4.18}$$

in which

$$k^2 = \left(\frac{\omega}{v}\right)^2 \frac{(1-\sigma^2) - (\omega/\omega_r)^2 - 1}{(\omega/\omega_r)^2 - 1}$$
$$\omega_r = \frac{v}{a} \tag{4.19}$$

Because $\sigma \approx 0.3$ for piezoelectric ceramic material, when $\omega/\omega_r > 1$, and $\omega/\omega_r > (1-\sigma^2)^{-1/2}$, there exists $k^2 > 0$. Therefore, Eq. (4.18) has a wave solution:

$$\xi_2 = A\sin(kz) + B\cos(kz) \tag{4.20}$$

Taking the center of the tube as coordinate original point, the node of vibration will be that for a symmetric vibration mode, when $z = 0$ and thus $\xi_2 = 0$, the constant $B = 0$ can be determined by Eq. (4.20) which becomes

$$\xi_2 = A\sin(kz)$$

Supposing both ends of the tube are free, ξ_2 has a maximum value at $z = \pm l/2$ at the resonance state. Thus, resonance condition is given by

$$k = \frac{m\pi}{l} \quad m = 1, 3, 5, \ldots$$

Substituting that into Eq. (4.19), we obtain

$$(1-\sigma^2)\left(\frac{\omega}{\omega_r}\right)^4 - \left[1 + \left(\frac{ma\pi}{l}\right)^2\right]\left(\frac{\omega}{\omega_r}\right)^2 + \left(\frac{ma\pi}{l}\right)^2 = 0 \tag{4.21}$$

Let

$$\frac{v\pi}{l} = \omega_l$$

Eq. (4.21) becomes

$$\left(\omega^2 - \omega_r^2\right)\left(\omega^2 - m^2\omega_l^2\right) = \sigma^2\omega^4 \tag{4.22}$$

This is the resonance frequency equation, provided that the thin-wall tube is at the vibration-free state.

Under the particular conditions of $m = 1$, and $\omega_l = \omega_r$, that is, $\pi a = l$, we get

$$\omega = \frac{\omega_r}{\sqrt{1 \pm \sigma}} = \frac{1}{a}\sqrt{\frac{Y_0^E}{\rho(1 \pm \sigma)}}$$

or

$$\omega = \frac{\pi}{l}\sqrt{\frac{Y_0^E}{\rho(1 \pm \sigma)}}$$

It is the angular frequency of coupling resonance with respect to a thin-wall tube with a length of one-half circumference. In this case, the severe coupling vibration between radial and axial directions will appear. Two corresponding resonance frequencies are as follows:

$$\omega_1 = \frac{1}{a}\sqrt{\frac{Y_0^E}{\rho(1+\sigma)}}, \quad \omega_2 = \frac{1}{a}\sqrt{\frac{Y_0^E}{\rho(1-\sigma)}}$$

The distribution of vibration displacements for the thin-wall tube at the coupling resonance state will thus be found when $\omega = \omega_l$, that is,

$$\xi_2 \approx 0, \quad \xi_3 = d_{31} a E_3 \frac{1+\sigma}{\sigma}$$

The amplitude distribution of the thin-wall tube is shown in Fig. 4.3. We see that the vibrating mode of the tube tends to vibrate radially, provided that $\omega = \omega_1$. Therefore, if this piezoelectric ceramic tube is used to form a transmitting transducer, to obtain the maximum

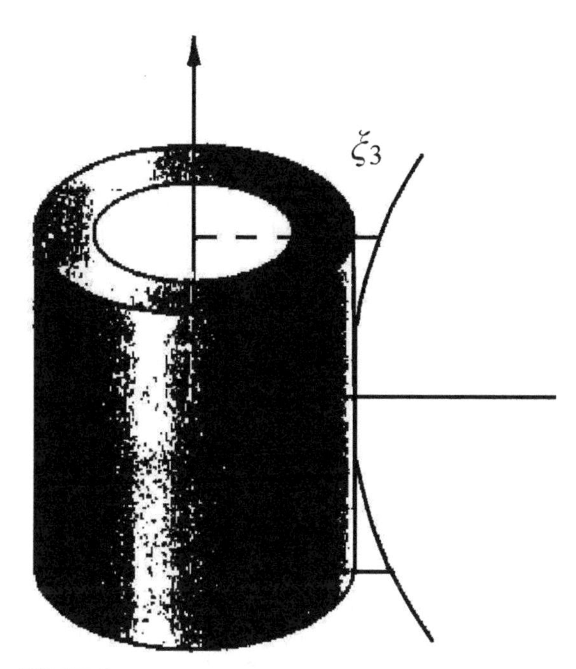

FIGURE 4.3 Amplitude distribution of a thin-wall tube.

energy of the radial vibration, the condition $\omega = \omega_1$ must be satisfied.

4.1.2.3 *Equivalent Electric Admittance for the Thin Wall Tubular Transducer*

According to Eq. (4.16), the electric current passing through a thin-wall piezoelectric ceramic tube polarized in the radial direction is given by

$$I = \frac{d}{dt}\int_{-l/2}^{l/2} a \int_0^{2\pi} D_3 dz d\theta$$
$$= \frac{d}{dt}\int_{-l/2}^{l/2} a \int_0^{2\pi}\left\{\varepsilon_{33}^* \frac{V}{b} + \frac{d_{31}Y_0^E}{1-\sigma}\left(\frac{\xi_3}{a} + \frac{\partial \xi_2}{\partial z}\right)\right\} dz d\theta \tag{4.23}$$

Moreover, $T_3 = 0$ at $z = \pm l/2$. Let

$$f(\omega) = (1-\sigma^2)\left(\frac{\omega}{\omega_r}\right)^2 - 1$$

According to vibration equations, we get

$$\frac{\partial \xi_2}{\partial z} = \frac{d_{31}E_3(1+\sigma)[f(\omega)+\sigma]}{[f(\omega)+\sigma^2]\cos\frac{kl}{2}}\cos(kz) \tag{4.24}$$

and

$$\frac{\xi_3}{a} = \frac{d_{31}(1+\sigma)E_3}{f(\omega)}\left\{\frac{\sigma^2 + \sigma f(\omega)}{[f(\omega)+\sigma^2]\cos\frac{kl}{2}}\cos(kz) - 1\right\} \tag{4.25}$$

Substituting Eqs. (4.24) and (4.25) into Eq. (4.23), we get

$$I = j\omega\frac{\varepsilon_{33}^* 2\pi al}{b}V + j\frac{d_{31}^2 Y_0^E(1+\sigma)2\pi al\omega}{f(\omega)(1-\sigma)b}V\left\{\frac{[f(\omega)+\sigma]^2}{[f(\omega)+\sigma^2]}\frac{2\tan\frac{kl}{2}}{kl} - 1\right\}$$

Because

$$\frac{[f(\omega)+\sigma]^2}{[f(\omega)+\sigma^2]}\frac{1+\sigma}{1-\sigma} = \frac{\left[(1+\sigma)(\omega/\omega_r)^2 - 1\right]^2}{(\omega/\omega_r)^2 - 1}$$

It can be written as follows:

$$I = jC_0\omega V + \frac{d_{31}^2 Y_0^E j\omega 2\pi al}{f(\omega)b} \times V\left\{\frac{\left[(1+\sigma)(\omega/\omega_r)^2 - 1\right]^2}{(\omega/\omega_r)^2 - 1}\frac{\tan\frac{kl}{2}}{\frac{kl}{2}} - \frac{1+\sigma}{1-\sigma}\right\}$$
$$= jC_0\omega V + \frac{j\omega k_{31}^2 C^T}{f(\omega)} \times V\left\{\frac{\left[(1+\sigma)(\omega/\omega_r)^2 - 1\right]^2}{(\omega/\omega_r)^2 - 1}\frac{\tan\frac{kl}{2}}{\frac{kl}{2}} - \frac{1+\sigma}{1-\sigma}\right\} \tag{4.26}$$

in which $k_{31}^2 = \frac{d_{31}^2 Y_0^E}{\varepsilon_{33}^T}$; $C_0 = \frac{2\pi al\varepsilon_{33}^*}{b}$ is "cut-off" capacitance; $C^T = \frac{2\pi al\varepsilon_{33}^T}{b}$ is "free," that is, low-frequency capacitance.

Therefore, the electric admittance of this tube at the vibration-free state is as follows:

$$Y = \frac{I}{V} = jC_0\omega + \frac{j\omega k_{31}^2 C^T}{f(\omega)} \times \left\{ \frac{\left[(1+\sigma)(\omega/\omega_r)^2 - 1\right]^2}{(\omega/\omega_r)^2 - 1} \frac{\tan\frac{kl}{2}}{\frac{kl}{2}} - \frac{1+\sigma}{1-\sigma} \right\} \tag{4.27}$$

The second term on the right side of Eq. (4.27) is dynamic electric admittance, which is reflected as an electric circuit when the tube is at vibration-free state. If $\omega \ll \omega_r$, that is, at low frequency, Eq. (4.27) can be simplified as follows:

$$Y_0 = jC^T\omega$$

Now the electric admittance has changed into electric susceptance generated by C^T at vibration-free state.

4.1.2.4 Analyses of Operating Characteristics for the Tubular Transducer Polarized in Radial Direction

The operating characteristics for a single short tubular transducer will first be discussed. The subscripts 1, 2, and 3 express the relative quantities of circumferential θ, axial z, and radial r directions, respectively, as follows.

A thin silver layer is applied on both the internal and external surfaces of the tube as two electric poles. If they are excited by a sine voltage V_3, the positive and negative electric charges will aggregate on the surfaces of both poles and form an electric field in piezoelectric ceramic material. Because the tube wall is assumed thin and only the radial vibration is taken, the electric field between the two surfaces can be considered uniform, thus its strengths are only the function of time t. The electric field strengths in the radial direction are given by

$$E_1 = E_2 = 0, \quad E_3 = E_3(t)$$

The mechanical vibrations due to piezoelectric effect excited by E_3 will be generated.

Let the thin-wall short tube (see Fig. 4.4) satisfy the following conditions:

$$b < \frac{a}{10}, \quad l < 2a$$

in which b, l, and a are the thickness, height, and average radius of the tube, respectively. If the frequency of exciting voltage can make the radial vibration achieve the mechanical resonance state, the modes of thickness and axial vibrations will be neglected because they are weak enough. In this case, only the radial vibration displacement ξ_r is considered. Moreover, only the stress T_1 in the tube is considered, and that is only the function of r and t, and independent of θ and z.

The stress T_1 will generate polarizing strength and electric displacement due to the piezoelectric effect. Now, piezoelectric equations are given by

$$\begin{aligned} S_1 &= s_{11}^E T_1 + d_{31}E_3 \\ D_3 &= d_{31}T_1 + \varepsilon_{33}^T E_3 \end{aligned} \tag{4.28}$$

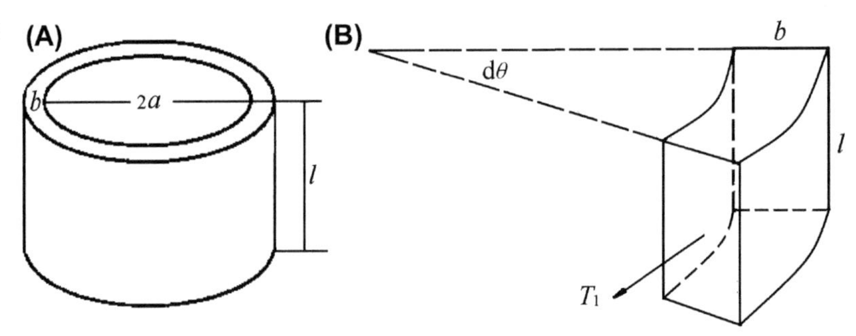

FIGURE 4.4 Vibration analyses for a short thin tube.

4.1.2.4.1 MECHANICAL VIBRATION EQUATIONS

A small volume element is taken in the tube as shown in Fig. 4.4B. Its stress of each flank is T_1, the area is lb, and the applied force at each flank is thus $T_1 lb$, their resulting force

$$2T_1 lb \sin\frac{d\theta}{2} \approx T_1 lb d\theta$$

Assume that the density of the ceramic material is ρ, thus the mass of the volume element is $\rho albd\theta$. If its displacement in the radial direction is ξ_3, according to Newton's second law, we get

$$T_1 lbd\theta = -\rho lbad\theta\frac{\partial^2 \xi_3}{\partial t^2}$$

That is,

$$\rho\frac{\partial^2 \xi_3}{\partial t^2} = -\frac{T_1}{a} \quad (4.29)$$

According to the first expression in Eq. (4.28), we obtain

$$T_1 = \frac{S_1}{s_{11}^E} - \frac{d_{31}}{s_{11}^E}E_3$$

while $S_1 = \frac{\xi_3}{a}$, therefore,

$$T_1 = \frac{\xi_3}{as_{11}^E} - \frac{d_{31}}{s_{11}^E}E_3$$

Substituting that into Eq. (4.19), we get

$$\rho\frac{\partial^2 \xi_3}{\partial t^2} = -\frac{\xi_3}{a^2 s_{11}^E} + \frac{d_{31}}{as_{11}^E}E_3 \quad (4.30)$$

When the tube is at the resonance vibration state, we have

$$\dot{\xi}_3 = j\omega\xi_3, \quad \ddot{\xi}_3 = -\omega^2\xi_3$$

Substituting them into Eq. (4.30), we obtain

$$\frac{2\pi lbd_{31}}{s_{11}^E}E_3 = \left(j2\pi a\rho lb\omega + \frac{1}{j\omega\frac{s_{11}^E a}{2\pi lb}}\right)\dot{\xi}_3$$

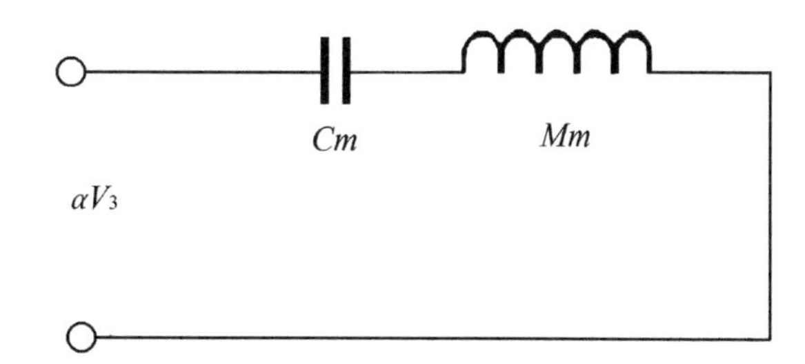

FIGURE 4.5 Mechanical equivalent circuit of the ceramic tube.

That is,

$$\alpha V_3 = \left(j\omega M_m + \frac{1}{j\omega C_m}\right)\dot{\xi}_3 \quad (4.31)$$

in which $V_3 = E_3 b$, $\alpha = \frac{2\pi l d_{31}}{s_{11}^E}$ is electromechanical conversion coefficient of the piezoelectric ceramic tube in radial vibration. $M_m = 2\pi a\rho lb$ is equivalent mass of the tube in radial vibration. $C_m = \frac{s_{11}^E a}{2\pi lb}$ is equivalent compliance coefficient of the tube.

According to Eq. (4.31) and the electromechanical analog, the mechanically equivalent circuit of the piezoelectric ceramic tube is shown in Fig. 4.5.

4.1.2.4.2 ELECTROMECHANICALLY EQUIVALENT CIRCUIT FOR THE SINGLE PIEZOELECTRIC CERAMIC TUBE

The electromechanical equivalent circuit for the single piezoelectric ceramic tube will be analyzed in the following pages. Circuit state equations will first be determined. According to the second expression in Eq. (4.28) and $T_1 = \frac{\xi_3}{as_{11}^E} - \frac{d_{31}}{s_{11}^E}E_3$, we have

$$D_3 = \frac{d_{31}}{as_{11}^E}\xi_3 + \left(\varepsilon_{33}^T - \frac{d_{31}^2}{s_{11}^E}\right)E_3$$

Moreover,

$$\varepsilon_{33}^{T} - \frac{d_{31}^{2}}{s_{11}^{E}} = \varepsilon_{33}^{S}$$

$$D_3 = \frac{d_{31}}{as_{11}^{E}}\xi_3 + \varepsilon_{33}^{S}E_3$$

in which ε_{ij}^{s} are clamping dielectric constants. The electric flux passing through the electric pole surfaces is

$$\Phi = 2\pi alD_3 = \frac{\varepsilon_{33}^{S}2\pi al}{b}V_3 + \alpha\xi_3$$

Thus, the corresponding electric current passing through the tube is given by

$$I = \frac{d\Phi}{dt} = j\omega\frac{\varepsilon_{33}^{S}2\pi al}{b}V_3 + \alpha\dot{\xi}_3 = j\omega C_0 V_3 + \alpha\dot{\xi}_3 = I_c + I_d \tag{4.32}$$

in which $C_0 = \frac{\varepsilon_{33}^{S}2\pi al}{b}$ is the steady capacitance of the tube. Eq. (4.32) is the circuit state equation, in which I_c is steady current and I_d is dynamic current.

If loss and radiant resistances are considered, the electromechanically equivalent circuit will be obtained as shown in Fig. 4.6, in which R_0 is electric loss resistance, R_m is mechanical loss resistance, R_s is radiant resistance, and M_s is same-vibrating mass.

The piezoelectric heap of the tubular transducer consists of several identical short ceramic tubes, which are parallel in the circuit.

4.1.2.4.3 OPERATING CHARACTERISTICS IN TRANSMITTING STATE

4.1.2.4.3.1 FREQUENCY OF MECHANICAL RESONANCE The condition of mechanical resonance is given by

$$\omega(M_m + M_s) - \frac{1}{\omega C_m} = 0$$

The corresponding frequency of mechanical resonance is

$$f_r = \frac{1}{2\pi\sqrt{(M_m + M_s)C_m}} \tag{4.33}$$

4.1.2.4.3.2 MECHANICAL Q-FACTOR

$$Q_m = \frac{\omega_r(M_m + M_s)}{R_m + R_s} \approx \frac{2\pi f_r M_m}{R_m + R_s} \tag{4.34}$$

4.1.2.4.3.3 BANDWIDTH

$$\Delta f = \frac{f_r}{Q_m} = \frac{R_m + R_s}{2\pi(M_m + M_s)} \approx \frac{R_m + R_s}{2\pi M_m} \tag{4.35}$$

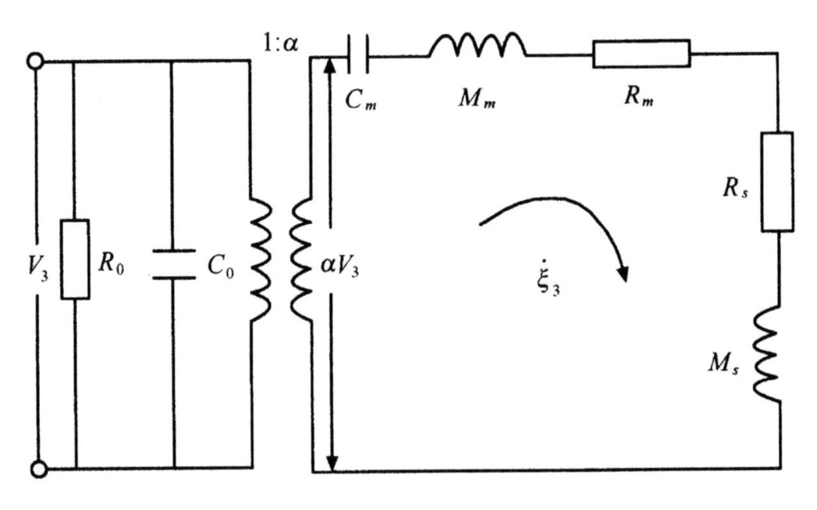

FIGURE 4.6 Electromechanically equivalent circuit.

4.1.2.4.3.4 POWER If n short tubes are parallel in the circuit, $\dot{\xi}_3 = \frac{\alpha V_3}{R_m+R_s}$ at the resonance vibration state, radiating power is given by

$$P_a = \frac{n}{2}\dot{\xi}_3^2 R_s = \frac{n}{2}\left(\frac{\alpha V_3}{R_m + R_s}\right)^2 R_s \tag{4.36}$$

Mechanical loss power is given by

$$P_{mn} = \frac{n}{2}\dot{\xi}_3^2 R_m = \frac{n}{2}\left(\frac{\alpha V_3}{R_m + R_s}\right)^2 R_m$$

Total mechanical power is given by

$$P_m = P_{mn} + P_a = \frac{n}{2}\frac{(\alpha V_3)^2}{R_m + R_s}$$

Electric loss power is given by

$$P_{en} = \frac{n}{2}\frac{V_3^2}{R_0}$$

and total input electric power is given by

$$\begin{aligned} P_e &= P_{en} + P_m = \frac{n}{2}\frac{V_3^2}{R_0} + \frac{n}{2}\frac{(\alpha V_3)^2}{(R_m + R_s)} \\ &= \frac{nV_3^2}{2}\frac{R_m + R_s + \alpha^2 R_0}{R_0(R_m + R_s)} \end{aligned} \tag{4.37}$$

4.1.2.4.3.5 EFFICIENCY Acoustic-mechanical efficiency is given by

$$\eta_{a/m} = \frac{P_a}{P_m} = \frac{R_s}{R_m + R_s}$$

Electromechanical efficiency is given by

$$\eta_{m/e} = \frac{P_m}{P_e} = \frac{\alpha^2 R_0}{R_m + R_s + \alpha^2 R_0}$$

and acoustic-electric efficiency is given by

$$\begin{aligned} \eta_{a/e} &= \frac{P_a}{P_e} = \eta_{a/m}\eta_{m/e} \\ &= \frac{\alpha^2 R_0 R_s}{(R_m + R_s)(R_m + R_s + \alpha^2 R_0)} \end{aligned} \tag{4.38}$$

4.1.2.4.4 ANALYSES OF RECEIVING STATE

The electromechanical equivalent circuit for a single piezoelectric ceramic tube at receiving state is shown in Fig. 4.7, in which v is the distortion coefficient of the sound field, S is receiving area of the transducer, and p_f is free-field sound pressure.

If the impedance Z_0 in the side of the electric circuit is converted into that of the mechanical circuit, we obtain the equivalent circuit as shown in Fig. 4.8. Therefore, the vibration velocity in the radial direction is given by

$$\dot{\xi}_3 = \frac{v p_f S}{R_m + R_s + j\omega(M_m + M_s) + \frac{1}{j\omega C_m} + \alpha^2 Z_0}$$

Obviously, the force applying to $\alpha^2 Z_0$ is as follows:

$$\begin{aligned} F &= \alpha^2 Z_0 \dot{\xi}_3 \\ &= \frac{v p_f S \alpha^2 Z_0}{R_m + R_s + j\omega(M_m + M_s) + \frac{1}{j\omega C_m} + \alpha^2 Z_0} \end{aligned}$$

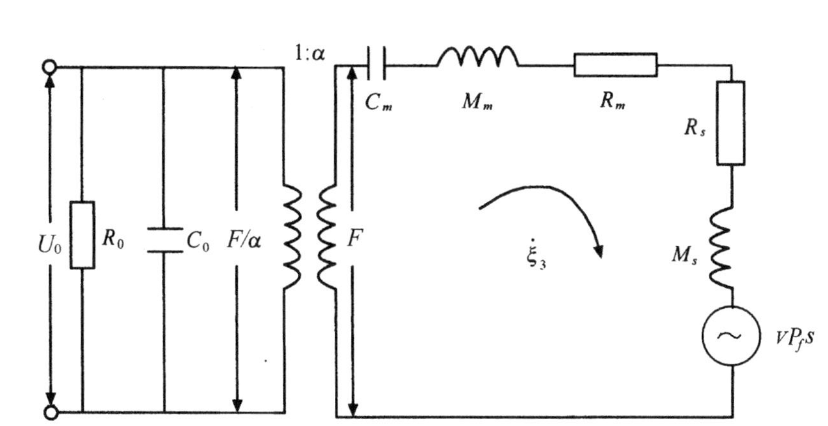

FIGURE 4.7 Electromechanically equivalent circuit for a single ceramic tube in the receiving state.

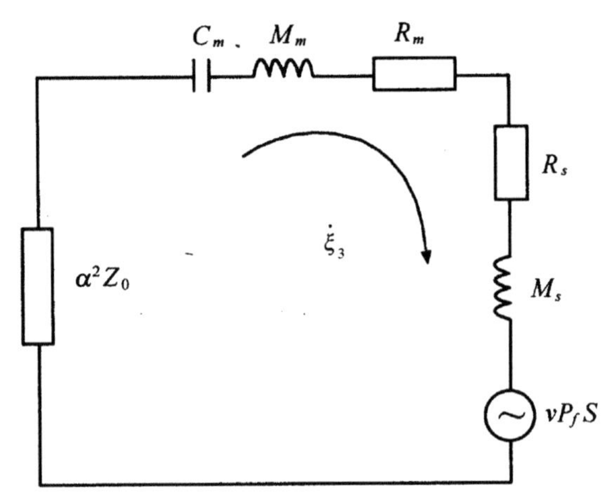

FIGURE 4.8 Equivalent circuit by passing through electric impedance converted to mechanical circuit.

Thus, the open circuit voltage of Z_0 is given by

$$\begin{aligned} U_0 &= \frac{F}{\alpha} \\ &= \frac{v p_f S \alpha Z_0}{R_m + R_s + j\omega(M_m + M_s) + \frac{1}{j\omega C_m} + \alpha^2 Z_0} \end{aligned}$$

Therefore, the receiving sensitivity of the open circuit voltage for each ceramic tube is given by

$$\begin{aligned} M_e &= \left| \frac{U_0}{p_f} \right| \\ &= \left| \frac{\alpha v S}{\alpha^2 + \left[R_m + R_s + j\omega(M_m + M_s) + \frac{1}{j\omega C_m} \right] \frac{1}{Z_0}} \right| \end{aligned} \tag{4.39}$$

When the tube is at the mechanical resonance state, because

$$j\omega(M_m + M_s) + \frac{1}{j\omega C_m} = 0$$

so that the receiving sensitivity for a piezoelectric ceramic tube may be expressed by

$$M_e = \left| \frac{U_0}{p_f} \right| = \left| \frac{\alpha v S}{\alpha^2 + \frac{R_m + R_s}{Z_0}} \right| \tag{4.40}$$

Because all tubes are parallel in circuit, the total receiving sensitivity at the mechanical resonance state may also be determined by Eq. (4.40).

To sum up, a tubular piezoelectric ceramic transducer may be considered as an electromechanically equivalent circuit, so that its parameters, including the mechanical resonance frequency, mechanical Q-factor, bandwidth, radiant power, and efficiency at the transmitting state, may be determined. Moreover, a power amplifier having a good impedance match with the transducer can be designed. Similarly, the receiving sensitivity and so on may also be found when the transducer operates at the receiving state. According to the equivalent circuit of the receiving transducer, the preamplifier of the matching receiver to that may also be designed, and the input SNR of the receiver will thus be improved.

In fact, the theoretical analyses and accurate performance predictions for an actual transducer are generally difficult; therefore, its chief specifications, including the directivity, are commonly determined by practical measurements. Some basic parameters, including transmitting response, impedance characteristic, receiving sensitivity, and directivity would be provided by relevant factory (referring to Figs. 4.39 and 4.40). It is better that the power amplifier–transducer set suitable to the requirement for a specific communication sonar can be provided by the corresponding factory (refer to Section 4.4).

4.2 UNDERWATER ACOUSTIC TELECONTROL COMMUNICATIONS BY USING DIGITAL TIME CORRELATION ACCUMULATION DECISION SIGNAL-PROCESSING SYSTEMS

We have developed underwater acoustic telecontrol communication sonar in which new signal-processing systems, namely digital time correlation (including autocorrelation and cross-correlation) accumulation decisions (TCAD), were used. This sonar had been employed in an underwater acoustic releaser as a remote-control instruction signal to control the release key. The experiments in situ had demonstrated that it performed excellently operating in shallow-water acoustic channels.

High robustness is a substantial specification for an underwater acoustic releaser. First, its false-alarm probability $P(f)$ must approach zero, even if the releaser is locked on the sea bottom for a long time, such as 3 months, which is a specification for the releaser developed, if the decision time is 2 s, the total decision numbers are up to 10^7. Then, the detection probability for the entire recovery process $P(d)$ must tend toward 1. Because the data rates for the telecontrol communication sonar employed in the releaser permit to be very low, the corresponding signal-processing time can greatly be increased. The detection probability for each accumulation decision $P_1(d)$ may thus be lower (as 0.3), because time diversity may be used for low data-rate communications to improve $P(d)$ and finally to make it approach 1. In fact, the remote-control instruction signal is continuously transmitted in the time interval of the entire recovery operation until the releaser is recovered.

The developed underwater acoustic releaser will be operated in shallow sea areas in which the depths are less than 200 m; moreover, required maximum telecontrol distance is only 4 km, so that operating frequencies may be selected to be as high as 20 kHz. The field experiments with respect to sound transmission characteristics will mostly choose that as a carrier frequency f.

4.2.1 Experimental Research on the Physical Characteristics of Shallow-Water Acoustic Channels

Understanding in detail the characteristics of sound transmission in shallow-water acoustic channels is necessary as a physical basis for designing a signal-processing system available to be deployed in telecontrol communication sonar. Its specific parameters can also be determined according to empirical data.

4.2.1.1 Statistical Characteristics of Sound Pulse Transmissions [3,4]

4.2.1.1.1 EXPERIMENTAL ARRANGEMENTS

The experimental research on the statistical characteristics of sound transmissions in Xiamen Harbor shallow-water acoustic channels has been performed. The experimental apparatus in situ is shown in Fig. 4.9A.

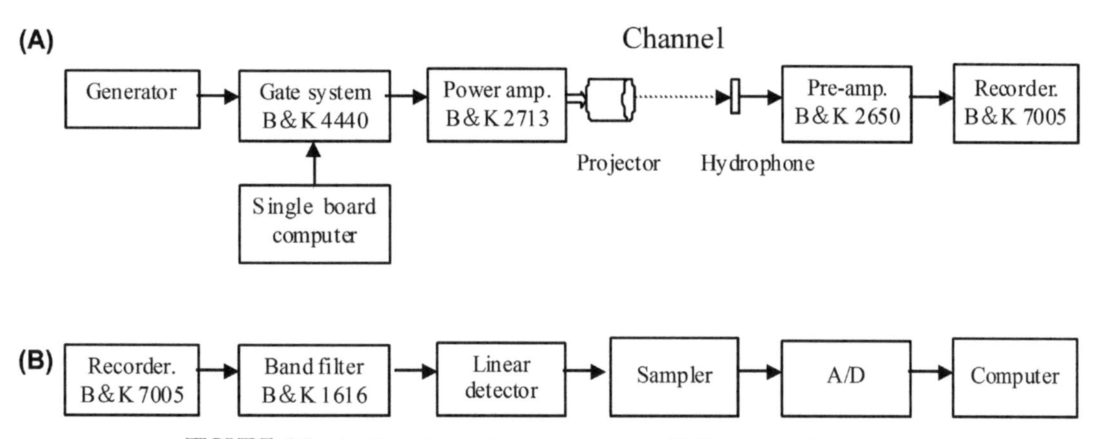

FIGURE 4.9 (A) Experimental apparatus in situ. (B) Data processing system.

FIGURE 4.10 Envelope of received waveform.

The projector was fixed against an extended wharf. The shallow-water channel is about 20 m deep, in which the distribution of sound velocity in the vertical direction is nearly homogeneous. Carrier frequencies, pulse, repetition, periods transmitting acoustic powers, and receiving sensitivities may be varied in field experiments.

The data processing system is shown in Fig. 4.9B. A one-third octave band filter (Model 1616) was used in that. Sampling and A/D were completed by means of a spectrum analyzer (model HP 3582A).

The statistical characteristics of direct pulses and multipath pulses (wave packets) were analyzed for a carrier frequency (20 kHz) pulse signal with a short duration of 0.5 ms to separate the direct and multipath pulses. A typical received waveform in an envelope is shown in Fig. 4.10.

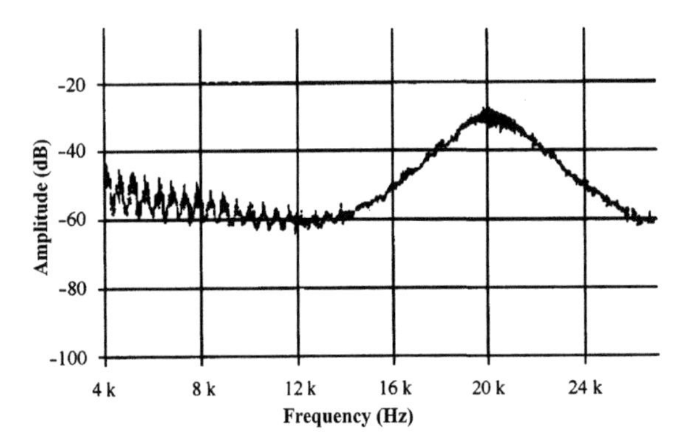

FIGURE 4.11 Spectrum of received signal.

The spectrum of received signal is shown in Fig. 4.11.

4.2.1.1.2 STATISTICAL CHARACTERISTICS OF DIRECT PULSES

By passing through an underwater acoustic communication channel, a transmitting-pulse signal will be changed into a random process; therefore, that must be described by some statistical quantities. To design a signal-processing system employed effectively in telecontrol communication sonar, some relative statistical quantities must be analyzed in detail.

4.2.1.1.2.1 PROBABILITY DISTRIBUTION OF THE AMPLITUDE OF THE DIRECT PULSES The probability density functions (PDFs) of the amplitude of the direct pulses at different distances are shown in Fig. 4.12, in which fluctuation rates are also indicated. The distances in Fig. 4.12A–D are 0.9, 1.8, 3.7, and 5.5 km, respectively, in which corresponding probability distribution curves are also drawn. PDFs of the amplitude of the direct pulses generally follow Rician distribution law, but they tend to obey Gaussian distribution law at shorter distances and Rayleigh distribution law at longer distances for which the fluctuation rate is up to 0.45. Empirical results are consistent with theoretical expectations.

4.2.1.1.2.2 FLUCTUATION IN REPETITIVE PERIODS The fluctuation in the repetitive periods reflects the time-variant property of underwater acoustic communication channels; this

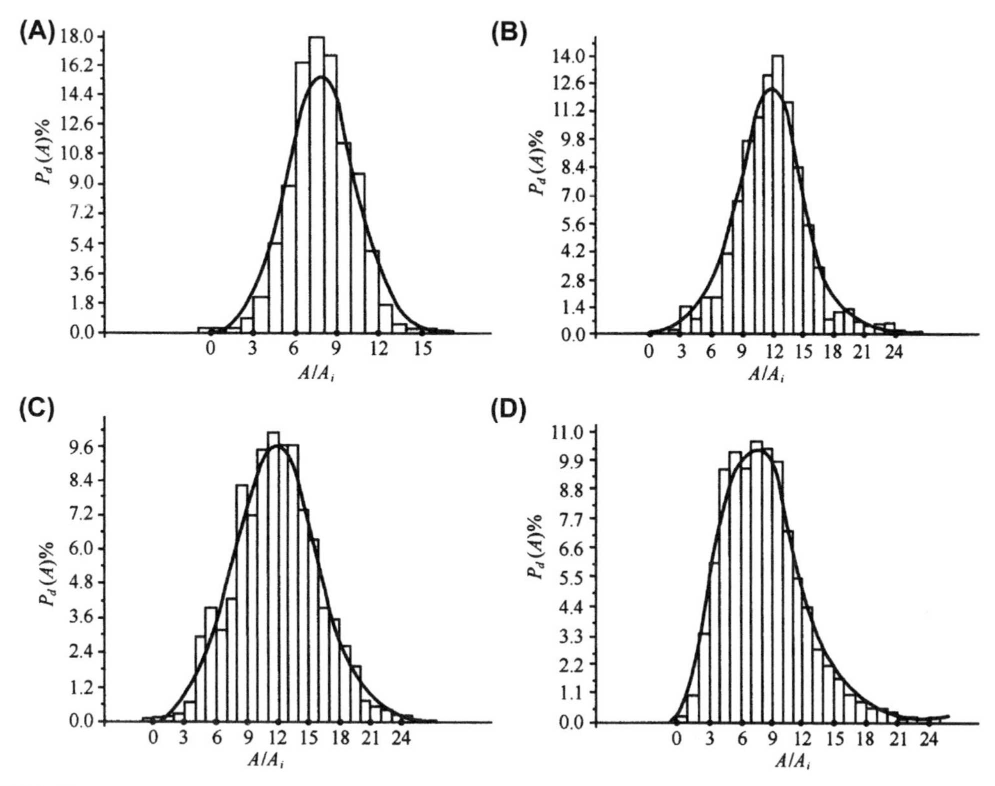

FIGURE 4.12 Probability density functions of the amplitude of the direct pulses. Fluctuating rates: (A) 0.171, (B) 0.156, (C) 0.296, and (D) 0.450.

property is an important parameter for designing robust digital underwater acoustic communication sonar.

The probability distributions of the repetitive period at different distances, which are the same ash those indicated in Fig. 4.12, are shown in Fig. 4.13. The repetitive period $T_0 = 10$ ms was taken in the experiments. It can be seen that most values are concentrated in the range of 10 ± 0.1 ms, although fluctuation deviations may increase with increasing distances.

Further experiments were performed in the sea outside the Jiao Zhou Bay where depths are about several tens of meters. The pulse duration $\tau_s = 8$ ms and $T_0 = 32$ ms were selected in the experiments. It had been proved that PDFs of the repetitive period obey the Gaussian distribution law as shown in Fig. 4.14A–B, in which the distances are at 1.8 and 5.5 km, respectively. We see that the fluctuation rates of T_0 range from 1.0% to 2.0% in these experimental conditions, and they will rise with increasing transmission distance r. So the fluctuation effect of T_0 would generally be considered for some relative signal-processing systems, in particular when R is higher (as in a developed image communication in which $R = 8$ kbp) and r is larger. This is a reason to adopt the total time-sampling processing for pseudo-random sequences instead of the conventional synchronical techniques.

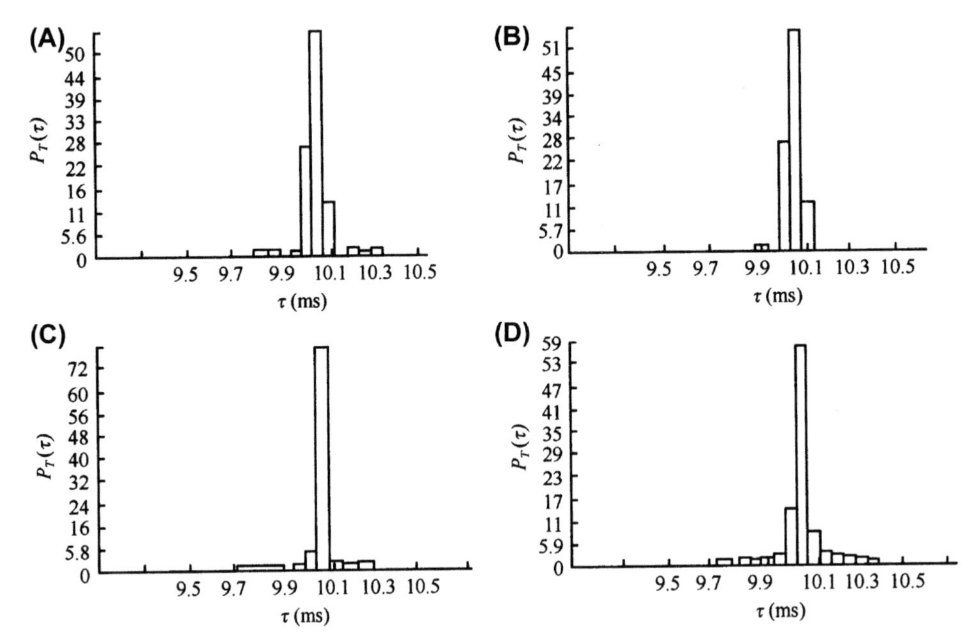

FIGURE 4.13 Probability distributions of repetitive period.

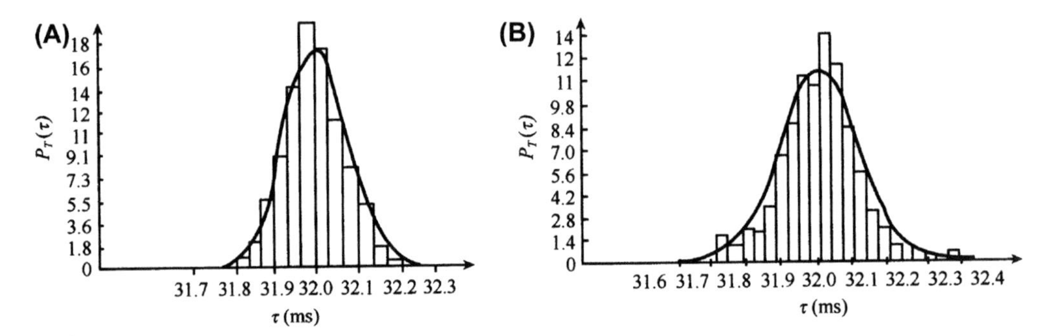

FIGURE 4.14 Probability density functions of repetitive period.

4.2.1.1.2.3 CORRELATIVITY OF DIRECT PULSES The experimental research on pulse-to-pulse correlativity was first performed. $T_0 = 50$ ms, that is, $R = 20$ bps, is selected, which is a main specification for the underwater acoustic telecontrol communication sonar. The pulse-to-pulse correlation coefficients at four different distances (as shown in Fig. 4.12) are given in Fig. 4.15 by A, B, C, and D. The pulse-to-pulse correlativity is strong enough, all correlation coefficients are larger than 0.85 under the experimental conditions. Of course, they will reduce with increasing r.

The experiments in relation to the correlation characteristics of pulse to pulse are further carried out at the sea area out of the Jiao Zhou

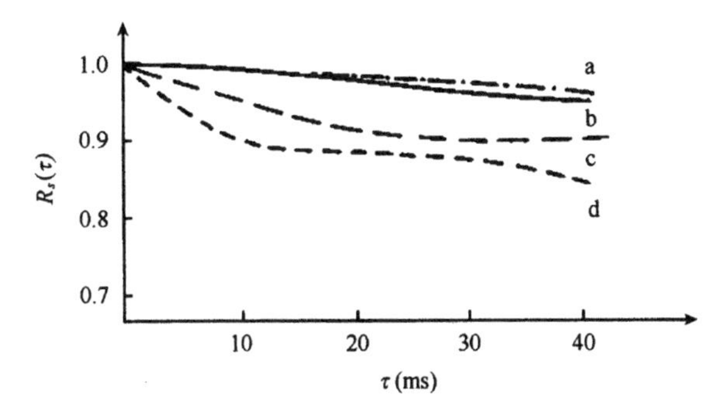

FIGURE 4.15 Pulse-to-pulse correlation coefficients.

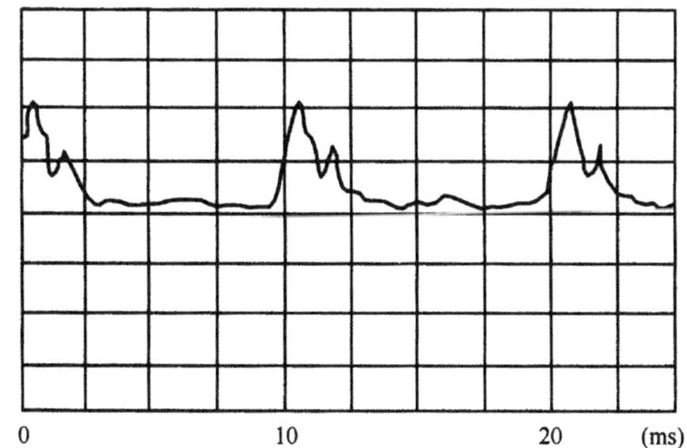

FIGURE 4.16 Received signal envelope.

Bay. $f = 20$ kHz, $\tau_s = 1$ ms, and $T_0 = 16$ ms are selected in the experiments. The correlation coefficients of pulse-to-pulse $R_d(\tau)$ for different τ at $r = 1.8$ km are listed in Table 4.1. The correlativity is strong enough, $R_d(\tau)$ are larger than 0.6 even if $\tau = 30$ s. Thus, if a scheme of pulse-to-pulse correlation detection is employed in underwater acoustic telecontrol communications, an excellent ability of antiinterference will be acquired.

Of course, the pulse-to-pulse correlativity is a randomly spatiotemporal–frequency variable, the relative data mentioned earlier were acquired under some specific conditions and are only for reference.

The correlativity for a copy correlator, in which a square wave is used as a reference signal, would be remarkably reduced even for a weak multipath underwater acoustic channel.

The received-signal envelope for a carrier-frequency pulse sequence transmitting in a shallow-water acoustic channel is shown in Fig. 4.16 [6], in which direct and multipath pulses are included. In this case, the operating result by means of a copy correlator is plotted in the lower part of Fig. 4.17, the correlation coefficient is only 0.256; that is to say, the remarkable decorrelation effect has been encountered; moreover, in which two side labels are present. If taking a received envelope as a reference signal, the pulse-to-pulse correlation coefficient will be remarkably raised as shown in the upper part of Fig. 4.17. We see that correlation coefficients are both equal to 0.996, and the side labels are absent for two operating results.

TABLE 4.1 Correlation Coefficients of Pulse to Pulse

τ(s)	2	4	6	8	10	15	20	25	30
$R_d(\tau)$	0.962	0.974	0.808	0.905	0.898	0.907	0.731	0.606	0.703

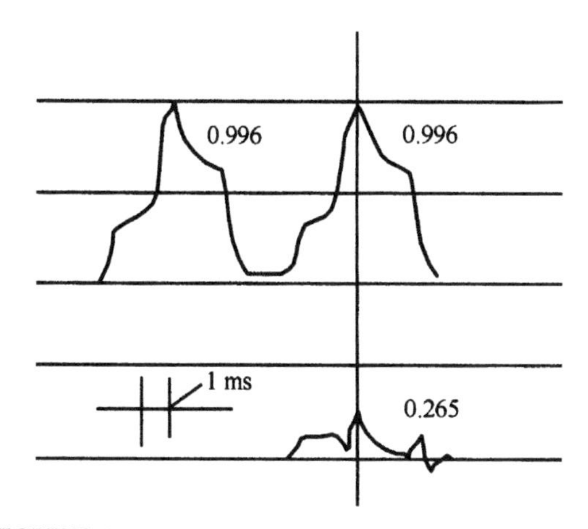

FIGURE 4.17 Correlation coefficients of both copy and pulse to pulse.

The multipath energy is also utilized in this operating scheme.

4.2.1.1.2.4 CHARACTERISTICS OF FREQUENCY EXCURSIONS The empirical research on the characteristics of frequency excursions Δf has been implemented in Xiamen Harbor.

The characteristics of Δf at a fixed distance (2 km) were first analyzed. PDFs of Δf at the carrier frequencies f of 16, 18.5, and 21 kHz are shown in Fig. 4.18A–C, respectively. We see that they follow the Gaussian distribution law and the variances for which will increase with rising frequency (refer to Fig. 4.19). PDFs of Δf at the different distances of 2, 4, 6, and 8 km are shown in Fig. 4.19A–D, respectively, and $f = 16$ kHz. They all obey the Gaussian distribution law, too. The maximum Δf is 25 Hz appearing in the experiments. Experiments proved that when the hydrophone is fixed on the sea bottom, PDFs of Δf are consistent with it being hung in the seawater. It means that the frequency excursions are mostly caused by turbulence in shallow-water areas with strong tidal current about 5 knots at empirical time interval.

4.2.1.2 Statistical Characteristics of Multipath Propagations in Shallow-Water Acoustic Channels

Studying the statistical characteristics of multipath propagation as well as direct-pulse signals, is necessary to improve the ability of antimultipath, or even to use multipath energy.

We have known that the laws of multipath propagation in shallow-water acoustic channels are complex and variable; that is, different conditions including different seasons, marine

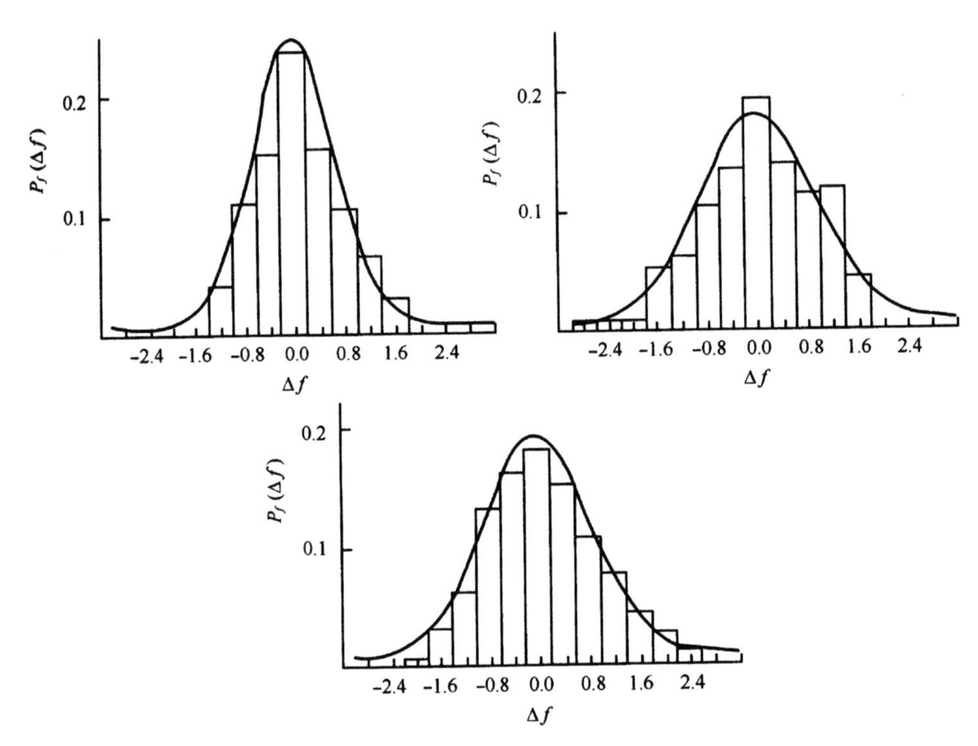

FIGURE 4.18 Probability density functions of frequency excursion at different frequencies.

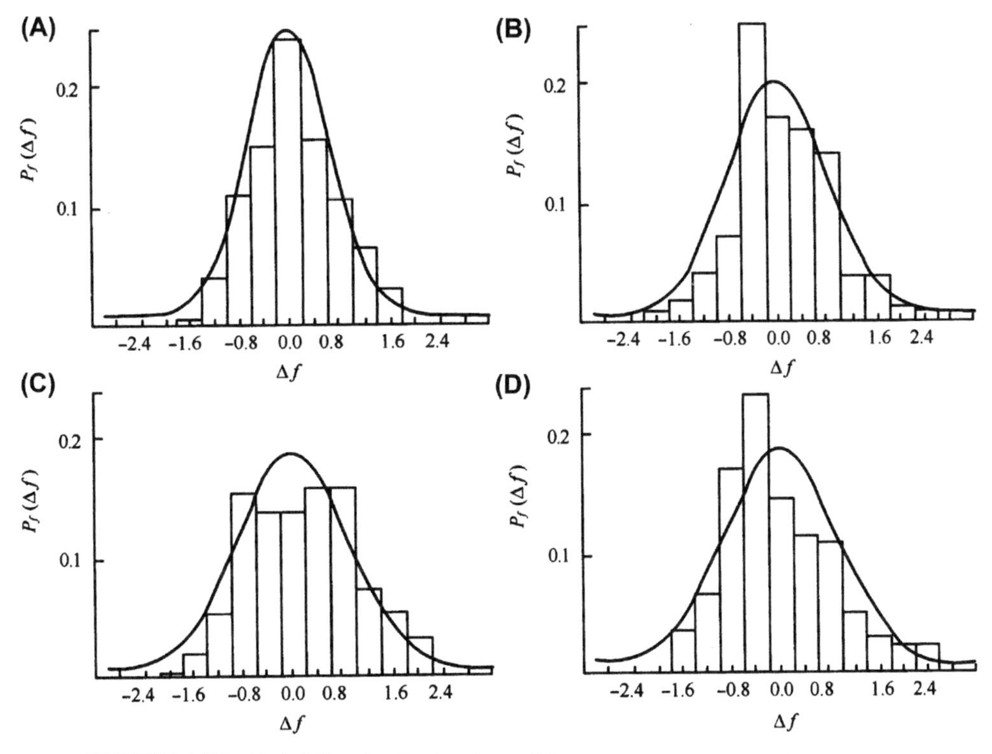

FIGURE 4.19 Probability density functions of frequency excursions at different ranges.

environments, acoustical parameters (such as carrier frequency), and transmission distances, etc., possibly have multipath structures with substantial differences. That is to say, they have severe random variability in spatiotemporal–frequency domains; therefore, theoretically analyzing those statistical characteristics is difficult. We carried out the experimental research with respect to some statistical characteristics of multipath propagations from the point of view of signal processing.

To analyze the characteristics of multipath propagation, a pulse sequence with shorter pulse duration is transmitted. Moreover, the hydrophone must be placed at a suitable depth, so that the multipath pulses (wave packets) can be separated with direct ones as shown in Fig. 4.10. We see that the same order multipath pulses have relative steady distributions, in particular in the case of low sea state.

4.2.1.2.1 PROBABILITY DENSITY FUNCTIONS OF THE AMPLITUDE OF MULTIPATH PULSES

PDFs of amplitude of the same-order multipath pulses at different distances (refer to Fig. 4.12) are given in Fig. 4.20, the fluctuating rates are also indicated in the figure. PDFs generally follow Rayleigh distribution law, because grazing angles are larger at shorter distances, thus the noncoherent-scattering effect becomes severe. Conversely, at longer distances, in which the Rayleigh parameter (see Section 2.4.2) becomes so small that the sea surface would be regarded as a smooth plane, PDFs still tend to follow Rayleigh distribution law due to the accumulation effect of sound fluctuations.

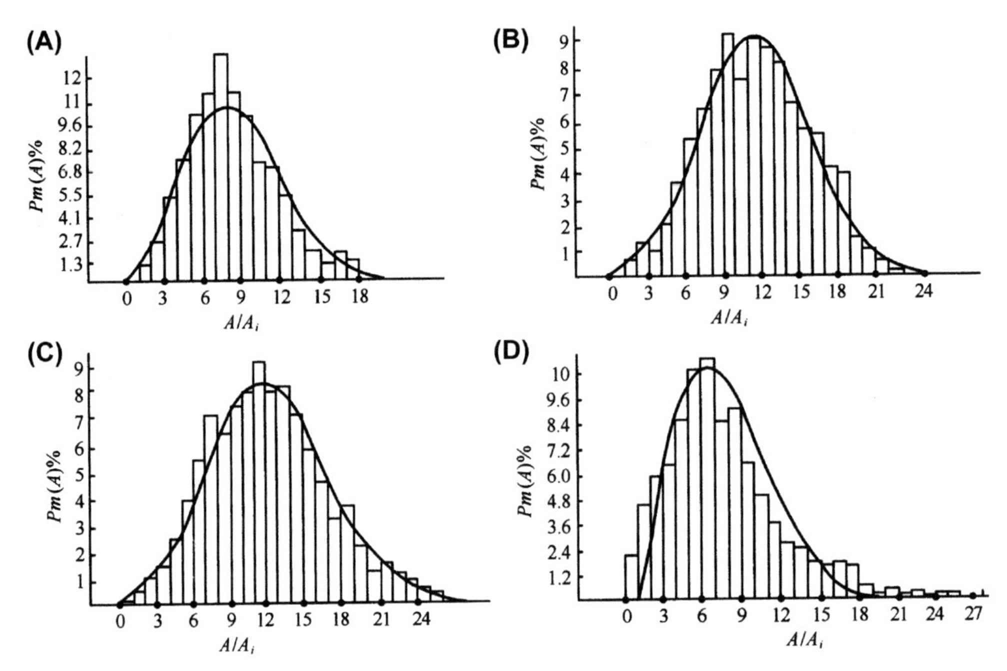

FIGURE 4.20 Probability density functions of multipath amplitude at different distances. Fluctuating rates: (A) 0.38, (B) 0.32, (C) 0.39, and (D) 0.58.

4.2.1.2.2 PROBABILITY DISTRIBUTION OF REPETITION PERIODS OF MULTIPATH PULSES

The probability distributions of repetition period of the same order multipath pulses T_m at different distances (refer to Fig. 4.12) are shown in Fig. 4.21. We see that T_m, as the repetitive period of direct pulsesT_0, has similar fluctuation characteristics. Making use of the multipath energy is operative, provided the digital time correlation accumulation decision system is employed in digital underwater acoustic communications.

4.2.1.2.3 THE CORRELATIVITY OF THE ENVELOPE OF THE SAME MULTIPATH PULSES

The typical correlation coefficients of the envelope of the same-order multipath pulses (wave packets) are shown in Fig. 4.22. The solid, dashed, and dotted curves in this figure represent the transmission distances at 0.55, 1.8, and 5.5 km, respectively. We see that the correlativity of the envelope of the same-order multipath pulses, like direct-signal pulses, is strong, the correlative coefficient is still about 0.9 at the processing time of 40 ms under the condition of a smoother sea surface. Of course, the correlative coefficients would reduce and disperse with further increasing transmitting distance, as well as increase in the roughness of the sea surface.

Because the correlativity of the same-order multipath pulses is strong, it would be expected that the multipath pulses, as well as direct-signal pulses, may be detected by means of the digital time correlation accumulation decision system; that is to say, the multipath energy is used by the system.

4.2.1.2.4 STATISTICAL CHARACTERISTICS OF DIRECT PULSES ADDED BY MULTIPATH PULSES

The direct and multipath pulses transmitting in a shallow-water acoustic channel cannot

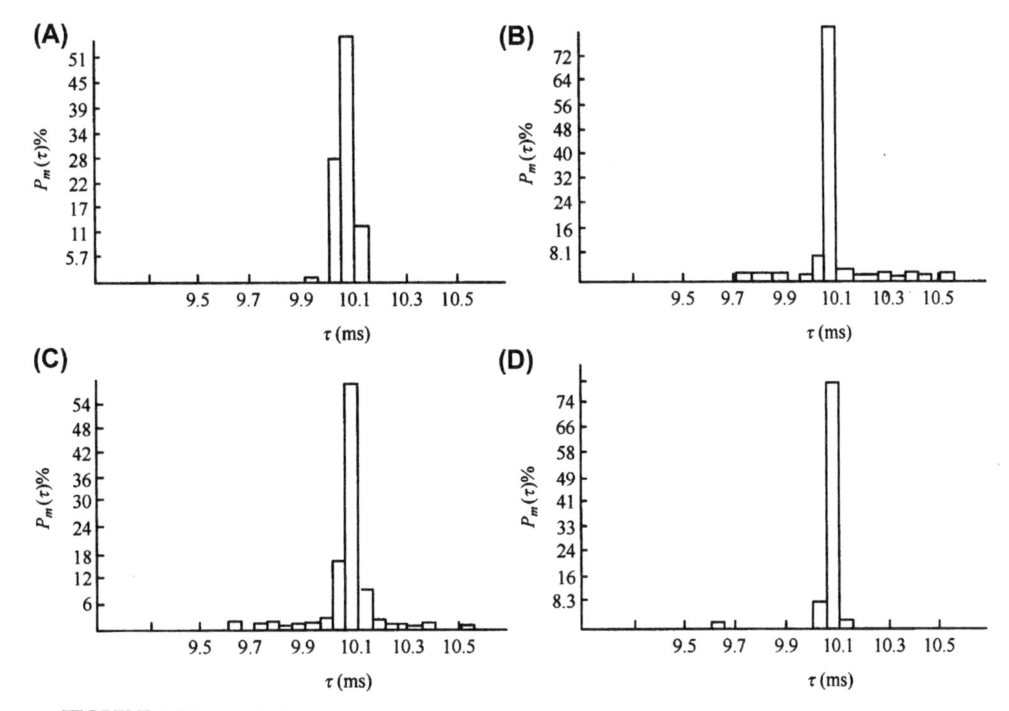

FIGURE 4.21 Probability distributions of repetition period of the same-order multipath pulses.

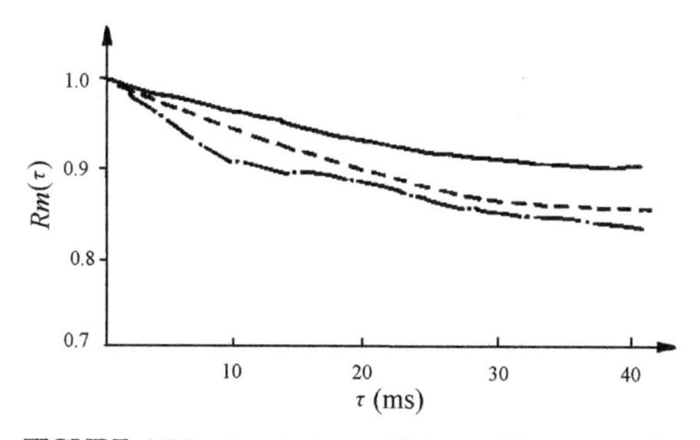

FIGURE 4.22 Correlative coefficients of the same-order multipath pulse.

generally be separated from each other when signal-pulse duration τ_s is wider, as 6.4 ms used in the underwater acoustic releaser developed. The statistical characteristics of the resulting signals will appear somewhat differently in comparison with those of direct or multipath pulses alone.

The probability distribution of the amplitude of direct pulses added to multipath ones is shown in Fig. 4.23. Double peaks are depicted in the figure, which mean that some low-frequency components are present.

The probability distribution of repetition period of direct pulses added to multipath ones is shown in Fig. 4.24 that disperses in a larger range than that of the direct or multipath pulses alone.

The correlative coefficients of the envelope of the direct pulses added to multipath pulses are shown by curve 3 in Fig. 4.25; transmission distance is 3.7 km. The curves 1 and 2 in this figure are those of direct and multipath pulses alone, respectively, under the same propagating conditions. We see that the correlative coefficients of the direct pulses added to multipath ones are less than those of direct pulses or multipath pulses alone.

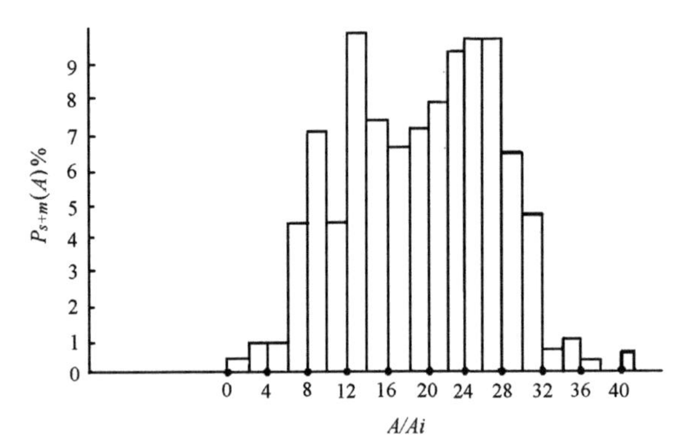

FIGURE 4.23 Probability distribution of amplitude of direct pulses added to multipath pulses.

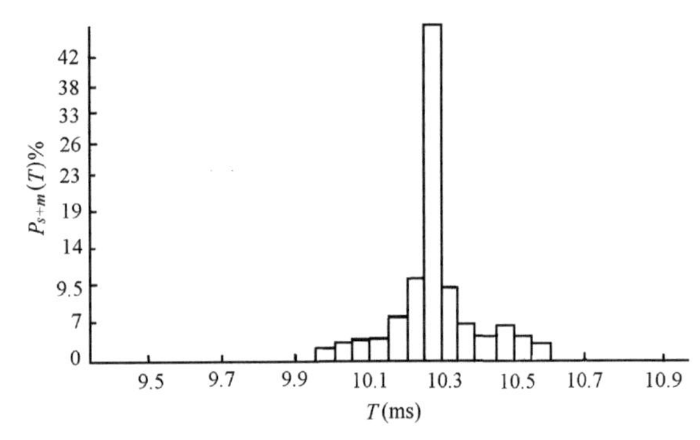

FIGURE 4.24 Probability distribution of repetition period of direct pulses added to multipath pulses.

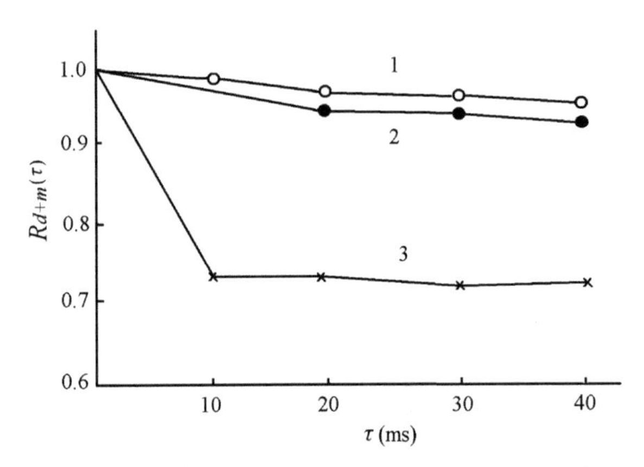

FIGURE 4.25 Correlative coefficients of direct pulses added to multipath pulses.

We should note that the empirical research on the statistical characteristics of sound propagation in shallow-water acoustic channels mentioned earlier follow the basic requirements for designing the releaser, in particular f (20 kHz), are higher and the correlation processing times (less than 50 ms) are shorter; the relative empirical data provided earlier are only for reference.

4.2.2 Development of the Underwater Acoustic Telecontrol Instruction Communication Employed in an Underwater Acoustic Releaser

We have analyzed the characteristics of sound propagation in shallow-water acoustic channels as a physical basis for designing the releaser; by combining its specific requirements, a suitable signal-processing system will be obtained.

4.2.2.1 Basic Principles Designing a Suitable Signal-processing System Employed in the Releaser

1. Because the pulse-to-pulse correlativity is much stronger than that of the noise, if a pulse-to-pulse correlation detection scheme is employed in underwater acoustic telecontrol instruction communication sonar, excellent antinoise ability would be expected.
2. The telecontrol key in a releaser is a single operation. In this case, to make $P(f) \to 0$ and $P_1(d)$ to have an expected value, the accumulation decision scheme may be adopted after correlation processing; that is to say, by increasing signal-processing times, the performance of the telecontrol communication may be improved. Moreover, the total pulse numbers of the accumulation decision and the pulse numbers of the accumulation decision threshold can be flexibly adjusted to adapt to variable-parameter underwater acoustic channels.

3. Because the required distances of telecontrol instruction communication are nearer, the operating frequency may be selected higher, say 20 kHz, thus the total delay spreads of multipath effects are shorter than 40 ms, which had been demonstrated by relevant experimental data. Therefore, once T_0 is chosen to be greater than 40 ms, intersymbol interference caused by multipath effects will disappear. Moreover, according to the strong correlativity among the same-order multipath pulses, particularly those repetition periods which are approximately equal to those of the direct-signal ones, the multipath pulses can be converted into signal pulses to be processed in the correlation accumulation decision system (refer to Fig. 4.30), and $P_1(d)$ will be greatly improved.
4. In situ experimental records have proved that some noise pulses with narrow durations but with high amplitudes are present in shallow-acoustic channels, as shown in Fig. 4.26. To combat them, PDFs of pulse duration for both the background noise in the sea and radiated noise of ships at the center frequency $f = 20$ kHz followed by a one-third octave filter have been analyzed. The results are shown in Figs. 4.27 and 4.28, respectively. We see that PDFs of both background noise and radiated noise follow Rayleigh distribution law, the fluctuation ratios for which are about 0.5 and average durations are about 0.5 ms. The duration of transmitting pulses is 6.4 ms in telecontrol instruction communication sonar, which is much larger than that of the former. Therefore, adopting a pulse-duration discriminator to combat the noise pulses with duration narrower than 6.4 ms (that is, "anti-narrow" processing) will be simple and effective. It should be noted that "anti-wide" processing to combat the durations of noise pulses that are larger than 6.4 ms is generally not suitable to use, because the signal pulses may be added to multipath and noise pulses; added pulse durations may be larger than 6.4 ms.

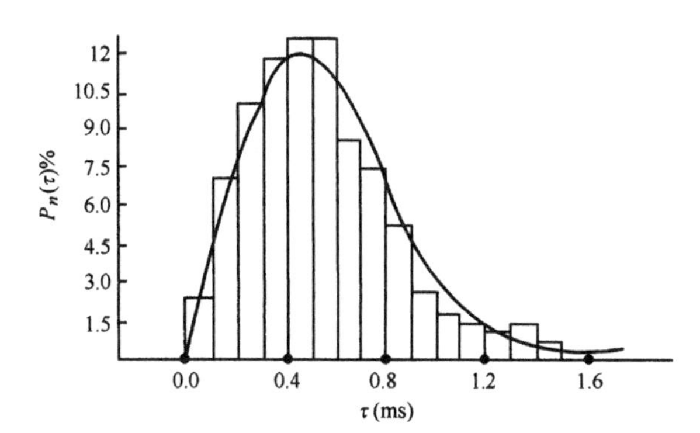

FIGURE 4.27 Probability density functions of the pulse duration of background noise.

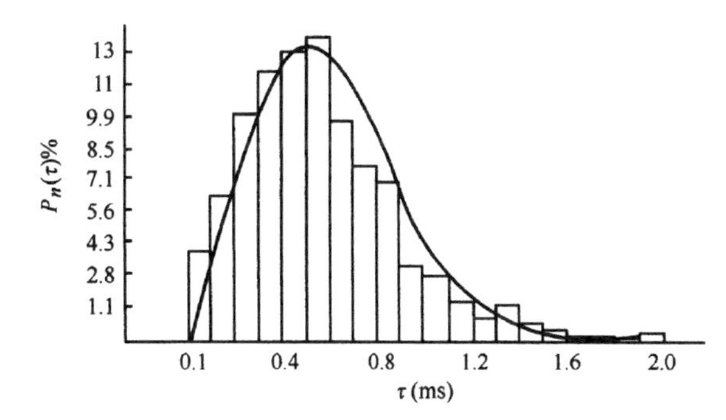

FIGURE 4.28 Probability density functions of the pulse duration of the radiated noise of a ship.

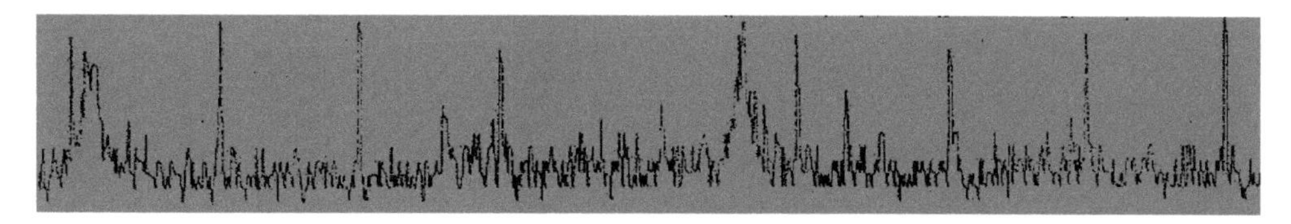

FIGURE 4.26 Record of telecontrol communication signals.

5. The fluctuation in amplitude of the sound signal pulses propagating in shallow-water acoustic channels is violent, the fluctuation ratios for which may be up to 0.5. Thus, PDFs will follow the Rayleigh distribution law at longer communication distances; moreover, the input SNR is lower at longer distances, so the $P_1(d)$ would thus decrease due to the fading of signal pulses. So the pulse numbers of accumulation decision threshold must be selected to remain some rooms to adapt to the signal fading. This is an effective approach to combat against low data-rate underwater acoustic communications using the time cross-correlative accumulation decision (TCAT) system.
6. The fluctuation of the repetition period of signal pulses generally follows Gaussian distribution law, which is considered in the operations of conformity in the correlation detection; however, the frequency excursions are much fewer in comparison with the operating frequency (20 kHz) and thus may be disregarded.

4.2.2.2 *Basic Principle and Structures*

Both size and power consumption for a releaser are limited. Moreover, the simplification of structures is commonly an effective approach to improve its robustness. They include two aspects: (1) The transmitted pulse sequence with the same f, τ_s, and T_0 will be selected in the telecontrol instruction communication, so that both transmitter and receiver can be greatly simplified. Moreover, the recovered instruction signal in situ is a pulse sequence in a continuously operating state, which is equivalent to time-diversity results and thus make $P(d) \to 1$. (2) Adopting one-bit quantization format, the pulse envelopes passing through a detector will be changed into spare waves with corresponding durations, and then accumulated and decided upon after the operation of polar registration correlation. Of course, the performance of antinoise will be reduced as the amplitude and phase messages of signal are lost by using the frame, but the structure of the correlator will be much simplified. We may utilize some other approaches to make $P(f) \to 0$, such as increasing the pulse numbers of the accumulation decision threshold.

4.2.2.2.1 DIGITAL TIME AUTOCORRELATION ACCUMULATION DECISION

The schematic block diagram of a digital time autocorrelative accumulation decision (TACAD) system telecontrol instruction communication receiver is shown in Fig. 4.29. When an instruction pulse sequence arrives at the receiver by passing through the peculiar underwater acoustic communication channels that will be changed into a random multipath structure in which direct signal and multipath pulses, as well as noise pulses, are involved, this constitutes the input signal of the receiver. Following this by amplification, filtering, detection, and analog-to-digital conversion (A/D) by means of one-

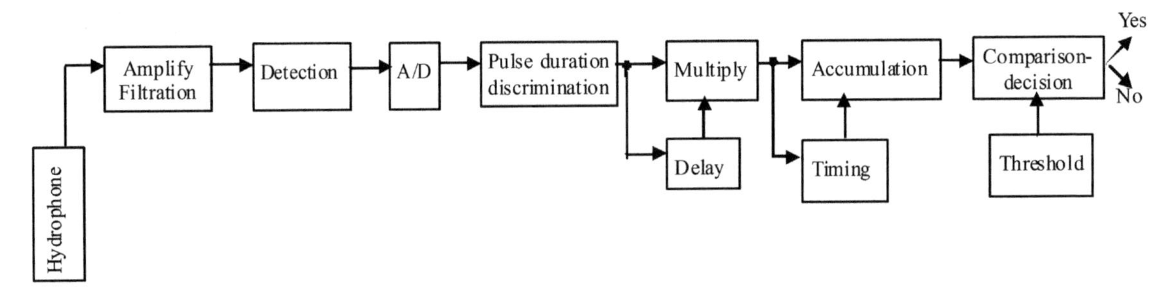

FIGURE 4.29 Schematic block diagram of the time autocorrelative accumulation decision (TACAD) receiver.

bit quantizing format, will form corresponding square waves if those amplitudes are greater than the preset-amplitude threshold. A pulse-duration discriminator is used to reject the noise pulses with the durations smaller than those of signal pulses and to limit the maximal numbers of noise pulses to be fewer than T_a/τ_s in one accumulation decision period T_a. The pulse-delay network for the square-wave sequence with the same T_0 can generally be formed by means of a monostable circuit; but consider the noise interference and to make use of the multipath energy; this delay circuit consists of an adaptive pulse distributor.

Of course, if the noise interference is neglected and using multipath energy is disregarded, the delay time will be a single T_0; such a signal-processing system principally belongs to the pulse-to-pulse correlation accumulation decision. However, experiments in situ have demonstrated that some noise pulses would conceal the pulse-to-pulse correlation detection, particularly in the case of lower-input SNR. Obviously, the multipath pulses have a similar effect on that. To overcome these problems, the adaptive pulse distributor is used; its operating model is similar to that of autocorrelation at a certain delay time T_0. Therefore, this signal-processing system is still called the TACAD system.

Logic multiplication is equivalent to a conformity circuit in a binary digital circuit. Total delay time $t = MT_0$, in which M is the total preset signal-pulse numbers of an accumulation decision. Let the pulse numbers of the accumulation decision threshold be M_0, thus even if $(M - M_0)$ signal pulses are lost due to the fading effect, the output of the accumulation decision is still "yes" in a statistical conception. We see that M and M_0 play important roles to determine $P_1(d)$ and $P(f)$, so they must be selected carefully according to the statistical characteristics of a certain channel. Generally speaking, $(M - M_0)$ would remain sufficient to satisfy the requirements for both $P_1(d)$ and $P(f)$. $M = 50$ and $M_0 = 45$ may be selected for complex and variable channels. Because $T_0 = 50$ ms, the accumulation decision time T_a is equal to 2.5 s.

It should particularly be noted that the TACAD system has excellent performance in making use of multipath energy [5].

A field record of a pulse sequence employed in the underwater acoustic telecontrol communication is shown in Fig. 4.30A, in which, besides the direct signal pulses, the first two orders of multipath pulses also appear. The horizontal dashed line plotted in the figure represents the level of the amplitude threshold. The direct signal pulses and the multipath pulses of the first order have amplitudes exceeding the threshold, they will be formed to corresponding square waves by following circuits. On the contrary, the multipath pulses of the second order have lower amplitudes than the level that will be eliminated, as shown in Fig. 4.30B. Because the

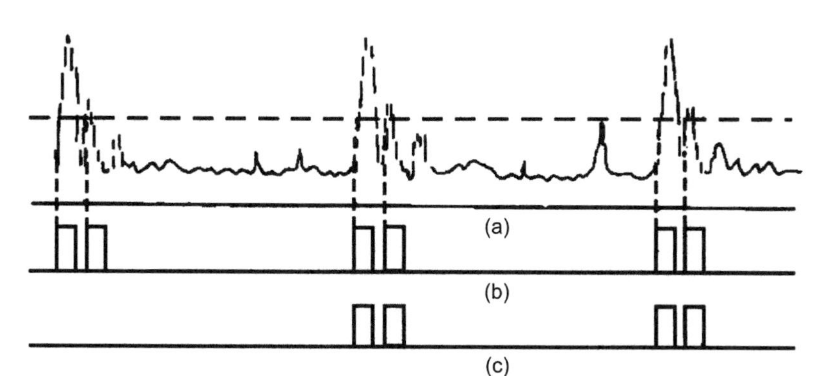

FIGURE 4.30 Schematic diagram for multipath energy to be used by the time autocorrelative accumulation decision (TACAD) system.

repetition periods of the same-order multipath pulses are approximately equal to T_0 as mentioned earlier, the output from the pulse-delay network is shown in Fig. 4.30C. Then the multipath pulses will be processed as direct signal pulses by the TACAD system, which is equivalent to a time diversity transmitting two pulses for the same information, in which the pulse numbers of accumulation would be two times greater in comparison with direct-signal pulses alone. $P_1(d)$ will much be improved.

Similarly, in the prerequisite of $P(f) \rightarrow 0$, whenever the two noise pulses occur with a space approximately equal to T_0, the accumulation numbers will increase one digit in the TACAD system. It means that "noise energy" would also be used in the system.

The theoretical analyses of performances and corresponding experimental results for the TACAD system will be introduced as follows.

We can see that (refer to Fig. 4.29) four thresholds: amplitude, pulse duration, repetition period, and accumulation decision are preset in the TACAD system receiver. The two pulses formed by the direct signal, multipath, or noise will form an output in multiple circuits as long as they have passed through the four thresholds.

In the case of telecontrol instruction pulses, the sound source level (*SL*) may be selected to be higher or the release distances nearer so that the receiver may operate at a state of high-input SNR, and thus the pulses have a high probability of passing through the amplitude threshold. The duration threshold is designed by referring to the duration of signal pulse τ_s; moreover, taking the delay time $T = T_0 \pm \Delta T$, in which ΔT is the maximum fluctuation quantity of the repetition period, can be predicted by examining the statistical characteristic of T_0, to realize the reliable conformity operation in the multiplier. Therefore, the probability of passing through the amplitude, pulse duration, and repetition-period thresholds would all have high values. Thus, the detection probability $P_1(d)$ of the TACAD system will be mostly determined by the threshold of accumulation decision that follows the binomial distribution law

$$P_1(d) = \sum_{J=M_0}^{M} C_M^J P^J (1-P)^{M-J}$$

in which P is the probability of adjacent signal pulses to form an output from the multiplier, which is higher as mentioned earlier.

In the cases of SNR $= 1$, $M = 50$, and $M_0 = 45$, the corresponding $P(d)$ are listed in Table 4.2.

The estimate of $P_1(d)$ mentioned previously does not consider the multipath pulses that will possibly be converted into signal pulses, thus actual $P_1(d)$ are greater than those listed in Table 4.2.

Now let us estimate $P(f)$ of the TACAD system.

Similarly, $P(f)$ follows the binomial distribution law

$$P(f) = \sum_{j=M_0}^{m} C_m^j p^j (1-p)^{m-j}$$

in which m is the total number of noise pulses arriving at the accumulation decision circuit for the decision time of 2.5 s, p is the probability of noise pulses forming an output from the multiplier.

Perhaps the noise pulses have a higher probability of passing through the amplitude threshold, that is, the receiver operates at a low-input SNR; but they would first be partially eliminated by the threshold of pulse duration, because those durations would generally be less (refer to Figs. 4.27 and 4.28) than preset pulse duration, say $(2/3)\tau_s$. In particular, the distributions of noise pulses are random in the time

TABLE 4.2 $P_1(d)$ of the TACAD System Employed in the Releaser

P	0.95	0.90	0.85	0.80
$P_1(d)$	0.95	0.60	0.21	0.05

domain; 45 pulses appearing in the decision time of 2.5 s, each with a space approximately equal T_0 (50 ms) are almost impossible, that is to say, P is much too small, and thus $P(f) \rightarrow 0$ under the conditions mentioned earlier.

The theoretical estimations had indicated that, provided the relative parameters, including SL and r, have been chosen in detail, the expected $P(f) \rightarrow 0$ and $P_1(d) \rightarrow 0.3$ would be realized, and then make $P(d) \rightarrow 1$ by means of the time-diversity technique. In fact, the telecontrol instruction pulses will be continuously transmitted until the releaser is recovered.

The performance of the releaser developed in which the TACAD is used as a telecontrol communication signal-processing system has been examined in situ.

4.2.2.2.1.1 ABILITY OF ANTINOISE The releaser was placed under the water near the shipping lane in Xiamen Harbor for 20 days to examine its ability of antinoise. It operated stably for the interval and was normally recovered at the distance of 3.7 km.

To examine its performance to combat wind noise due to great wind speeds, the releaser was locked in a shallow-water area in Xiamen Harbor for a full typhoon process, in which force 13 wind (on the Beaufort scale) appeared. The releaser still operated normally and was recovered after the typhoon had passed at an expected distance.

4.2.2.2.1.2 OPERATING STATE AT A DEPTH OF 200 M The releaser had been locked in a lake where the depth is 200 m for 30 days. The expected results were obtained.

4.2.2.2.1.3 EXPERIMENTS IN SHALLOW-WATER AREAS IN THE SOUTH CHINA SEA The experiments had been done at four different positions in the sea areas where the depths are 29, 44, 75, and 105 m, respectively. The releaser was locked on the sea bottom. The water temperature was approximately homogeneous in the vertical direction, but the sea state of 4 to 5 had been encountered.

The experimental results had indicated that the release distances at the depths of 29, 44, and 75 m are above 5 km; whereas that is up to 7.5 km at the depth of 105 m. The release distances are much larger than those in shallow-water areas in Xiamen Harbor. The releaser normally floated to the sea surface for the intervals of 8–30 s at the different depths after the telecontrol instruction signal pulses had been transmitted.

Based on the theoretical analyses and experimental results, the developed releaser had satisfied expected specifications. The TACAD system telecontrol communication sonar has excellent performance, including simple construction, small size, and low energy consumption, which are suitable to employ at low data-rate underwater acoustic communications; in particular, it is suitable to be placed under the sea for a long time.

4.2.2.2.2 DIGITAL TIME CROSS-CORRELATION ACCUMULATION DECISION SYSTEM

The TACAD system and its applications to underwater acoustic telecontrol communication have been described earlier. The TACAD system is proven useful in an underwater acoustic channel with severe spatiotemporal variant parameters, violent multipath effects, and high noise levels. Obviously, a digital time cross-correlation accumulation decision (TCCAD) system can also be employed in the channel, and excellent results could be obtained. Moreover, the ability of antinoise may be higher than that of the TACAD system.

The schematic block diagram of a simplified TCCAD system receiver is shown in Fig. 4.31. A pulse sequence with the same f, T_0, and τ_s as a telecontrol instruction signal is received in the hydrophone; and then the one-bit quantizing format in A/D is adopted, as well as with the TACAD system. The chief differences between them are that the TCCAD system needs a

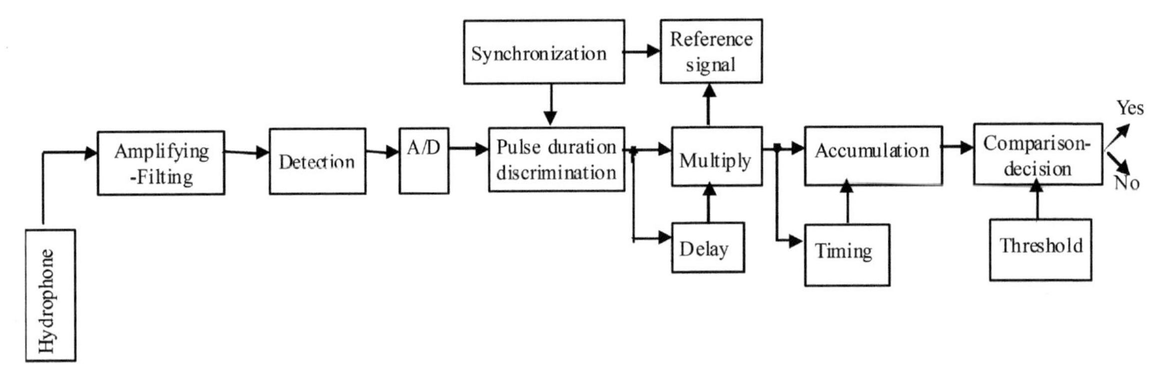

FIGURE 4.31 Schematic block diagram for a time cross-correlation accumulation decision (TCCAD) system receiver.

reference signal to perform cross-correlation operations; this signal may be composed of a preset square-wave sequence or introduced directly from the transmitter. Moreover, to realize the conformity operation in the cross-correlator, a synchronizing signal is required, which may be acquired from the first several pulses of the transmitting pulse sequence, and then terminal outputs will be determined by means of a comparison–decision circuit. Therefore, if the relative parameters, in particular M and M_0, have availably been selected, $P(f) \to 0$ and $P_1(d)$ with an expected value would be obtained, as well as with the TACAD system.

4.2.2.2.3 EXTENSIONS

We have analyzed simple the performances of the digital autocorrelation and cross-correlation accumulation decision signal-processing systems. It is worthwhile to point out that the data rates of the systems are spent for obtaining such high indexes as $P(f) \to 0$ and $P(d) \to 0$. However, the applied underwater acoustic telemetry and multiaddress telecontrol communication equipment, such as multiplex transponders, multiplex releasers, and oceanographic factor-integrated telemeters, require higher data rates. Because the relative parameters of the digital TCAD system, such as f, τ_s, T_0, M, and M_0, may easily be selected, their data rates may be raised, especially in the cases of nearer-range communications. On the other hand, multibit quantizing schemes may be used to improve those antinoise performances.

Some telecontrol and telemetry underwater acoustic communications based on the TCAD system by combining with different signal-modulation schemes will briefly be introduced as follows.

A multiple frequency-shift keying (MFSK)-TCCAD system [7] will first be introduced.

As mentioned previously, MFSK modulation is one of the more appropriate schemes employed in underwater acoustic communications.

We adopted the merits of both MFSK and TCCAD and developed a new MFSK-TCCAD signal-processing system employed in low data-rate communications. Theoretical estimations and experimental results in situ have showed its excellent performance, in particular rejecting ISI and adapting to the violent signal fluctuations, therefore a high $P_1(d)$ and very low $P(f)$ under the condition of a low SNR can be obtained.

To examine the performance of the MFSK-TCCAD system, only six frequencies are taken in our prototype; the first frequency acts as a synchronizing pulse, the others act as coded pulses.

A phase-locked loop (PLL) scheme is employed to demodulate the MFSK signal, and then transferred to a microprocessor. The operations of digital time cross-correlation accumulation decision and decoding are accomplished in that. To reject ISI caused by multipath effects, each frequency f_i is transmitted only once in each block of coded pulses. Once a pulse with the frequency f_i is received, this frequency will automatically be rejected in the duration of one block of the pulses. In addition, to adapt to signal fluctuations, the appropriate M and M_0 can be selected, so that satisfactory $P(d)$ and very low $P(f)$ could be attained by means of the MFSK-TCCAD signal-processing system.

The field tests were conducted in Xiamen Harbor. Communication range is 4.7 km. Entire experimental time lasted approximately 1 h; no error was detected, that is, $P(f) \to 0$.

$P_1(d)$ of theoretical estimations at input SNR was 8 dB and acquired from experiments in situ with SNR $\cong$ 8 dB as listed in Tables 4.3 and 4.4, respectively. We may use a time-diversity scheme for lower data-rate underwater acoustic communications to make $P_1(d) \to 0$ as mentioned earlier.

PDCM (pulse-wide encode modulation)-TACAD signal-processing system [8] employed in underwater acoustic multiaddress instruction telecontrol and lower data-rate telemetry has been developed, in which different pulse durations and corresponding coding express one-bit or multibit information. However, each pulse duration will be repetitively transmitted M times and then an accumulation decision is made after autocorrelation operations. Fine performance have proved system by in situ experiments.

TABLE 4.3 Estimated $P(d)$ at Different M_0

M_0	5	4	3	2
$P_1(d)$	0.34	0.87	0.94	0.95

TABLE 4.4 Measured $P_1(d)$ at Different M_0

M_0	5	4	3	2
$P_1(d)$	0.83	0.98	0.98	1.0

A PPM (pulse-position modulation)-TCCAD signal-processing system [9] has also been explored, in which the different pulse positions and carrier frequencies express one-bit or multibit information, but they are repetitively transmitted M times and then correlation operation and accumulation decision processes are performed. This system has better performance than the usual PPM system, if it is employed in lower data-rate underwater acoustic communications.

The TCAD signal-processing systems can be extended to be employed in some active sonars to improve their performance of both $P(d)$ and $P(f)$.

Active sonar using several TCCAD signal-processing elements to be organically combined may be applied to some complicated fields, such as to shallow-water target telemetry sonar, which may be employed in exploring new shallow-water routes and so on. To design active telemetry sonar operated in a sea area several meters deep, the essential difficulty is how to resist the multipath interferences caused by sound reflection and scattering from both the sea surface and the sea bottom in target echoes. Solving the difficult problem is possible with a spatiotemporal division scheme supported by a specific underwater acoustic transducer operating via a particular model, and if the TCCAD signal-processing elements are properly combined.

A schematic block diagram of the target telemetry sonar based on the TCCAD system is shown in Fig. 4.32. It organically consists of three TCCAD elements. The first transmitter emits a high carrier-frequency pulse train to one side of the ship. If the echoes reflected from a target at this side are decided as "yes," an output is formed from the first TCCAD element, and then the second transmitter is triggered by that and emits a pulse train to another side of the ship; the echo signals are decided by the second TCCAD element. The width of the shallow-water route is therefore acquired by the third

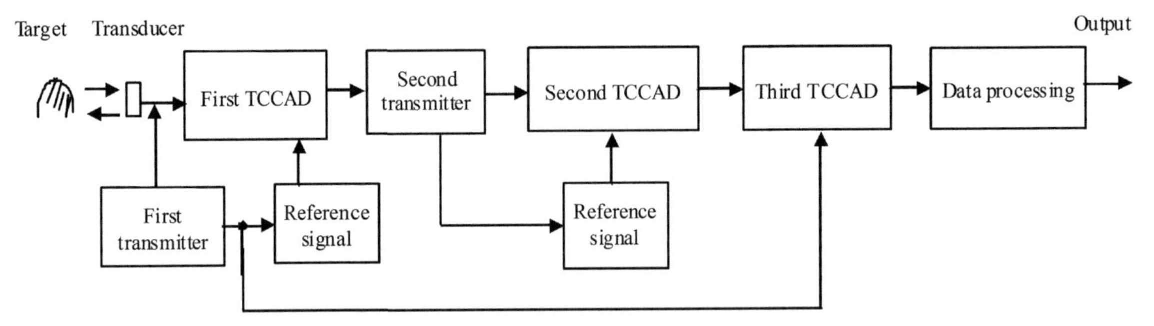

FIGURE 4.32 Schematic block diagram of a target telemetry sonar.

TCCAD element. The telemetry sonar is operated in sailing state, so that noise levels may be quite high. To improve the ability of antiinterference, the scheme of accumulation decision was thus used.

Some active sonars, such as depth sounder, only need to know the message of ranges; but others still need to acquire the echo intensity, waveforms, and so on for the reference to identify target properties, such as the abundance of schools of fish in a fish-finding sonar.

The schematic block diagram for an exploring prototype with double functions of both sounding the depth and finding the schools of fish by means of the TCCAD system is shown in Fig. 4.33 [10].

Considering that large dynamic ranges exist in the amplitude domain for an echo signal and background noise, an adaptive amplitude threshold may be adopted in the prototype. The first pulse of a transmission pulse sequence is taken as a standard to set the adaptive threshold of amplitude and the time of signal arrival; moreover, that is also introduced to a register and used as an initial reference signal. By passing through the TCCAD element, the digital messages of both depth and the abundance of the schools of fish would be extracted by means of following fish–bottom identification network based on the differences between their echo signals in intensity, waveform structure, and so on.

We used the prototype to conduct an experiment in Xiamen Harbor, in which two plastic balls with different sizes were hung at different depths as simulated targets. It had been proved

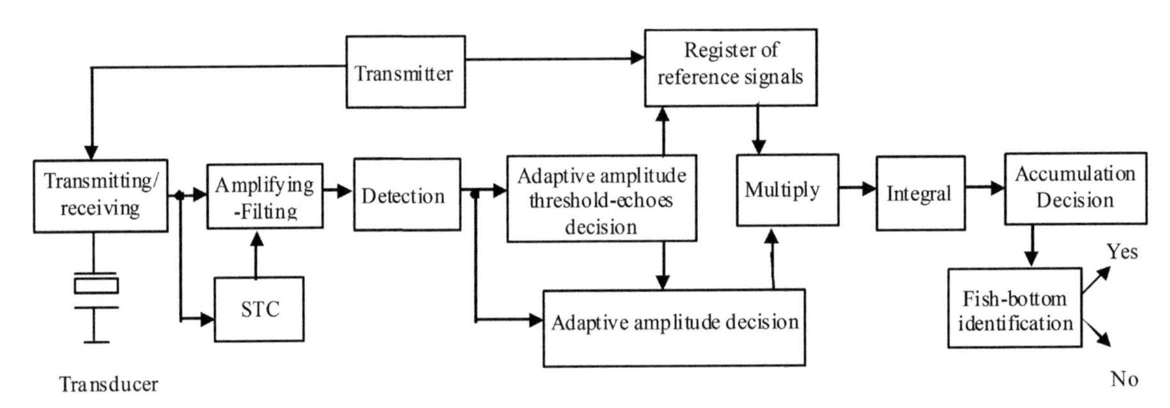

FIGURE 4.33 Schematic block diagram of a double-function prototype.

that the echoes from the targets can be eliminated, and the probability to detect the depth message is equal to 1. By changing programs to suppress bottom echoes, the prototype can correctly provide the relative digital messages of the targets, as the depths and echo intensity. When the ship sailed at the speeds of 2 and 8 knots, the digital messages of depths were steadily acquired, which proved that the prototype has a high ability of antinoise.

An experiment had been done in comparison with the ability of the antiinterference between the prototype and a usual fish finder against the same background noise. The experimental results are shown in Fig. 4.34. The record in the right-hand side of this figure was acquired by the fish finder; many obvious interference points appear. The left one shows the experimental results for the prototype in which almost all interference points were suppressed. The first echo and the secondary echo from the sea bottom were also recorded as shown. Because the $P(d)$ of the first echo is higher than 0.9 in such severe background noise, it is consistent with the theoretical estimation.

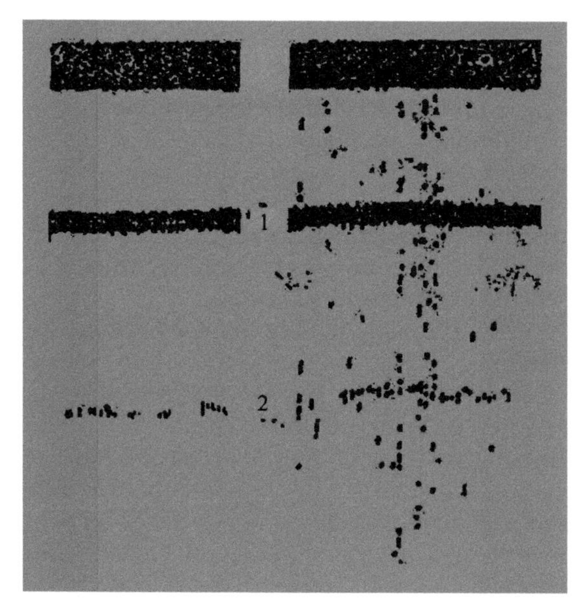

FIGURE 4.34 Experimental records of the contrast of antinoise.

4.3 DEVELOPMENT OF ADVANCED FH-SS SYSTEM DIGITAL MULTIMEDIA UNDERWATER ACOUSTIC COMMUNICATION EQUIPMENT

4.3.1 Expected Developing Target

The expected target is to develop an underwater acoustic data communication prototype having moderate communication ranges and moderate data rates. It can be extended to digital multimedia underwater acoustic communications and established with compatible multimedia communication equipment. Alternatively, we can select one or two media to form corresponding systems to adapt to the requirements of exploring and using the natural resources of the ocean at present and in the foreseeable future.

Main specifications of the experimental prototype are as follows. They have been approximately predicted by the communication sonar equation mentioned in Section 1.4.

1. sound source level SL: 198 dB;
2. center frequency: 15 kHz;
3. bandwidth: 3 kHz;
4. communication range r: 15 km;
5. communication data rates R and corresponding bit error rate (BER)
   $R = 600$ bps, BER $= 3 \times 10^{-2}$;
   $R = 300$ bps, BER $= 5 \times 10^{-2}$;
   $R = 200$ bps, BER $= 10^{-2}$.

4.3.2 Chief Key Techniques

The basic principle, specific structures, and performance analyses for frequency-hopping spread spectrum (FH-SS) system radio communications have been briefly introduced in Chapter 3. Because the system has some outstanding advantages, such as simple structures and better channel adaptability, etc., it would first be considered as a signal-processing system employed in civil underwater acoustic communications. Of course, to adapt to the peculiarities of underwater acoustic communication channels, making some essential modifications for present FH-SS systems employed in radio communications is necessary, in particular how to adapt to the complex and variable multipath effects that must be considered.

Four-frequency shift keying (4FSK) modulation has been adopted in the underwater acoustic data communication prototype. As in two-frequency shift keying (2FSK) modulation, frequency points are fully used in 4FSK; moreover, its duration of frequency codes is two times greater compared to 2FSK modulation; it is available for following relative signal-processing parts. The scheme of Chirp-Z transform (CZT) spectrum analysis (it will be introduced shortly) is employed in the demodulator that has a higher frequency distinguishability than usual Fast Fourier Transform (FFT) ones, so that the whole bandwidth B of the prototype may be reduced.

Some key techniques that will be encountered in designing the prototype and how to solve them will briefly be discussed in the following section.

4.3.2.1 Design of Frequency Code Compression Scheme to Adapt to the Great Intersymbol Interference Caused by Multipath Effects

This is a key technique for FH-SS system digital underwater acoustic communications. In the case of low data-rate (as several 10 bps) underwater acoustic communications, it is possible to adopt the simple scheme: let $n\tau_s > T_M$ (refer to Section 3.2.2) to overcome ISI caused by multipath effects. However, the data rate $R = 600$ bps in this prototype, thus $\tau_s = 1.67$ ms and frequency distinguishability is 600 Hz, provided the scheme of FFT spectrum analysis is used. According to present experimental data, the maximum delay spread of multipath T_M is possible up to 40 ms in most underwater acoustic channels even if carrier frequencies are above 15 kHz. Therefore, the numbers of the frequency codes without repetitive use $n = 24$, and the whole bandwidth $B = 15$ kHz. Obviously, it is only adaptable to short-range (as several kilometers) underwater acoustic communications [11].

To solve this problem, a scheme of adaptive frequency code compression to combat against ISI due to multipath effects is employed in the communication prototype, which is based on distinguishing the sparse characteristics of multipath structures. This scheme can remarkably reduce n, such as $n = 8$ for $T_M = 40$ ms. Provided FFT spectrum analyses is used, $B = 4.2$ kHz. That will be 2.5 kHz if CZT spectra analysis is used as subsequently mentioned. Realizing medium ranges (as 15 km) and medium data-rate (as 600 bps) digital underwater acoustic communications are a basic premise.

4.3.2.2 Discriminating Frequencies by Means of CZT Spectrum Analyses Scheme

Underwater acoustic channels have strict band-limited peculiarity, whereas the basic request for FH-SS system communications is spread spectrum. Therefore, to find a scheme of discriminating frequency having the distinguishability Δf higher than that of the usual FFT scheme is an available approach.

CZT arithmetic has the higher flexibility, in particular, higher Δf than that of FFT one can obtain. It has been proved that by elemental research Δf using the former has a ratio of 2/5 higher than that of the latter. That is to say, CZT arithmetic may discriminate the two adjacent frequencies with a space of $(2/5) \times (1/\tau_s)$, in which τ_s is the duration of signal pulse. Fig. 4.35A and B shows the discriminating frequency diagrams by means of FFT and CZT

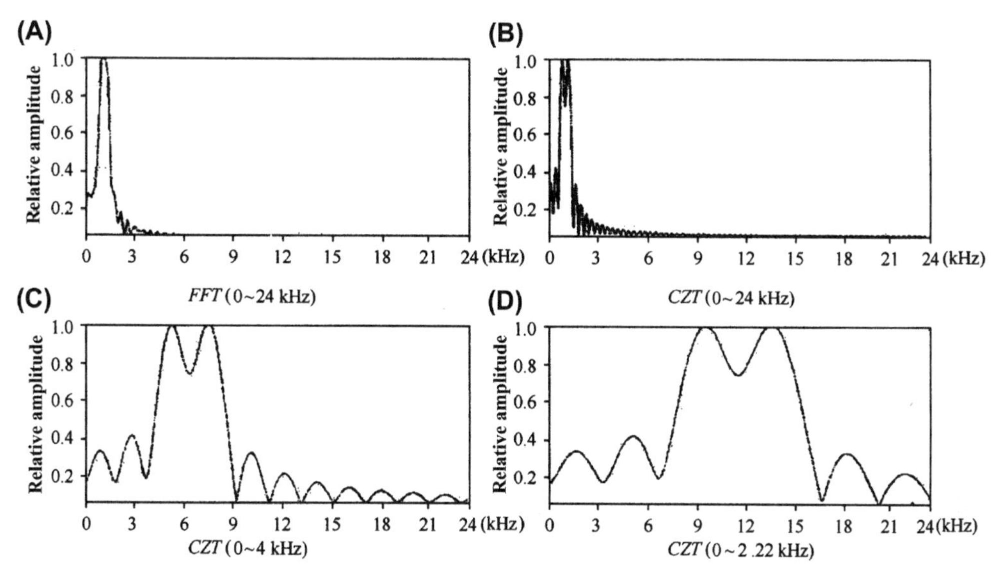

FIGURE 4.35 Diagrams of discriminating frequencies by means of Fast Fourier Transform and Chirp-Z transform arithmetic, respectively.

arithmetic, respectively, with respect to the two carrier frequencies of 1 and 1.2 kHz, and $\tau_s = 3$ ms. Fig. 4.35C and D shows the fining diagrams in which the frequencies ranging from 0 to 4 kHz (fining four times) and 0 to 2.22 kHz (fining 11 times) are selected, respectively. We see that although the FFT scheme cannot discriminate the two frequencies that are separated by a frequency gap less than $1/\tau_s = 333$ Hz; the CZT scheme can precisely discriminate a frequency gap of 222 Hz. It is significant for underwater acoustic communication channels having strict band-limited peculiarity.

4.3.2.3 *Selections of Synchronous Techniques*

In the case of weak multipath channels, as in shallow-water areas of Xiamen Harbor, field experiments have demonstrated that adopting synchronous head combining with matched filter is effective [12].

Of course, adopting total time-sampling processing is preferable as mentioned previously for the underwater acoustic communication channels having violent signal fluctuations in time–amplitude domains.

4.3.2.4 *Explorations Employing Turbo Code in Digital Underwater Acoustic Communications*

The convolutional code and interleaver are excellently combined in Turbo code; the idea of random encoding has thus been realized. Moreover, soft input–soft output interactive decoding is adopted to approach maximum-likelihood encoding. Excellent detection performance has been obtained under the condition of low SNR. Therefore, the application of Turbo code to underwater acoustic data communication is receiving much attention at present.

For a communication system based on Turbo code, the performances of BER and real-time operation are chiefly determined by the modes to acquire soft-decision messages of code element sequences at the decoding end, and to select and optimize iterative arithmetic.

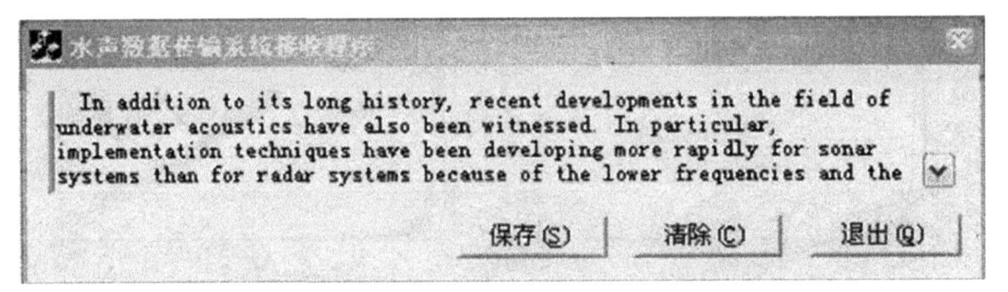

FIGURE 4.36 Computer interface for the Turbo decode system.

A Turbo decode system was constructed, in which using model No.6711 digital signal processing (DSP) as a core arithmetic processing chip, employing Goertzel arithmetic to realize soft dehop, and soft output Viterbi algorithm (SOVA) as an iterative decoding core arithmetic.

Numerous tests have been carried out in lab pool for the communication system. Transmitted messages are text data. Computer interface is showed in Fig. 4.36; the transmitted text data had been correctly detected.

The typical experimental data received and corresponding outputs of every iterative decodes are shown in Table 4.5. The bit ordinal numbers express the positions of data lined up after encoding. The messages of soft decision express those calculated after real-time spectrum analyses at the received end. We see that by comparing three-time iterations, the messages being decided gradually become indistinct. The transmitting bit data have been correctly reestablished after iterative decoding processing in a real-time mode.

The tests in the lab pool have demonstrated that the Turbo code performs well against interference. It may be expected to efficiently lower BER in digital underwater acoustic communications. The conclusions had been proved by the experiments in the sea (refer to Section 4.3.4).

TABLE 4.5 Output of Iterative Decodes

Bit ordinal number	2	8	9	22	32	52	95
Transmitting bit	1	1	1	1	0	1	0
Soft-decision messages	−3	−3	0	−3	3	1	−2
Output before decoding	0	0		0	1	1	0
First iterative	7.7	6.7	6.7	7.2	−7.5	8.3	−8.1
Second iterative	9.3	9.2	9.5	9.0	−8.8	9.0	−8.9
Third iterative	9.7	9.8	9.8	9.7	−9.5	9.6	−9.6
Output after decoding	1	1	1	1	0	1	0

4.3.3 Experimental Explorations for the Advanced FH System Digital Underwater Acoustic Data Communication Prototype

4.3.3.1 *Experiments of the Simulations for Adaptive Frequency Code Compression at Violent Multipath Circumstances*

In the case of moderate data-rate (as 600 bps) digital underwater acoustic communication, the numbers of frequency codes must be compressed to adapt to the channels with longer total-delay spreads T_M due to the multipath effects and thus remain a suitable bandwidth B in the communication. A basic premise is to realize the developing target as mentioned earlier.

Because multipath structures have the spatiotemporal–frequency variability, making the adaptive matched experiments in situ via varied underwater acoustic channels is more difficult. One of the available methods is that the experiments of simulations may first be implemented, and then we may select some typical channels to accomplish corresponding tests and then compare with simulative results; some advances and improvements for the developed prototype would be obtained from the practice processes.

Based on the analyses for the sparse characteristics of the multipath structures, the simulative results of the adaptive frequency codes compression are shown in Fig. 4.37.

Fig. 4.37A shows a selected frequency-hopping (FH) pattern. Every frame consists of 14 FH code elements.

Fig. 4.37B expresses the impulse response for simulative channel 1 in which $T_M = 225$ ms.

Fig. 4.37C shows the transmitted frequency codes corresponding to the data of [11001110010101]. The duration of pulses can be changed at a certain range according to actual data rates. Spaces in adjacent code elements show their differences.

Fig. 4.37D shows the output of a frame signal from the preamplifier, which is a complex waveform mixed with the multipaths and severe noise.

Fig. 4.37E shows a frame output from dehop processing, which will be passed through the following comparison–decision circuit. To adapt to high background-noise levels, a weak signal-detection scheme is used. Obviously, the accurate reestablishment of the transmitted data can be realized for such a high-input SNR in the decision circuit.

Fig. 4.38 shows another simulative channel 2 and corresponding experimental results. Transmitted data are the same as the simulative channel 1. Fig. 4.38A shows the impulse response for channel 2 that belongs to a slowly attenuating model, which is more difficult to adapt for actual underwater acoustic communications, because the violent random fluctuation in multipath structures must be considered, and the decision in the amplitude domain is thus ineffective.

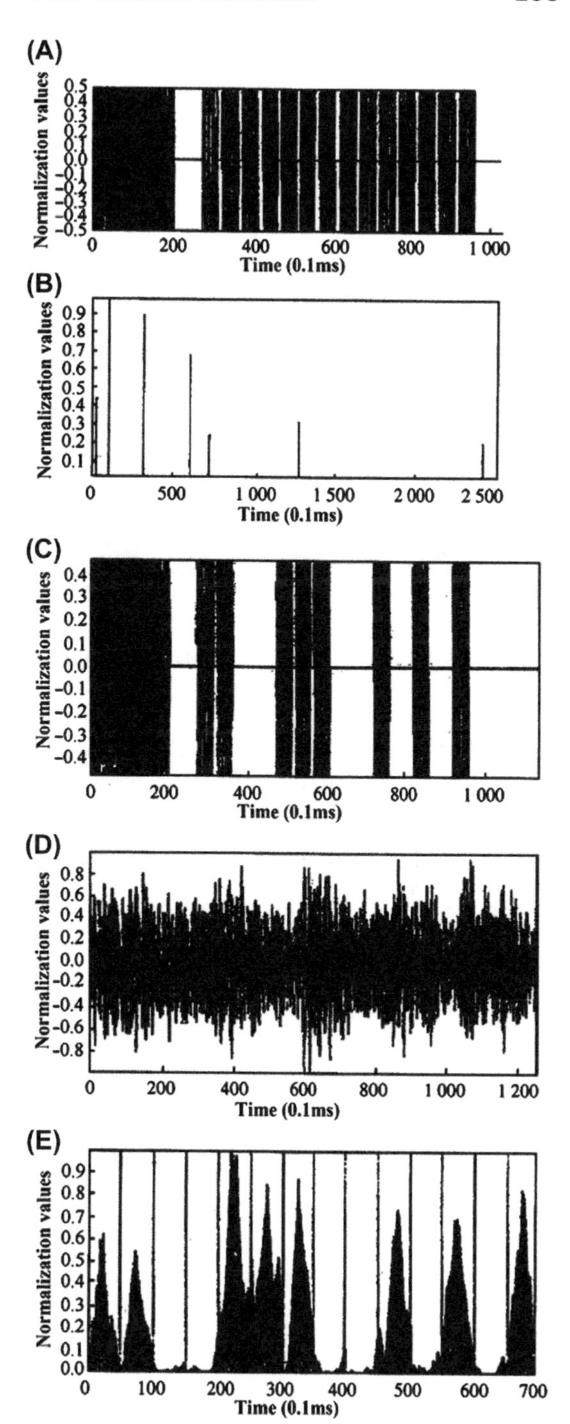

FIGURE 4.37 Simulative experiments of adaptive frequency code compression (channel 1).

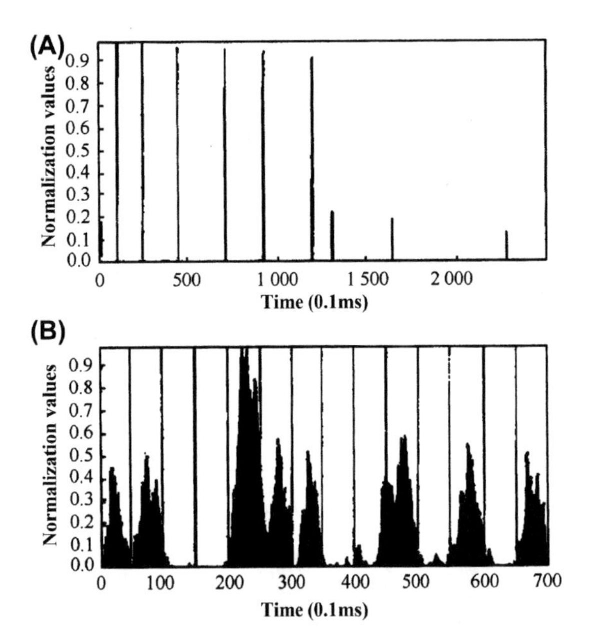

FIGURE 4.38 Simulative experiments of the adaptive frequency code compression (channel 2).

Fig. 4.38B shows the output of the dehop processor for the simulative channel. As well as channel 1, accurate data detections can be realized by means of a suitable comparison–decision circuit.

The results of simulative experiments had proved that the multipath structures with larger T_M may suitably be matched by means of the scheme of the adaptive antimultipath frequency code compression; moreover, the compressing ratios are greater. It provides an available condition to design communication sonar having the expected moderate communication ranges (as 15 km) and moderate data rates (as 600 bps).

Although the adaptive frequency codes compression employed in this prototype cannot use the energy of multipath propagations, that may simply realize the compression of the bandwidth with greater ratios. It would be expected to provide technical support to efficiently transmit the data information with moderate ranges and data rates in strict band-limited underwater acoustic communication channels.

4.3.3.2 Experimental Explorations in the Sea

Some parameters of the underwater acoustic data communication prototype are as follows: $SL = 190$ dB, data rate $R = 600$ bps; spherical piezoelectric ceramic transducers are used, the transmitting response and receiving sensitivity for which are shown in Figs. 4.39 and 4.40.

One of the spherical transducers was located on a sea beach during low tide; experiments were carried out during high tide having depths of about 3 m, as shown in Fig. 4.41. Moreover, a sharply directional transducer that consists of a cylindrical piezoelectric ceramic element combined with a reflecting cover is used to examine the directional effect on digital underwater acoustic communications. A spherical receiving transducer that is the same as the transmitting one was hung by a drifting ship at several meters depth in different ranges. Experimental sea area is the same as that for the image communication (see Section 3.2.1).

The waveform in time domain, signal spectrum, and autocorrelation function received at the range of 6 km are shown in Figs. 4.42, 4.43 and 4.44, respectively.

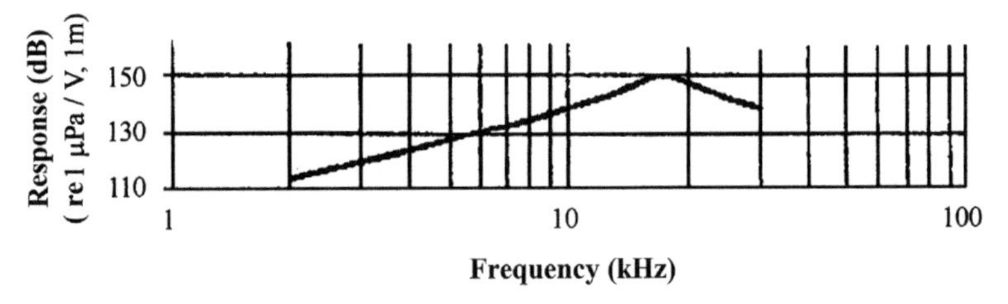

FIGURE 4.39 Transmitting response.

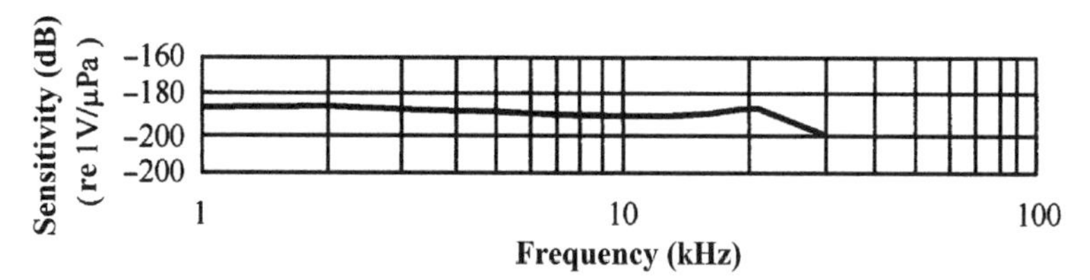

FIGURE 4.40 Receiving sensitivity.

FIGURE 4.41 Transmitting transducers located on a sea beach.

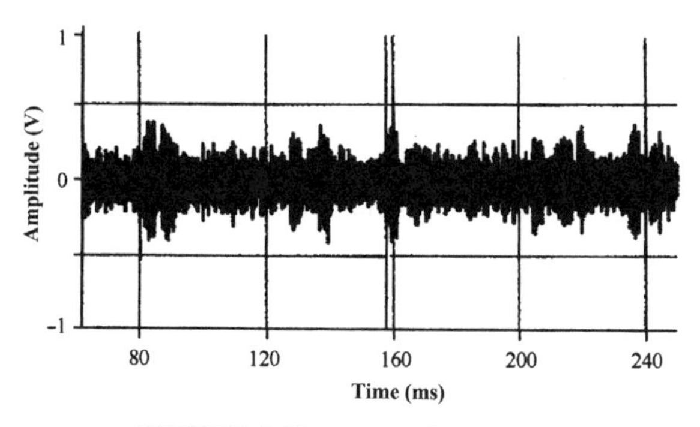

FIGURE 4.42 Received waveform.

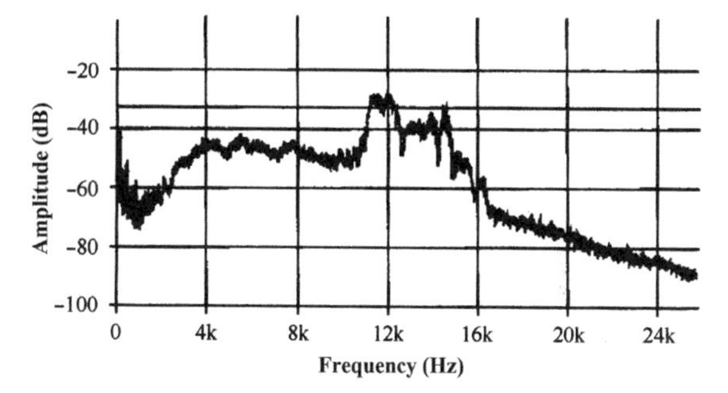

FIGURE 4.43 Received signal spectrum.

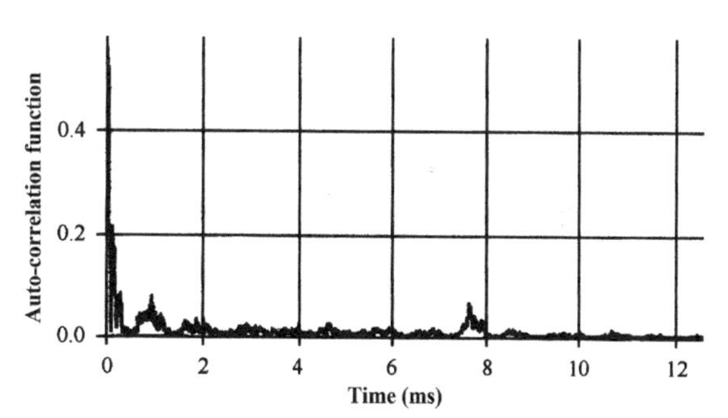

FIGURE 4.44 Autocorrelation function of received signal.

Fig. 4.45 shows a typical record in situ for the data communication prototype at 6 km, and BER $\cong 10^{-3}$. Experimental results had demonstrated that the prototype has robust communication performance in complex and variant shallow-water acoustic channels.

According to the prediction by means of the communication sonar equation (see Section 1.4), when the prototype is operated in deeper underwater acoustic channels and the *SL* for which is appropriately raised, moderate communication ranges (as15 km) would be realized. The underwater acoustic releaser developed having communication ranges much greater in the South China Sea than those in the Xiamen shallow-sea area (refer to Section 4.2.2) provides a typical example.

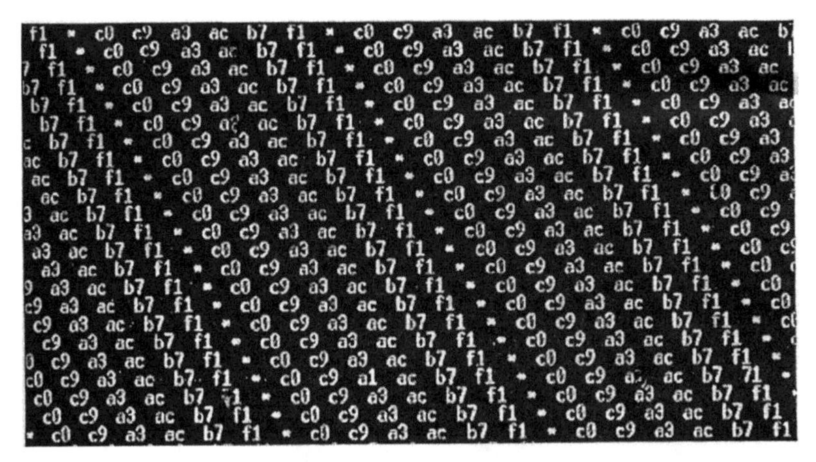

FIGURE 4.45 Received data.

4.3.4 Analyses of Feasibility Extending the Prototype to Underwater Acoustic Multimedia Communications

The expected target for the data communication prototype is to establish the multimedia communication equipment that can be applied to varied civil underwater acoustic communication fields. Of course, developing single or two-media communication equipment according to practical requirements is suitable.

4.3.4.1 Image and Image Text Communication

Information content is voluminous in image communications, thus adopting an image data-compression technique for the expected $R = 600$ bps is necessary. For example, for a frame image that consists of 100×100 (image elements) $\times$ 16 (gray scales), if compression ratios are 8:1, transmitting time is about 9 s. It would be expected to adapt to some civil underwater image communications.

The experiments of image communication have recently been performed in a shallow-water area in Xiamen Harbor; typical experimental results are shown in Fig. 4.46. Fig. 4.46A is the originally transmitted image. The records for the image communications not using channel encoding and using a Turbo soft decision decode are shown in Fig. 4.46B and C, respectively. Communication range $r = 5$ km. The quality of the image communication has been remarkably improved by using the Turbo decode; the images received agree with the original one.

Adopting the image data compression for the image text communication provided that $R = 600$ bps is unnecessary, because 250-bit text information content may form a Chinese character, thus corresponding to a ratio of 2.4 Chinese characters per second, which extends to speech communication in real time.

Because received image text as Chinese characters, digits, and so on, may be distinguished by prior knowledge, image text communications have an extremely high performance with which to combat against both signal fluctuations and background noise. Fig. 4.47A and B shows field records with respect to digits and Chinese character messages acquired in the Xiamen shallow-water acoustic channels. We see that transmitted image texts can still be distinguished by the corresponding prior knowledge under the conditions of violent fluctuations of signals and high background noise. The similar conclusion had been pointed out

FIGURE 4.46 Records of image communications using and not using Turbo decode.

for the "falling out" of synchronous signals, as shown in Fig. 1.7, in which the text messages can still be distinguished.

4.3.4.2 Voice Communication

Provided the scheme of voice recognition code communication is selected, and BER is loweredto less than 5×10^{-4}, data rates R are also lowered; voice communication would be realized under the low-R condition. Therefore, robust and intelligible underwater acoustic voice communication would be achieved. Provided a speech synthesizer is adopted to play back received Chinese characters, the output of double terminals for speech and Chinese characters may also be realized [13].

4.3.4.3 Underwater Acoustic Data Communications

The main specifications of this prototype, such as $r = 15$ km, $R = 300$ bps, and BER $= 5 \times 10^{-4}$ are generally available for civil underwater acoustic data communications. Of course, they may be changed according to specific requirements.

Some transmitted data are directly acquired from underwater relative sensors. In this case, their output waveforms may be directly transmitted as a image as shown in Fig. 4.46C. This scheme has extremely high channel adaptability.

4.3.4.4 Code Text Communications

Because R is so low for this communication model, we can lower R to improve robustness. Realizing this design thinking at the present condition of $R = 600$ bps is easier, including longer communication ranges and possibly hidden digital underwater acoustic communications.

Otherwise, it would be expected that the underwater acoustic multimedia communication prototype mentioned earlier may be employed in underwater acoustic networks, including underwater e-mail transmission [12], because it has a code division multiple access (CDMA) function.

4.4 EXPLORATIONS DEVELOPING AN INNOVATIVE APNFM SYSTEM DIGITAL UNDERWATER ACOUSTIC COMMUNICATION EQUIPMENT

The basic principles, specific structures, and analyses of performance with respect to the APNFM signal-processing system have been described in Chapter 3. By organically combining the APNFM system receiver and corresponding transmitter, the APNFM system digital underwater acoustic communication equipment can be established.

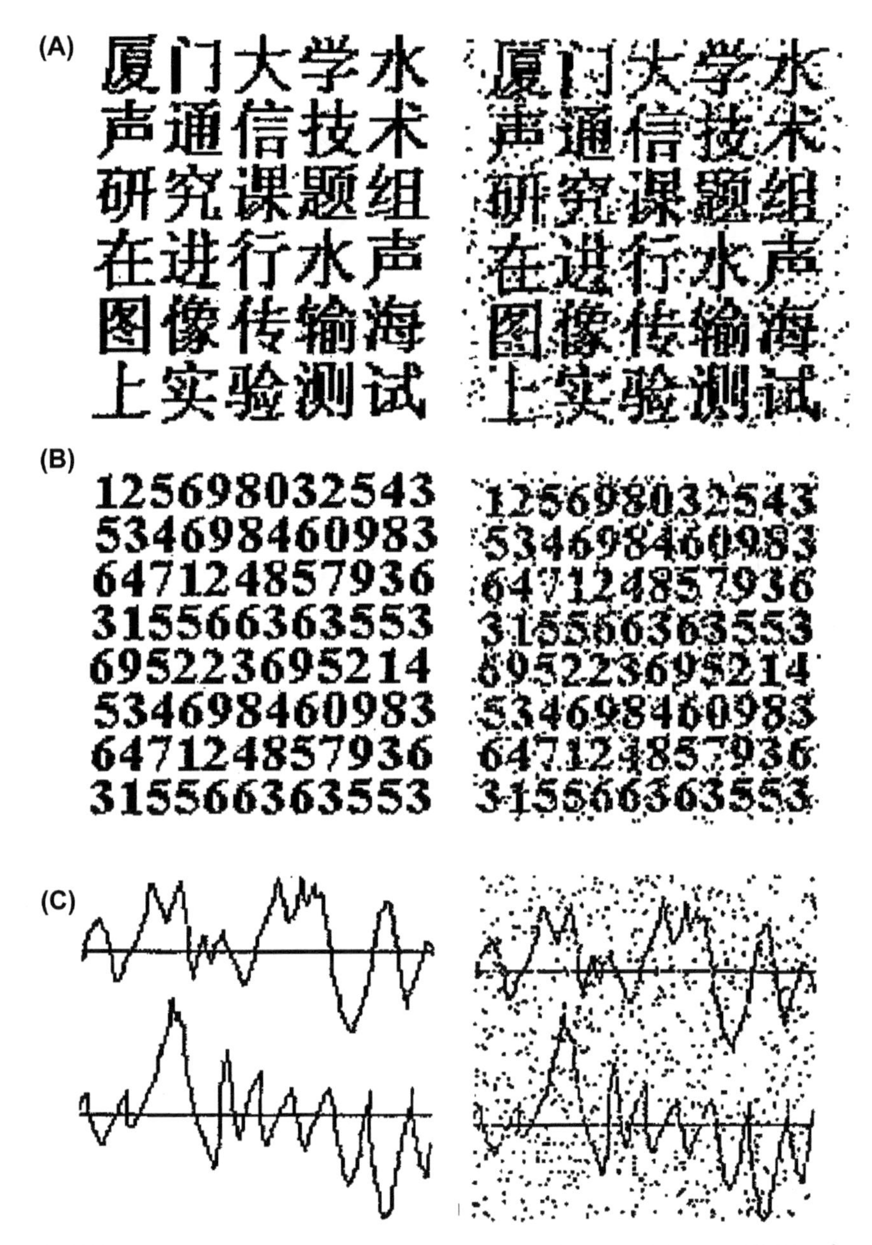

FIGURE 4.47 Image text communications (A) Chinese characters (B) Digits (C) Waveform.

4.4.1 Outline of the Overall Structure and Operating Processes for the APNFM System Digital Underwater Acoustic Communication Equipment

As mentioned earlier, two specific implementing approaches for the APNFM signal-processing system are employed in digital underwater acoustic communications. One is the perfectly adaptive processing approach, which is more suitable to employ in military communications. The other is a more firm approach in which the field inputs of a communication receiver, which is multipath structures mixed with background noise in the sea, is first acquired. This approach is more suitable to employ in civil underwater acoustic communications, because they generally operate in fixed points, low speeds of ships, or shorter-range communications; therefore, the communication circumstances are more stable, in which a higher bandwidth efficient may thus be obtained. The second approach will briefly be introduced as follows.

The schematic principal block diagram of civil digital underwater acoustic communication equipment using the APNFM system is shown in Fig. 4.48. In this communication equipment, the channel parameters, include multipath structures. Doppler frequency excursion, and so on, are firstly acquired in situ and processed by means of the preset parts (refer to Fig. 1.10), and then the detected parameters will provide to all adaptive parts as preset data by using feedback loops, as shown in Fig. 4.48. Both transmitter and receiver are included in the communication equipment, transmitting or receiving states will be changed by a converting key, and then two-way underwater acoustic communications will implement.

The operating processes of the APNFM system communication sonar will briefly be described as follows:

4.4.1.1 *Call and Preset*

The side requiring communication sends a suitable call signal and lets the receiving side respond. This call signal is also used as a channel-exciting source to obtain field channel parameters, and thus a PNFM pattern adaptively matching to the channel without prior knowledge will be preset on this side.

Once the receiving side receives the call signal, it sends as soon as a responding signal agreeable to the calling one. The transmitting

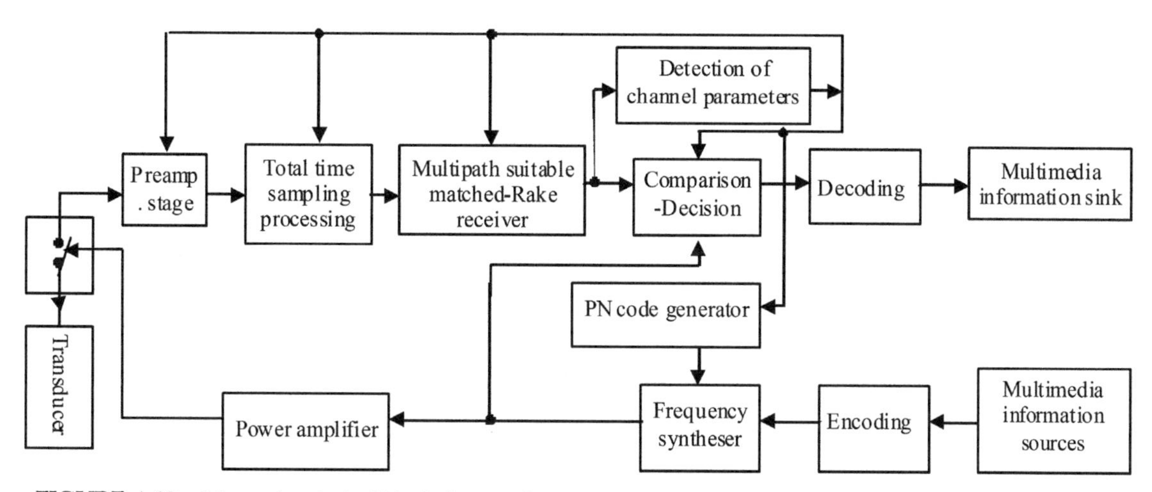

FIGURE 4.48 Schematic principal block diagram for an APNFM system underwater acoustic communication sonar.

side will acquire corresponding channel parameters and make corresponding signal processing like the receiving end.

4.4.1.2 *Formal Communication*

Multimedia information sources, such as voice, image, text, data, and so on, will be converted to the bit streams without essential differences by applying a quantization format, following by information source encoding and channel encoding, they are made use of to control frequency synthesizer by combining with PN codes generator, which is preset according to acquired channel parameters in situ to form PNFM pattern that suitably match to field underwater acoustic channels. Then passing through power amplifier and transducer, corresponding information signals are transmitted in the channels. Because an adaptive Rake receiver is included in the APNFM system, which has a function of path diversity, channel encoding is generally unnecessary to remain higher communication data rates. Although in some communication conditions, such as the operating frequencies are higher for some shorter-range underwater acoustic communications, the multipath structures will reduce to single pulse (wave packets) with considerable amplitude, the function of paths diversity is absent. In this case, in orders to lower BER, channel encoding would thus be utilized. Of course, the bandwidth efficient is higher in this conditions, adopting efficient channel encoding schemes to raise the robustness of communication sonar is possible. For some communication media having large information contents, in particular for image communication, to raise frame rates, data compression is necessary.

Once the communication sonar is converted to the receiving state, the transducer, now as a hydrophone, receives the channel outputs, which are the multipath structures mixed by the background noise in the sea. By passing through adaptive preamplifying part, the multipath structures will be preprocessed and the background noise will partially be suppressed, and then the transmitted PNFM sequence codes will be correctly detected at a permitting BER by the total time sampling processing part. In these processes, some violent fading frequency codes may be distinguished and the noise would further be suppressed.

Based on distinguishing the spare characters of the multipath structures by passing through field studies and analyses, the lobe numbers of "rake," the space between adjacent lobes and raking rates in the adaptive Rake receiver may adaptively be adjusted to adapt to the spatiotemporal variability of multipath structures without prior knowledge. As a result, significant multipath pulses (wave packet) may be "raked up" by the adaptive Rake receiver. If an optimal combiner is selected, expected to obtain optimum detecting result against the multipath propagations. Thus the output PNFM sequence from Rake receiver and local PNFM pattern preset are made corresponding processing in following comparison−decision part. Finally, by means of information source decoding and channel decoding, multimedia information can be reinstated at an expected BER in the information sink.

4.4.1.3 *Adaptive Correction*

As mentioned above (refer to Section 2.3), multipath structures have highly stationary distributions in some underwater acoustic channels, therefore high robust underwater acoustic communications would performed in the channels, in particular in civil fixed-point communications.

Of course, the stationarity of multipath structures would be changed for different underwater acoustic channels. Provided the changes of channel parameters have a considerable impact on the communication performances, an adaptive channel corrective part must be used to adapt to that. The adaptive adjusting times to suitably match to the changing parameters underwater acoustic channels depend not only on communication ranges and data rates, but also on channel characteristics; whereas they would be less than 1 s in most communication circumstances. So that real-time communication would be realized.

As mentioned previously, to develop the underwater acoustic communication equipment that may availably employed in peculiar underwater acoustic channels, and can be compatible with different applied fields by adopting corresponding communication media, it is seen that the APNFM system is an appropriate signal-processing one. However, there exist weak multipath propagations in some underwater acoustic channels. In this fine communication circumstance, utilizing such a complex signal-processing system is unnecessary. Fig. 4.49 shows a weak multipath structure acquired in Xiamen Harbor shallow-water acoustic channel in a certain direction.

In this case, we can use a matched filter or cross-correlator to detect the output signal of the channel, which is corresponding to a weak multipath structure, as shown in Fig. 4.50. The outputs of the preamplifier and cross-correlator are shown at the lower and upper parts in the figure, respectively. We see that direct signal pulses can correctly be detected, multipath and noise interferences are perfectly eliminated by the correlator.

4.4.2 Experimental Explorations of an Innovative APNFM System Digital Underwater Acoustic Communication Prototype

The two typical communication media: image (including image text) communications having higher data rates and code text communications

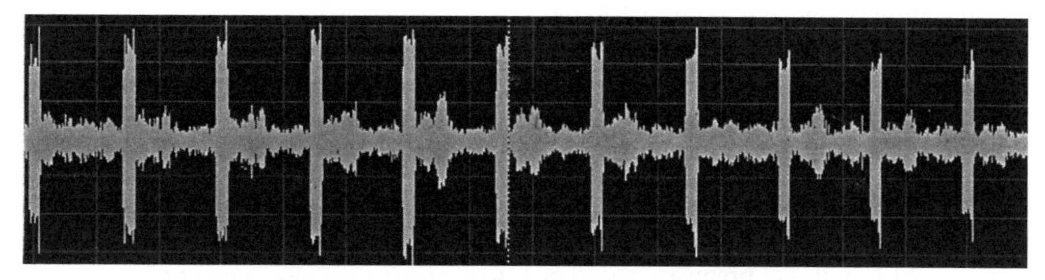

FIGURE 4.49 Signal structure for a weak multipath channel.

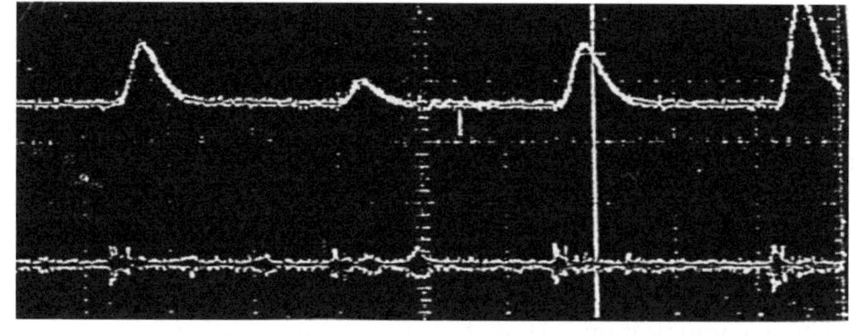

FIGURE 4.50 Signal processing using a cross-correlator.

that permit lower data rates are included. They can be extended to other communication media.

As mentioned earlier, the compression of the image data under the condition of the higher BER is a key technique in the image communication, which will first be briefly introduced.

4.4.2.1 *Primary Experimental Research on the Image Data Compression*

Because the information contents is voluminous for image communications, the scheme of the image data compression would generally be adopted, in particular for longer range (as above 20 km) image communications in which data rates have to be reduced as operating frequencies must correspondingly be lowered. For example, for a simple picture having the structures of 140×140 (picture elements) $\times$ 8 (gray scales), its information contents are up to 3.9×10^4 bit. For the data rate to be 600 bps as mentioned above, the time transmitting this picture is up to 65 s. Obviously this frame rate is too low for actual applications. If we can realize the image data compression having compressing ratios of 8:1, that will be reduced to 8 s; this may be acceptable to some civil underwater acoustic communication engineering.

The compression techniques of image data in radio communications are normalized and efficient, moreover, the compressing ratios are also great enough, but a basic premise: requires extremely low BER to perform the compression. We have known that by analyzing the peculiarities of underwater acoustic communication channels relative to radio ones, BER for the former are much greater than that of the latter; moreover, the greater the compression ratios, the more sensitive the image quality with respect to BER, or even BER are still moderate, image messages may perfectly be lost. So that the present schemes of image data compression in radio communications cannot directly be employed in underwater acoustic communications. How to realize the image data compression having a better quality and higher bandwidth efficient under the condition of larger BERs (as 3×10^{-2}) is quite difficult. Therefore, it has become a key technique in underwater acoustic image communications.

The research on image data compression based on the wavelet transformations had been implemented. In the case of lower compression ratios, the primary experimental results for underwater acoustic image data compression had been obtained in laboratories and in the sea, which will be introduced as follows. In additions, the experiments to compare image communications not using image data compression with using that will also be included.

4.4.2.2 *Experiments in Laboratories*

4.4.2.2.1 SIMULATIVE EXPERIMENTS FOR ADAPTIVE MULTIPATH SUITABLE MATCH PROCESSING

Adaptive Rake receiver plays a key role in the APNFM signal-processing system. The adaptive multipath suitable match processing is a basic technical support to that.

Because multipath structures have spatiotemporal—frequency variability, accomplishing the field experiments in varied underwater acoustic communication channels is difficult, therefore simulative experiments in laboratories would be available for the explorations to design the APNFM signal-processing system.

Fig. 4.51 shows a typical multipath structure formed by a random multipath generator, which consist of eight ideal multipath pulses, because the numbers of multipath pulses having significant effect on underwater acoustic communications are generally fewer than eight for most communication channels provided they are first processed by an adaptive preprocessing circuit in the preamplifying stage. The multipath generator may form simulative multipath structures having the spreading time of 1 s and varied distributing modes, including the different

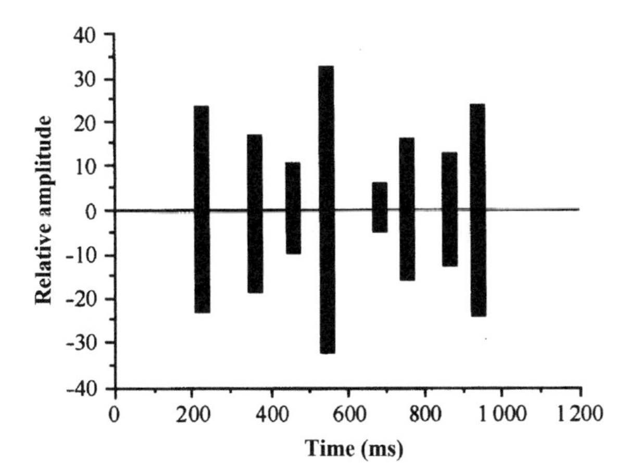

FIGURE 4.51 Typical outputs of multipath simulation generator.

numbers of multipath pulses, whose appearing times and relative amplitudes.

Fig. 4.52 shows the two parts of output waveforms from the preamplifier for the planar near-field measurement false random frequency modulation (PNFM) pattern that suitably matches the simulative channel shown in Fig. 4.51. The starting section of the output waveform is shown in Fig. 4.52A, which belongs to an equal-amplitude wave before multipath superposition, that is, transmitting the PNFM pattern itself, and then following with the waveforms superposed by two to three multipath pulses. Fig. 4.52B shows a section of output waveforms superposed by eight multipath pulses. By passing through the adaptive multipath suitable-match processing, and then implementing comparison–decision with local PNFM pattern preset in the corresponding part, a received PNFM pattern perfectly agreeing with the transmitted one will be obtained. Using the different distributing modes of multipath structures and the corresponding input PNFM patterns suitably matching for them, we may get consistent results. As viewed from the simulations, it means that the adaptive multipath suitable-match processing has performance matching varied multipath structures, which is an essential support to following the Rake receiver. Moreover, adapting to complicated and variable multipath structures by means of fewer frequency codes is possible; thus, bandwidth efficiency will be remarkably improved. It is also a basic premise to realize longer-range underwater acoustic communications.

As viewed from the compression of the bandwidth, they are basically consistent between adaptive multipath suitable match processing here and adaptive frequency codes compression in the advanced FH system (refer to Section 4.3.2). Of course, the former must overall be designed combining with a Rake receiver, the

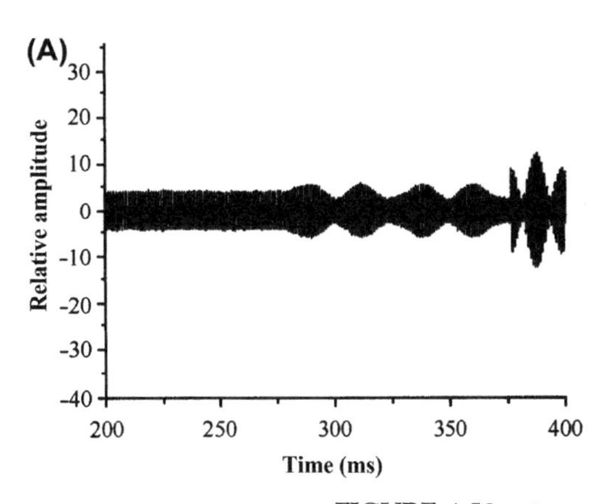

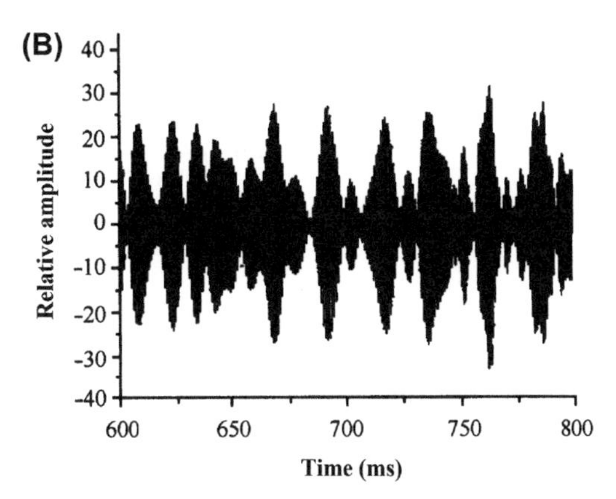

FIGURE 4.52 Output waveforms of preamplifier.

bandwidth compression ratios for which are thus slightly less than those in the latter.

In the case of weak multipath communication circumstances, the Rake receiver would not be adopted, and perhaps by using a correlator instead as previously mentioned, the communication sonars will remarkably be simplified.

4.4.2.2.2 TESTS IN EXPERIMENTAL POOLS

Before carrying out the tests in the sea, we performed many tests in an experimental pool for the developed communication prototypes to examine their basic performance. The experimental pool does not process the absorbing sound property, therefore, in which severe multipath effects occur, including the signal fading due to the interference of multipath propagations and spatiotemporal–frequency variability. Typical waveforms recorded are shown in Fig. 4.53A–C that are relative to an identical transmitted pulse at different times with great spaces of time, different depths located transducers, and different carrier frequencies, respectively. We see that narrow pulses with approximately equal amplitude occur at the starting part of every waveform, which correspond to the direct sound pulse without superposing the reflected sound pulses from the boundaries of the pool, and then by following complicated and varied multipath structures.

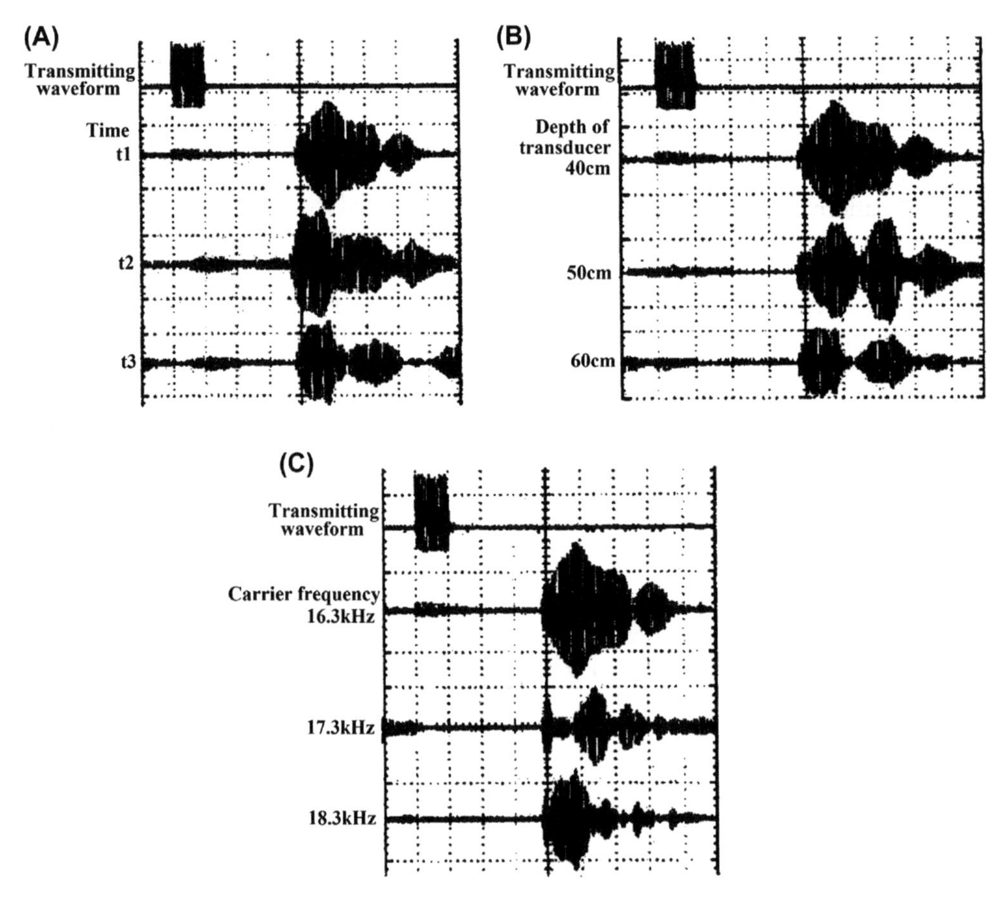

FIGURE 4.53 Spatiotemporal–frequency variability for the multipath structures in an experimental pool.

Obviously, the experimental pool belongs to a nonlinear system for the transmitted pulse with a wider duration.

4.4.2.2.2.1 CODE TEXT COMMUNICATIONS [14] Because the low BER must be satisfied for the code text underwater acoustic communications, channel error correction encoding scheme is employed in the communication prototype. Excellent experimental results had been obtained. Fig. 4.54A shows the transmitted code text messages; (B) shows the results for them traveled in the pool, and remarkable errors occur in the complex and varied communication conditions. Fig. 4.54C shows the results having a traveling condition like (B), while the channel-encoding scheme is employed. In this case, a robust code text communication had been realized.

4.4.2.2.2.2 IMAGE TEXT COMMUNICATIONS The image text (the image with two gray scales) communications have high performance combating violent interference. That is to say, we can still distinguish the text messages from prior knowledge under the condition of a high BER. The records for the image text communications at the different BER are shown in Fig. 4.55. We see that transmitted Chinese characters can still be identified even if BER = 5%.

4.4.2.2.2.3 IMAGE COMMUNICATIONS NOT USING THE IMAGE DATA COMPRESSION Image communications not using the image data compression have the similar property as image text ones. Of course, the gray scales are 16 in the developed prototype, thus the information contents are four times in comparison with that of the latter. The records for the image communications not using image data compression at the different BERs are shown in Fig. 4.56. We see that the image communications not using the data compression may adapt to the decision condition with a greater BER, above 10^{-2}. Therefore, the scheme not using image data compression would be adopted at shorter range (several kilometers) image communications, and the higher data-rate, as 8 kilobits per second (kbps) can be realized [18].

4.4.2.2.2.4 IMAGE COMMUNICATIONS USING THE DATA COMPRESSION Information contents being processed are much greater for image communications not using data compression. To increase frame rates, the image data compression has to be used. Of course, image resolution will

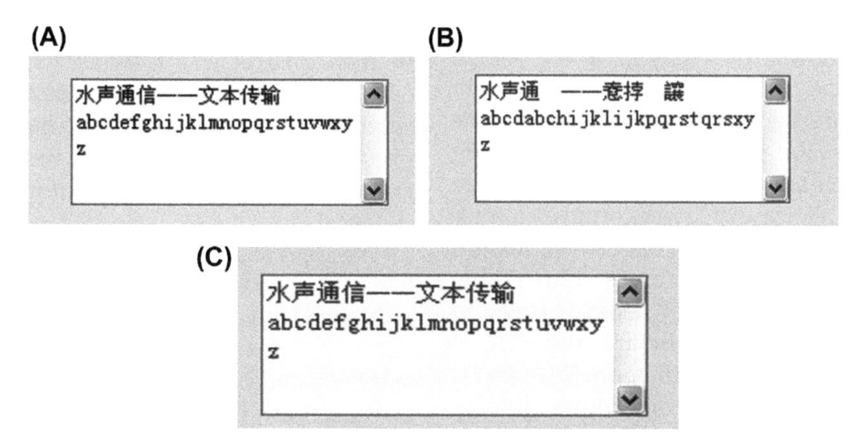

FIGURE 4.54 Records for code text communications.

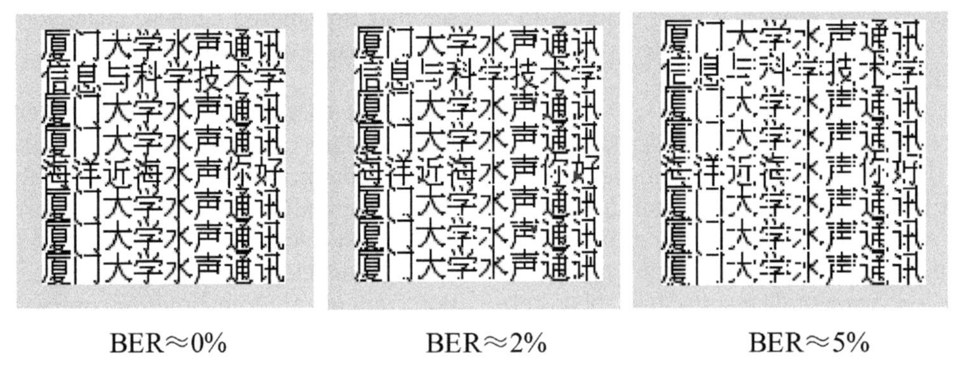

FIGURE 4.55 Records for image communications not using data compression at the different bit error rates (BERs).

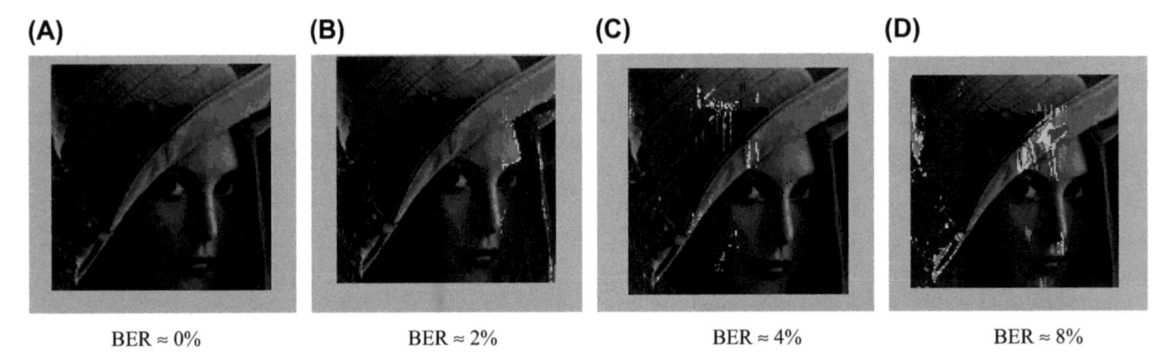

FIGURE 4.56 Records for image communications not using data compression at the different BERs.

be reduced, or even some image messages will be lost. That is to say, the price for compressing the data is to increase the request for BERs.

Fig. 4.57 shows the records of the image communications by using the image data compression in the experimental pool at the different BERs. We see that the images having basic resolution are obtained, even if BERs are up to 2–4% in the case of lower compression ratios. Of course, the resolution of received images using the data compression is remarkably reduced relative to ones not using it (refer to Fig. 4.56). According to primary research, it seems that controlling BER to about 2×10^{-2} is preferable in image communications having lower data-compression ratios. This request is lower.

4.4.2.3 Primary Experiments in Shallow-Water Acoustic Channels

Source level $SL \cong 200$ dB, and the operating frequency band was selected within the range of 4.5–7 kHz in the prototype developed. Corresponding transducer–power amplifier set was provided by a relative factor. Realizing longer-range underwater acoustic communications is expected.

The expected in situ experimental results have been obtained. Typical records for the code text and image communications are shown in Fig. 4.58. Fig. 4.58A and E was transmitted original texts and images, respectively. The field records include the code text (b), the image text, (c) regular, and (d) backscript Chinese characters; moreover, the images not using and using

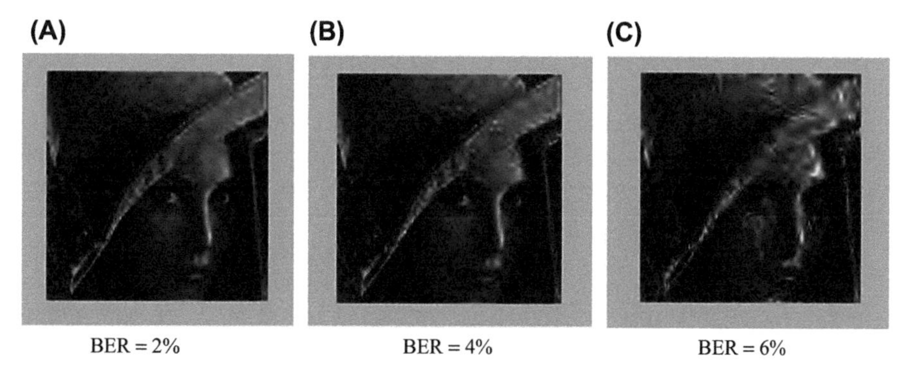

FIGURE 4.57 Records for image communications using the data compression at the different BERs.

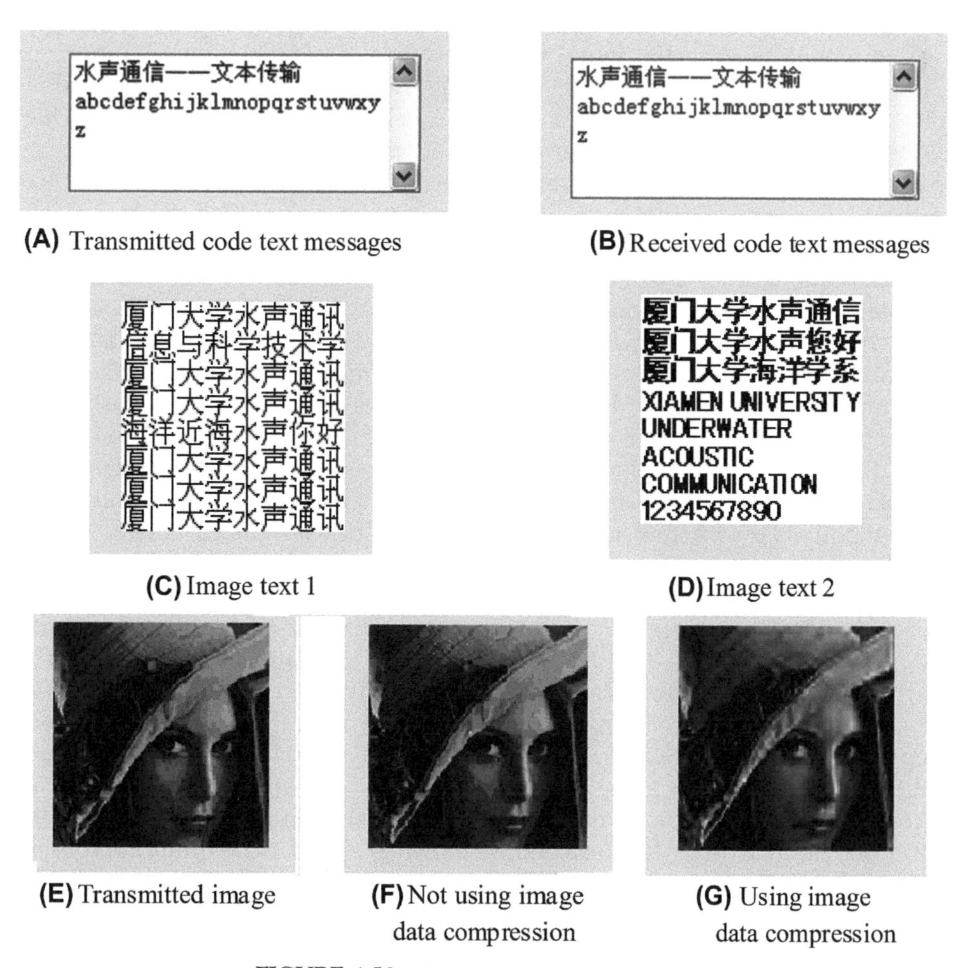

FIGURE 4.58 Experimental recodes in the sea.

data compression are also shown in (f) and (g), respectively. We see that the expected experimental results of both the code text and the image communications in the shallow-water acoustic channels has been obtained, which are consistent with those in the experimental pool in which violent signal fading and ISI caused by multipath effects appeared.

It could be expected that underwater acoustic communication equipment in which the APNFM signal-processing system has been employed, just as the underwater acoustic data communication prototype developed by using the advanced FH-SS system may be extended to the multimedia communications and has a prospect of establishing compatible multimedia digital underwater acoustic communication equipment. This equipment would adapt to some peculiarities of underwater acoustic channels in a certain range, in particular the multipath effects even if using its energy.

Of course, in the case of civil underwater acoustic communications, adopting compatible modes to adapt to different communication fields at the same time is generally not necessary. That is to say, we may develop a series of communication sonars, including different communication media and specifications, such as different communication ranges, data rates, operating depths, fixed or mobile points, and compatibility with two media, as speech and image, to adapt to varied users. It would be expected to elaborate the active effects to the exploitations and uses of marine resources.

It seems that the APNFM signal-processing system is one of the available signal-processing schemes employed in civil underwater acoustic communication equipment. Based on that, a series of communication products may be provided to different civil users. In particular, for shorter range and thus higher frequency underwater acoustic communications, the APNFM system will reduce essentially to the advanced FH-SS system, the corresponding communication equipment will remarkably be simplified, or even can be used by divers.

As mentioned previously, the matched filter or cross-correlator may be used to detect transmitted signals in a weak multipath underwater acoustic communication circumstance.

To examine the relation between the output of the cross-correlator and the frequency shifts of reference signals, understand the channel adaptability of autocorrelation and cross-correlation signal-processing schemes, and make the exploration-raising bandwidth efficient, corresponding cross-correlative analyses have been performed as shown in Fig. 4.59 by connecting with the development of the APNFM system underwater acoustic communication prototype. The frequency of transmitted signal is 5.5 kHz. To analyze the frequency response characteristic of cross-correlation, the reference signal frequencies are shifted to be 25, 50, and 100 Hz relative to the transmitted frequency of 5.5 kHz, respectively. Fig. 4.59A shows the signal acquired in a shallow-water acoustic channel with a weak multipath structure, as shown in Fig. 4.49. Fig. 4.59C–E shows the outputs of the cross-correlator corresponding to the reference signals having the frequency shifts of 25, 50, and 100 Hz, respectively. We see that although the received waveform has a remarkable distortion, the cross-correlativity is still strong enough if the frequency shifts are absent, as shown in (b); whereas the cross-correlation values will be remarkably reduced with increasing the frequency shifts of reference signal. That is to say, the cross-correlator lacks frequency shift adaptability, and thus cannot adapt to high-speed mobile communications having greater Doppler frequency shifts. Perhaps it would point out a direction to improve the bandwidth efficient, which is significant for strictly band-limited underwater acoustic communication channels.

In the case of weak multipath underwater acoustic channels, an autocorrelation detection scheme may also be employed in underwater acoustic communications, although

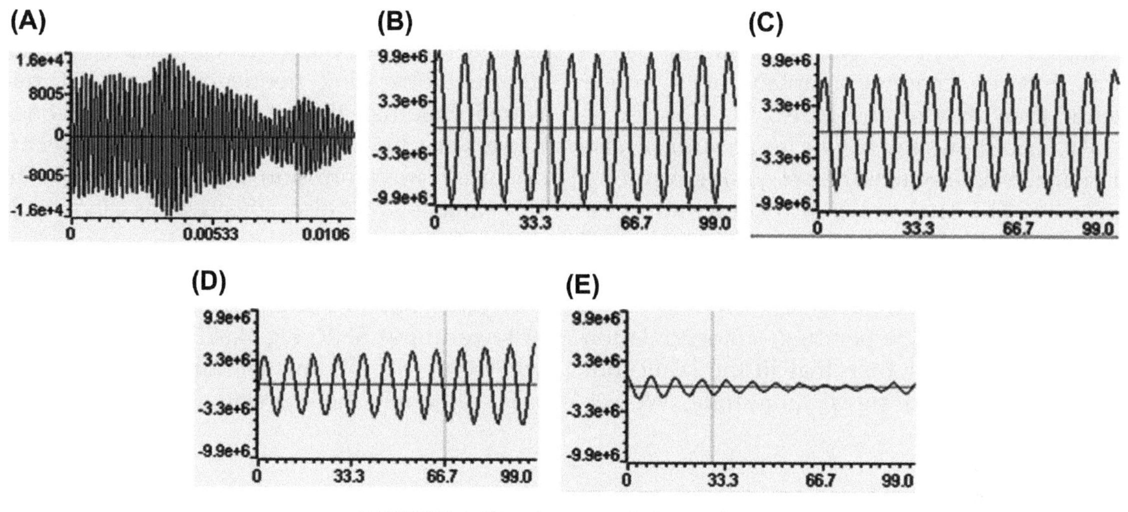

FIGURE 4.59 Cross-correlation analyses.

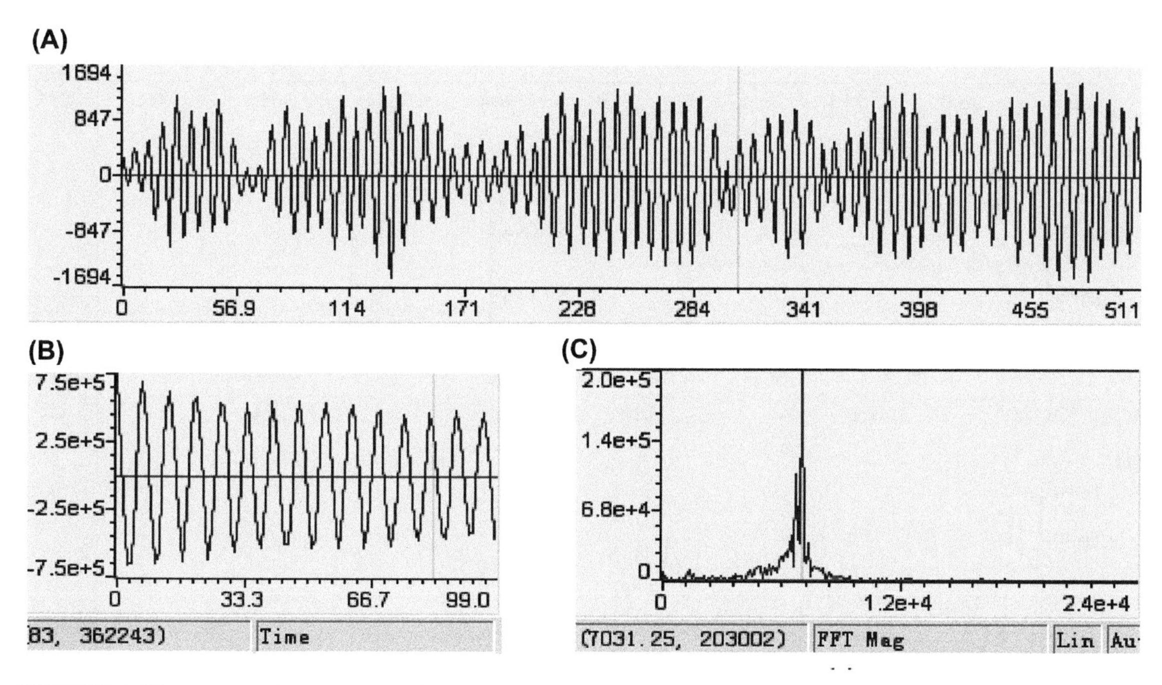

FIGURE 4.60 Autocorrelation analyses for experimental data in the sea. (A) Experimental data. (B) Autocorrelation analysis. (C) Fast Fourier Transform analyses.

its performance combating interference is weaker than that of cross-correlation detection, but its channel adaptability is much better than that of the latter.

Fig. 4.60 shows the results for an autocorrelation analysis, in which the processing result by means of FFT analyses is also shown to make the comparison between them. The input signal amplitude here is less and has more severe fluctuations than that shown in Fig. 4.59; therefore, the corresponding autocorrelation value is also lower than that in the latter; but it still has xpected signal amplitude. We see that the autocorrelation signal-processing schemes have a better channel adaptability for different operating frequencies and distorted waveforms than that of cross-correlation ones; moreover, they also have an ability to rectify distorted waveforms in a certain range. The FFT analysis is also effective as shown in Fig. 4.60C.

Fig. 4.61 shows the analysis results for autocorrelation and FFT schemes under the condition of lower-input SNR. Fig. 4.61D is the amplified waveform of (b) smaller amplitude. In this case, although the correlativity is remarkably

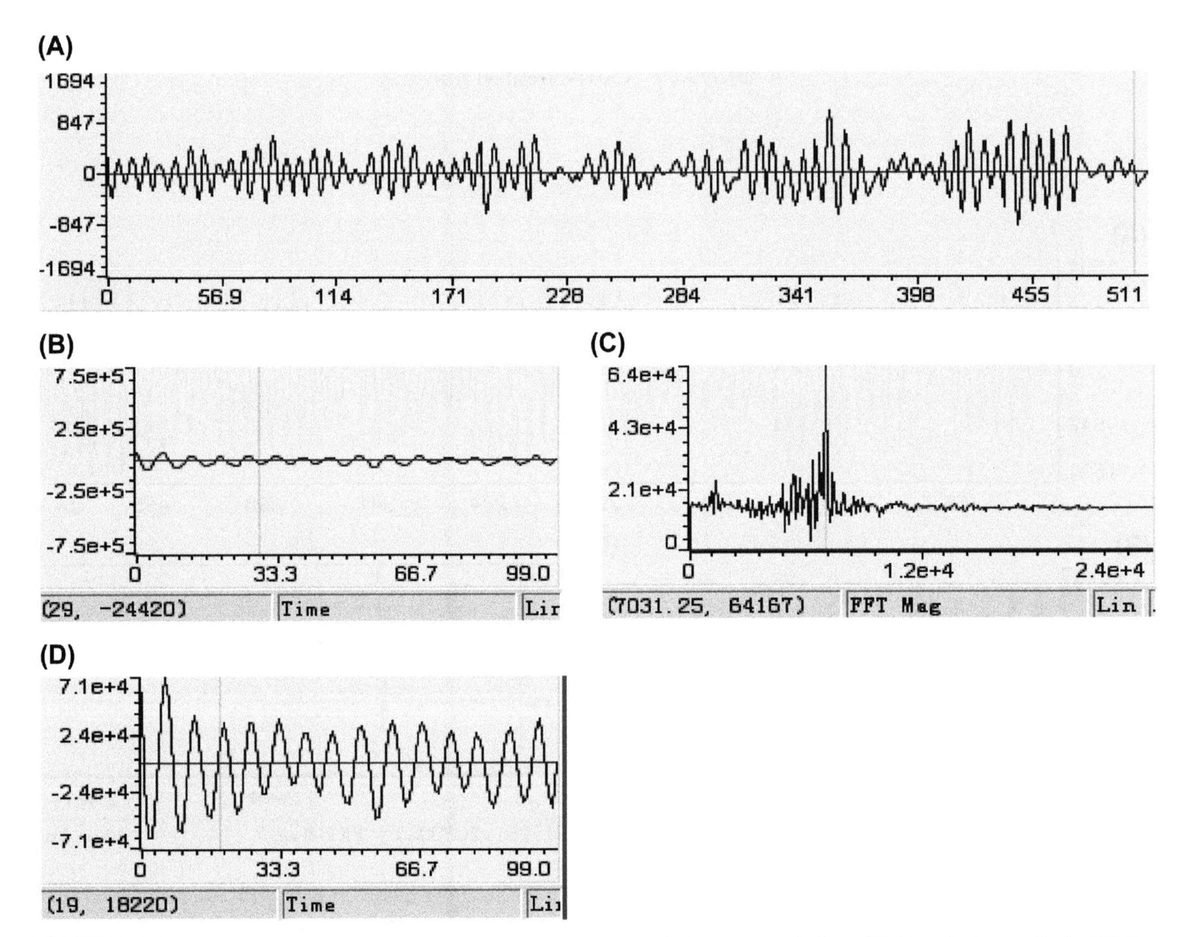

FIGURE 4.61 Autocorrelation analysis for a lower-input SNR: (A) Experimental data. (B) Correlation analysis. (C) Fast Fourier Transform analysis. (D) Amplification with respect to (B).

reduced, and the correlation component of noise in the sea occurs, the correlativity of signals still occupies the main component. The discriminating frequencies may also be realized by using an FFT scheme, as shown in Fig. 4.61C.

Obviously, the structure for an underwater acoustic communication equipment by using the correlation detection schemes will remarkably be simplified, including establishingequipment small in size, which can be used by civil divers at shorter (as several kilometers) range communications.

Before ending the discussion of the main body of this book, it must emphatically be pointed out that developing an innovative APNFM signal-processing system and establishing new digital communication equipment based on the system are only at the state of initial explorations. Some key techniques have yet not been examined in practice, in particular testing in varied peculiar underwater acoustic channels. Authors expect that the shortcomings will be resolved in further research. The signal-processing system and corresponding communication equipment may be improved and raised by passing through practical examinations.

Moreover, maintaining secrecy for some key techniques is necessary, because our research group is preparing to undertake relevant research tasks and perhaps develop some relevant civil underwater acoustic communication equipment. Provided that conditions are ripe, we should apply relative patents, such as an advanced FH-SS system and APNFM system digital underwater acoustic communication equipment, including the total-time sampling processing with multifunctions, adaptive Rake receiving technique with an optimum combiner, and adaptive rapid channel adaptor.

As a result, discussions in the book for some key techniques are simplified or even neglected.

Finally, it should be noted that the digital time correlative accumulation decision system and its applications in underwater acoustic active detecting equipment, such as the echo sounder, fishing sonar, and target telemetry sonar have been discussed in Section 4.2, the excellent detection results have presaged that the APNFM system could also be employed in some active sonars. It would be expected to improve their total performance, because we can think of the echo signals from targets as a PNFM sequence and make corresponding signal processing as mentioned earlier in APNFM system active sonar.

References

[1] L. Mingde, Principle of Underwater Acoustic Transducers, China Ocean University Press, 2001.

[2] H. Zhou, Underwater Acoustic Transducer and Array, National Deference Industry Press, Beijing, 1984.

[3] T. Xu, A digital time correlative accumulation (I)—auto-correlative accumulation, Chinese Journal of Acoustics 4 (1990) 365–371.

[4] T. Xu, A digital time correlative accumulation (II)—cross-correlative accumulation, Chinese Journal of Acoustics 1 (1991) 86–91.

[5] T. Xu, A digital time correlative accumulation(III)—the anti-multipath ability, Chinese Journal of Acoustics 2 (1991) 149–155.

[6] Y. Liang, Y. Fu, Characteristics of random time-variant sound channel in shallow sea, in: Proceedings of International Workshop on Marine Acoustics, Beijing, 1990, pp. 233–236.

[7] T. Ming, X. Tianzeng, H. Ximing, MFSK-DTCCA and its applications to underwater acoustic telemetry, in: IAC 14, BP-9, Beijing, 1992.

[8] X. Xiaomei, X. Tianzeng, The application of PDC-DTACAA to signal detection in shallow water acoustical cannel, Journal of Xiamen University (Natural Science) 34 (2) (1995) 243–248.

[9] N. Baoqing, H. Yanzhen, J. Zhidi, Study on anti-noise scheme for detecting PPM information in shallow water channels, Chinese Journal of Oceanology and Limnology 17 (3) (1999) 278–280.

[10] S. Xu, T. Xu, Application of digital time correlative accumulation to echo sounder and fish finder, in: In Proceedings of International Workshop on Marine Acoustics, China Ocean Press, Beijing, 1990, pp. 165–170.

[11] A.D. Waite, Sonar for Practicing Engineer, third ed., National Deference Industry Press, Beijing, 2004.

[12] C. En, E-Mail Transmission in Shallow Water Acoustic Channels, The Collection of Thesis for First Oceanic Development Forum, 2004, pp. 234–238.

[13] K. Xu, T. Xu, Underwater acoustic wireless communications, Journal of Xiamen University (Natural Science) 40 (2) (2001) 311–319.

[14] L. Xu, A research on the text data transferring in underwater acoustic channel, Ocean Technology 25 (3) (2005) 46–48.

Appendix: Ultrasonic Sensing Systems in the Air Medium

The rules with respect to ultrasonic radiation, transmission, reflection, scattering, and reception in the air medium will first be briefly described in this appendix. They are the physical fundamentals for designing ultrasonic sensing systems employed in an air medium. Next, the three new, important ultrasonic sensing systems developed: ultrasonic-ranging and bearing-sensing systems employed in concrete-jetting manipulators, ultrasonic terrain obstacle-sensing systems employed in mobile robots, and ultrasonic navigation-sensing systems employed in automatic guided vehicles (AGV), will be discussed in this appendix.

Ultrasonic surveying and positioning systems would generally be considered as the particular examples of underwater acoustic ones in which high operating frequencies (corresponding to short surveying distances) are used. Their basic principles or even some signal processing schemes are basically agreeable. For example, surveying demolished basic faces is similar to the telemetry of shallow-water targets and the digital echo sounder to survey the sea bottom (refer to , Chapter 4). The thinking involved in measuring the heights of terrain obstacles and the widths of ditches comes from the "white area" (corresponding to the sound shadow region) appearing in the echo wave of the fishing sonar (refer to , Chapter 1). Moreover, underwater acoustic positioning principles can be directly introduced in the air medium. This is why we develop these ultrasonic sensing systems for employment in the air medium.

Of course, because there exist some essential differences in sound propagation media between the air and water that would indicate the directions toward solving peculiar problems that arise in ultrasonic sensing systems in the air medium.

1. ULTRASONIC SENSING SYSTEM EMPLOYED IN CONCRETE-JETTING MANIPULATORS

In modern constructions for rapid building, concrete is directly jetted onto demolished basic faces. Experiments have proved that the concrete reflected from basic faces to splash down in a construction process will be least when the distance between jetting mouth and the basic faces has a suitable value for a certain jetting manipulator, as 80 cm for our manipulator, and the direction of jetting flow is vertical to the basic face. How to realize the automatic operation adapting to these conditions instead of a human-manipulated one is a problem urgently necessary to solve.

In principle, this problem would be solved by means of the ultrasonic ranging and bearing sensing system, but it is quite difficult to implement that, because the operating conditions of jetting manipulators and the properties of targets being sensed (as demolished basic faces) are complex and variable. Therefore, it is necessary to carry out careful analyses and research in relation to ultrasonic wave transmission characteristics in the air medium and to design a corresponding ultrasonic ranging and bearing sensing system suitably employed in concrete-jetting manipulators.

1.1 Specifications and Principles

According to the specific requirements of field construction, the expected specifications of the ultrasonic ranging and bearing sensing system are as follows:

1. Sampling area of basic faces: $\geq 25\ \text{cm}^2$;
2. Ranging regions and precision: 0.7–3 m, ± 0.5 cm;

3. Bearing region and precision both in vertical and horizontal directions: 0–45 degrees, ±5 degrees; larger than 45 degrees that will be processed as 45 degrees in actual operation;
4. Surveyed distances of obstacles: less than 70 cm in the direction of jetting flow; less than 50 cm in left or right horizontal direction relative to the jetting flow.

To survey the distance between jetting mouth and a basic face, it is only necessary to measure the time difference between transmitting ultrasonic pulses by the transducer mounted on the jetting mouth and receiving corresponding echo waves reflected from the basic face under the condition of sound velocities in the air medium being known. To measure the inclined angle α of the jetting flow relative to the basic face, it is necessary to adopt a transducer array in which every element has a sharp directivity, as 2–3 degrees, and measure the differences of sound ways in both horizontal and vertical directions, as shown in Fig. 1.

For a definite transducer array shown in this figure, the distance values of a, b, and c are known; therefore, provided the distances BD and CH can be measured, α can thus be determined. The distance r between jetting mouth and basic face can also be found by the average value measured by every transducer in the array. Sometimes α may be large. To adapt to this condition, ranging regions must be increased, as up to 3 m. In actual operations, if r is larger or less than 80 ± 5 cm, the message of advance or retreat will be sent to control part in the manipulator. Similarly, once $\alpha > 5$ degrees, the rotational message of the jetting mouth will also be sent to that. Because the specific values being measured for r and α are unnecessary to the control part, this control mode is thus easily realized.

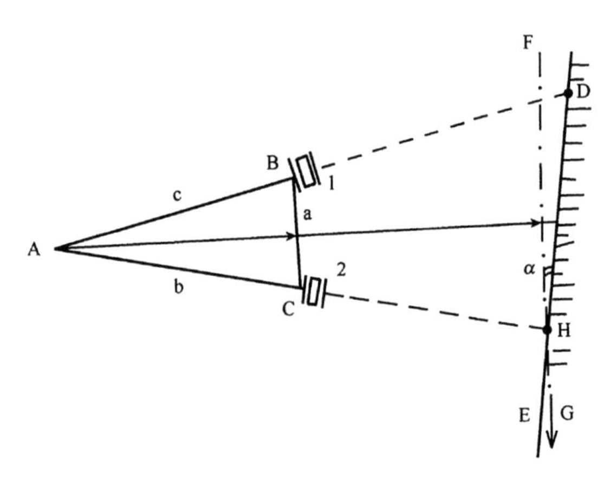

FIGURE 1 Diagram of geometrical relations for ultrasonic ranging and bearing.

1.2 Some Key Techniques Necessary to Solve

Although operating principles are simple for the ultrasonic ranging and bearing sensing system, some key techniques must be carefully solved in actual developing processes.

1. Design of transducer
   The ultrasonic sensor is mounted on wet jetting manipulators and the surface of the transducer will be lashed and piled up by the concrete and stones; thus, it is impossible to use the usual air-medium ultrasonic transducers. The piezoelectric ceramic element bound by a steel layer will first be considered, which can obtain different vibrating modes and operating frequencies. However, its sound impedance is very large and thus difficult to match with the air medium. To solve this problem, besides raising sound source levels (SLs) and receiving sensitivity, the characteristics of transmission sound of the coupled layer would definitely be investigated. This coupled steel layer is also used as a lash-proof and etch-proof one.

   The characteristics of transmission sound versus the thickness of the coupled layer are shown in Fig. 2. The curves 1 and 2 express

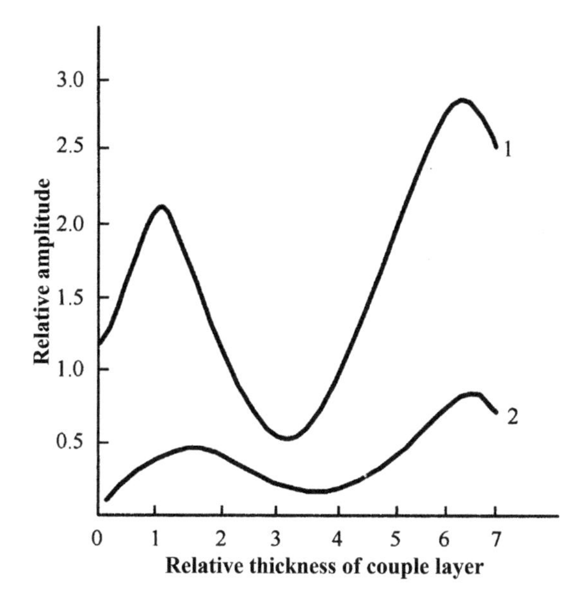

FIGURE 2 Acoustic characteristic curves for couple layer.

the first and second echo signals from a ceiling, respectively. We can see that larger echo signals will be obtained if coupled layer is at a "matched state." That is to say, it has higher sound transmission efficiency and receiving sensitivity in matching thicknesses than other ones, and thus is suitable to be employed in the sensing system.

2. To overcome the lingering vibration of the transducer array

It is better to adopt transmitting–receiving mode for the transducer array employed in the ultrasonic sensing system. Because the load of transducers is quite light in an air medium, the receiving circuit with high gains will have to be used, and the coupling effect among all transducers in the array will appear, the lingering vibration having long spreading times will thus be encountered. In particular, stronger sound power must be used in the sensing system. Therefore, it is impossible to measure the targets at near distances. This is the surveying "blind area" for a transmitting–receiving transducer system. Provided some integral approaches, including selecting suitable vibrating mode, adequately reducing Q value, and adopting an adaptive self-tuning controller (STC), are employed in that, the lingering vibration will remarkably be suppressed and thus the array will adapt to the requirement measuring the targets at near distances.

3. In the case of a basic face having large fluctuations, it is possible that ultrasonic wave will be incident on them at small grazing angles; echo signals are weak enough and thus difficult to detect. Besides raising SL, operating frequencies must also be raised to enhance backdirectional scattering components.

Of course, transmission loss will correspondingly be increased versus increasing operating frequencies; thus, it is necessary to raise sound power. However, the lingering vibration will also be strengthened with that; therefore, these parameters must be carefully adjusted in experimental processes.

4. When the transducer is operated upward, its surface will be piled up by concrete and stones up to several centimeters that severally impacts the performances of both transmitting and receiving ultrasonic waves, which must be cleared away in situ.
5. To suppress false echo signals reflected from splashing down concrete, stones, and electromagnetic interference.

The interference echoes caused by the reflection of sound from splashing down concrete, in particular the stones, severely impact the correct detection of echo signals reflected from basic faces. The signal-processing scheme of the digital time correlation accumulation decision employed in the digital echo sounder (refer to , Section 4.2.2) will be efficiently introduced in the ultrasonic sensing system. Because the distributions of interference echoes reflected from splashing down concrete and stones are random, the false alarm probability caused by them will tend to zero if the relative parameters in this scheme are suitably selected.

Obviously, the scheme of the accumulation decision has a function combating against the electromagnetic interference, because its distributions in time domain are also random.

6. The echo signals from the basic faces have a fluctuating character, because they are changed versus the grazing angles and the roughness of basic faces. The problem would be solved by adopting the approaches of adaptive amplitude threshold and a time correlation accumulation decision. The parameters of total accumulation decision numbers M and threshold numbers M_0 in the latter must suitably be selected (refer , Section 4.2) in practice.

1.3 Experimental Results

1. Experiments tracking the fluctuations of demolishing basic faces

   Fluctuating demolishing of a stone wall in a tunnel is selected to integrally examine the performance tracking of the fluctuations of the basic faces by using different operating frequencies. Based on that, the relative parameters of the sensing system, as operating frequencies, sound power, and so on, can also be determined.

   Tracking results by using different frequencies of 48, 78, 96, and 110 kHz are shown in Fig. 3A–D, respectively. The real lines in the figures express the fluctuating tendency of the stone wall having a length of 2 m; the imaginary lines show the tracking curves that are plotted according to experimental data.

   We see that, provided the operating frequencies are selected to be higher, that is, by using the transducer with sharper directivity, the changing tendency of fluctuation of the basic face is agreeable to tracking one. It means that the ultrasonic sensor would realize the ranging and bearing with high precision and high distinguishability.

2. Quantitative experiments for ranging and bearing implemented at a static state

   The typical experimental data with respect to the ranging and bearing using the ultrasonic sensing system are shown in Tables 1–3.

   Main experimental results are as follows:

   1. Ranging precision: relative error is about 1% in the experimental region of 165 cm.

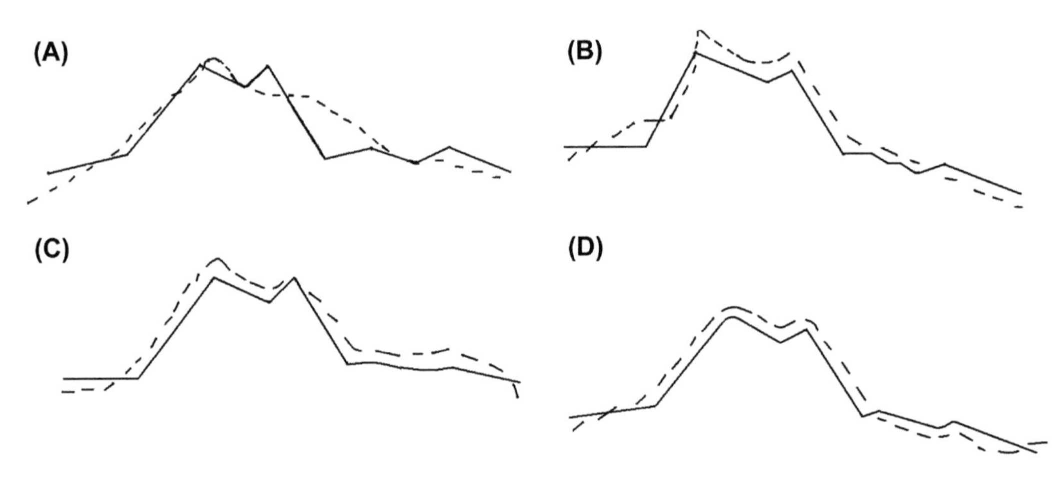

FIGURE 3 Tracking experiment for a fluctuating basic face.

TABLE 1 Ranging

Ruler Reading (cm)	165	154	100	80
Measured Data (cm)	164	153	101	79

2. Bearing precision: ±5 degrees both in vertical and in horizontal bearing when the grazing angles are less than 45 degrees; provided they are larger than 45 degrees that will be processed as 45 degrees in actual constructions.

The specifications mentioned previously accord with designing demands.

Based on the experiments at static state, the ultrasonic sensing system had been mounted on a jetting manipulator and carried out in field ranging and bearing operations in a building tunnel. High robustness had been demonstrated by practical constructions for a long time. The fully automatic operations of a concrete-jetting manipulator had been realized.

The ultrasonic sensing system had obtained high appraisal: to be in the lead in the world by two relative evaluations by the Ministry of Construction, etc., in 1989 and 1991 respectively.

2. ULTRASONIC TERRAIN OBSTACLE-SENSING SYSTEM EMPLOYED IN MOBILE ROBOTS

Ultrasonic ranging and bearing sensors have widely been employed in varied fields, including mobile robots. However, their ranging regions are generally nearer, and bearing precisions are lower. In particular, the ultrasonic transducers employed in them generally adopt the membrane-vibrating mode, although the mode has higher transmitting efficient and receiving sensibility, whereas they cannot be used in the adverse circumstances of the wild.

TABLE 2 Horizontal Bearing

Bearing Apparatus Reading (degree)	0	4	8	11	15	18	22	28	31	37	47
Measured Data (degree)	3	3	5	7	11	16	25	28	30	37	48

TABLE 3 Vertical Bearing

Bearing Apparatus Reading (degree)	0	3	8	11	15	19	26	30	34	40
Measured Data (degree)	4	6	9	9	16	19	22	36	31	37

2.1 Specifications and Principles

There are two functions in the ultrasonic terrain obstacle-sensing system: one is ranging and bearing for terrain obstacles; the other is surveying some parameters of the obstacles to determine whether the mobile robots can stride across them. The main specifications are as follows:

2.1.1 *Ultrasonic Ranging and Bearing for Terrain Obstacles*

1. Minimum distinguishing area of obstacles: 35×35 cm^2 plane at the distance of 18 m;
2. Longitudinal ranging precision and range: Relative errors are less than 5% in the range of 3–18 m;
3. Transverse angle distinguishability: ±4 degrees.

2.1.2 *Ultrasonic Terrain Obstacle Parameter Sensor*

1. The measured range of the heights of obstacles: 0.4–0.6 m is the criterion whether robots can stride across those; relative errors are less than 10%;
2. The widths and depths of ditches: Minimum distinguishing width is 0.5 m with relative errors less than 15%. The measuring region of depths is 0.4–0.6 m; relative errors are less than 10%. They are the criteria for striding across them.

Surveying processes are as follows: the existence of obstacle and its distance and direction are first surveyed, and then let the robot stops in front of that having the distances about 5 m to survey the relative parameters of terrain obstacles in the static state. They are the criteria whether robots can stride across the obstacles.

The usual method of echo detection is employed in ranging. Direction finding may adopt several transducers having directivities 3–4 degrees arranged side-by-side or in beamforming schemes.

The principles for surveying the parameters of obstacle targets are inspired by fishing sonar. Provided there are some submerged reefs or deep ditches in its surveying direction, corresponding echo signals would perform a "white zone" (correspondent with sound shadow region), as shown in Fig. 4. We see that, if the arrival time of the echo signal from obstacle target t_1 and that from a rough surface (like the reverberation of the sea bottom for an active sonar) t_2 are found, the height for which will thus be determined.

Similarly, we can find the widths and depths of ditches by means of the method of sound shadow.

2.2 Some Key Techniques Necessary to Solve for Designing the Sensing System

1. Selections of relative parameters
   The ultrasonic sending, traveling, and receiving laws, in particular the backscattering characteristics from the obstacle and the ground, versus operating frequencies, grazing angles, and the roughness of the ground have experimentally been investigated. Based on them, the optimum operating frequency, SL, and so on will be determined.
2. Development of ultrasonic transducers
   The ultrasonic transducer here is similar to that employed in the concrete-jetting manipulator. Of course, some relative parameters, such as SL, operating frequency, and directivity for the former must be modified.

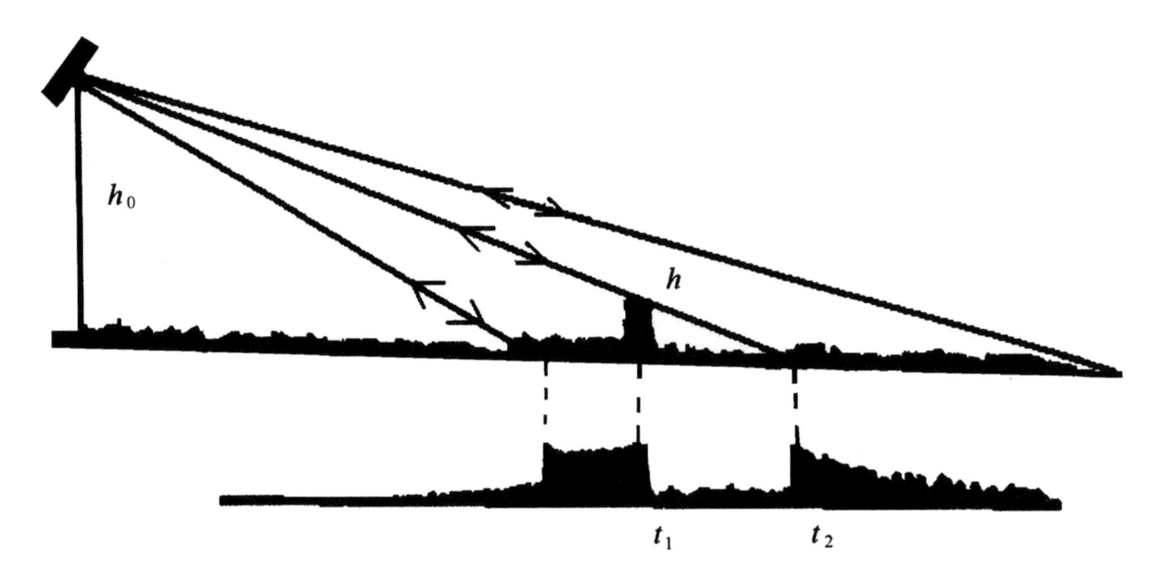

FIGURE 4 Principal diagram to measure the height of an obstacle by using the method of sound shadow.

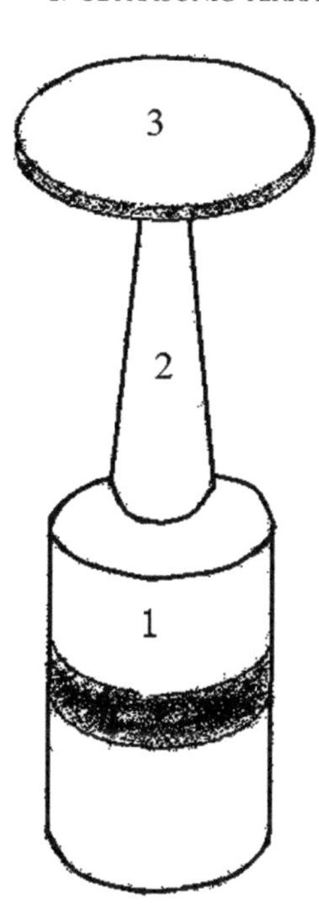

FIGURE 5 Structure of ultrasonic transducer.

Another type of transducer employed in the ultrasonic sensing system has been developed; its specific structure is shown in Fig. 5. It consists of three parts: longitudinal compound vibrating part 1, cone changing amplitude pole with the length one-half wavelength 2, and the ladder disc to produce winding vibration 3, which is excited by the vibrating part 1.

The acoustic–electric parameters for the two similar transducers mentioned are shown in Table 4.

We see that this type of transducer may adapt to the requirements of the ultrasonic terrain obstacle-sensing system, in particular its side maxima are reduced up to 16 dB that is a basic premise combating the echo interference from the ground just below the sensor.

Field experiments have also proved that this type of transducer has a good matching property with the air medium; therefore, it has higher transmitting efficiency and receiving sensitivity. The surveying "blind area" due to lingering vibration can also be suppressed by applying suitable processing schemes as mentioned in Section 1 in this appendix.

3. Combating against electromagnetic interference and signal fading
The ultrasonic sensing system is mounted on mobile robots, so the background noise levels are high. It has been proved that the fading of echo signals may reach 15 dB in the case of strong airflow. To improve the robustness of the sensing system, the approaches of adaptive signal processing and statistical decisions would be employed.

2.3 The Basic Structures of the Ultrasonic Sensing System

The principal block diagram for signal-beam ultrasonic terrain obstacle-sensing system is shown in Fig. 6. Surveying signal pulses are amplified by power amplifier and then applying on the transducer, ultrasonic signals are thus transmitted in the air medium. Then a converting key will be locked at receiving state. Once

TABLE 4 Acoustic-Electric Parameters for the Two Transducers

Numbers\ Parameters	Resonant Frequencies	Equivalent Resistances	Static state Capacitances	Beam Angles	Bandwidths	Side Levels
1	48.13 KHz	836 Ω	1756 pF	3.6 degrees	2.07 KHz	−16 dB
2	47.50XEz	840 Ω	1787 pF	3.7 degrees	2.15 KHz	−16 dB

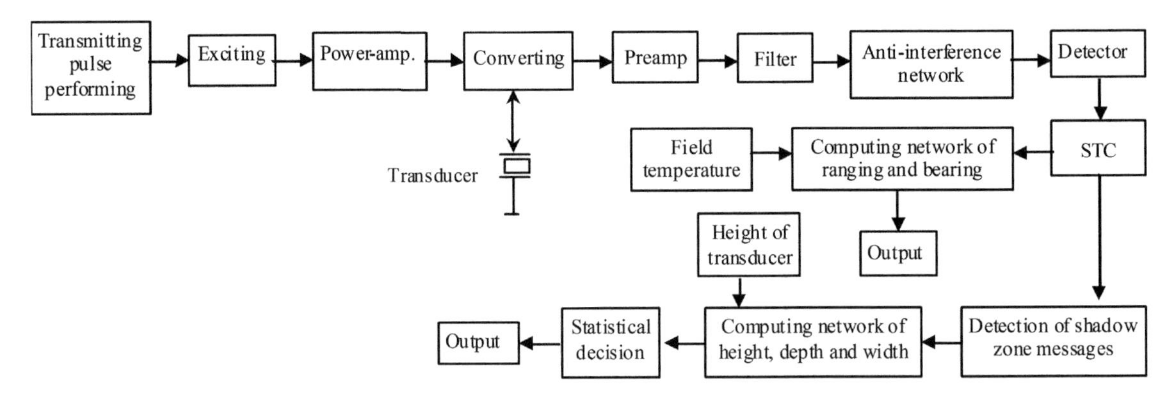

FIGURE 6 Principal block diagram for a signal-beam ultrasonic sensing system; STC, self-tuning controller.

echo signals received pass through a preamp-filter and an antiinterference network, the interference would be suppressed and the fading of signal would partly be compensated by STC. One of the outputs from STC is to make use of the messages of the ultrasonic ranging and bearing. Because the temperature has a remarkable effect on sound velocity when modified by the field temperature, ranging precision may achieve 1%.

The other output from STC is introduced to surveying the network of the terrain parameters. The parameters of t_1 and t_2 shown in Fig. 8 will first be detected; next, the corresponding heights, widths, and depths will be found according to definite geometric relations. They are relative to the heights of the transducer array; therefore, they are also introduced into the computing network. Lastly, the signals are processed by statistical decision circuit in which n times echo signals are taken to make a statistical average to combat fading and interference. The robustness of the ultrasonic sensing system will thus be remarkably improved.

The developed prototype of the ultrasonic terrain obstacle-sensing system is shown in Fig. 7. One of its transducers is mounted on a support, the heights and oblique angles for which can be changed.

2.4 Experimental Results for the Principal Prototype

The principal prototype of the sensor had been examined under different conditions. Typical experimental results are shown in Tables 5–8.

1. Longitudinal ranging.
2. Heights of terrain obstacles, which are simulated by the planks with different heights.
3. Depths of ditches.
4. Widths of ditches.

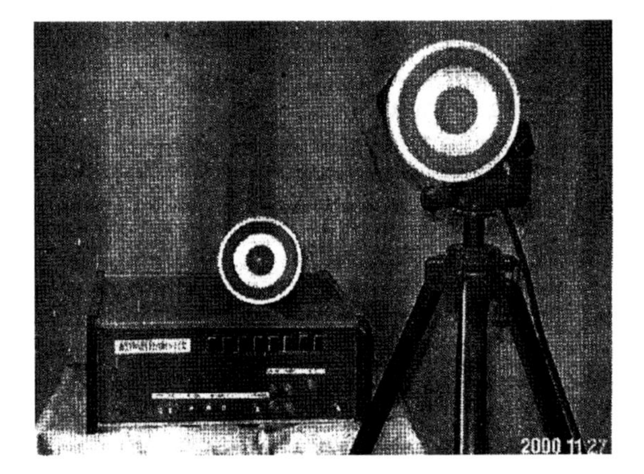

FIGURE 7 Prototype of ultrasonic terrain obstacle-sensing system.

TABLE 5 Ranging Data

Ruler Reading (m)	5	10	15	20	23
Measured Data (m)	4.99–5.01	9.95–9.97	14.95–14.98	19.88–20.12	22.85–23.13
Maximum Relative Error (%)	0.2	0.4	0.33	0.60	0.65

TABLE 6 Measured Data of the Heights

Ruler Reading (cm)	35	40	50	60	70
Measured Data (cm)	33–37	37–30	46–48	64–65	67–68
Maximum Relative Error (%)	5.7	7.5	8	8.3	4.3

TABLE 7 Measured Data of the Depths

Ruler Reading (cm)	42	54	60	70
Measured Data (cm)	43–45	51–53	55–57	68–71
Maximum Relative Error (%)	7.1	5.6	8.3	2.9

The experimental results pointed out that the ultrasonic sensor has performancesensitivity and reliably surveys obstacles at farther distances and with higher precisions. Its bearing precisions are the same with the directivities of transducers at 3–4 degrees. Moreover, it is possible to measure the heights of terrain obstacles, depths, and widths of ditches in a static state.

The prototype does not carry out the tests combined with mobile robots. It seems that, by adopting a signal-beam surveying scheme, it is difficult to adapt to the complex and variable parameters of terrain obstacles. Perhaps by means of multibeam integral sensing and merging techniques it would adapt to practical surveying circumstances. To decide whether robots can stride across the terrain obstacles encountered in situ, it is only necessary to provide the criteria of corresponding heights, widths, and depths of terrain obstacles in actual operations (as well as the decision mode in the ultrasonic sensing system employed in the concrete-jetting manipulator), it would be expected to more easily detect target parameters.

3. ULTRASONIC NAVIGATING SENSING SYSTEM EMPLOYED IN AUTOMATIC GUIDED VEHICLES (AGV)

Material handling is a significant part of the manufacturing process, both in terms of cost and time. Indeed, statistics show that the processing time only occupies 5% of the manufacturing time of a typical job, the

TABLE 8 Measured Data of the Widths

Ruler Reading (cm)	50	70	90	110	130	150	170	190	200
Measured Data (cm)	50–57	60–77	77–96	99–111	116–131	135–158	175–182	192–210	202–219
Maximum Relative Error (%)	14	14	14	10	11	10	7.1	9.5	10.5

remainder is spent in storage and transportation processes. With advantages such as mobility, flexibility, efficiency, and so on, automatic guided vehicles (AGVs) are playing a more and more important role in modern manufactory systems. Thus, the research of AGV technology makes important sense for the industrialization of automatic handling systems.

The main task of AGVs is to transport materials from one workbench to another in an industrial environment. During the stage of running, AGVs need to locate their two-dimensional position; when an AGV runs near a workbench, it enters docking mode.

3.1 Main Functions and Features for the AGV Ultrasonic Navigation Sensing System

1. Positioning
   It is necessary to determine the real-time positions of an AGV to control that by a control center. In this case, when a positioning range includes the full operating area, positioning precision is set to medium.
2. Path guiding
   To guide the AGV moving in its operating paths, the positioning ranges are two sides of the paths, positioning precision is also medium.
3. Stopping and docking at workbenches
   An AGV will be guided in front of workbenches and then the materials will be loaded and/or unloaded. In this case, the operating ranges are small, but docking precisions are quite high. It is a key technique in the ultrasonic navigating sensor.
4. Avoiding obstacles
   Surveying obstacles (including the human body) around the AGV at a certain range as several meters would carry out in the full operating process. The method of the echo detection is employed in that. The positioning precision is permitted to be low.

3.2 Properties of Ultrasonic Beacon Navigation-Sensing System

1. There are some advantages by using an ultrasonic beacon navigating sensing system in comparison with traditional methods using electrical wires buried in floors, which is necessary when merging with a laser guide and ultrasonic avoidance of obstacles.
   a. The sensing system has the properties of multifunctions, including AGV positioning, guiding, docking, and avoiding obstacles, and does not request merging of multisensors.
   b. The similar operating mode may be adopted to realize fixed or free-path guiding.
   c. Electromagnetic interference may easily be suppressed by using suitable signal-processing schemes (as correlator); moreover, smoke screen, fog, usual sound, light, and so on cannot affect the sensing system. That is to say, it has high circumstance adaptability.
   d. The beacons may be located or hung conveniently and flexibly according to actual operating circumstances.
   e. There are some outstanding advantages for the sensing system, such as simple in structure, reliable in performance, and low in cost.
2. There are some advantages using an ultrasonic beacon navigation scheme in comparison with using ultrasonic echo detecting:
   a. The beacon-navigating scheme is a single-way operating state; the requirement for SL is much lower than for an echo detector.
   b. The waveforms received by beacons are independent of the properties of targets; therefore they are steady and more suitable for processing following circuits.
   c. Provided the system is availably designed and arranged, high navigating precision

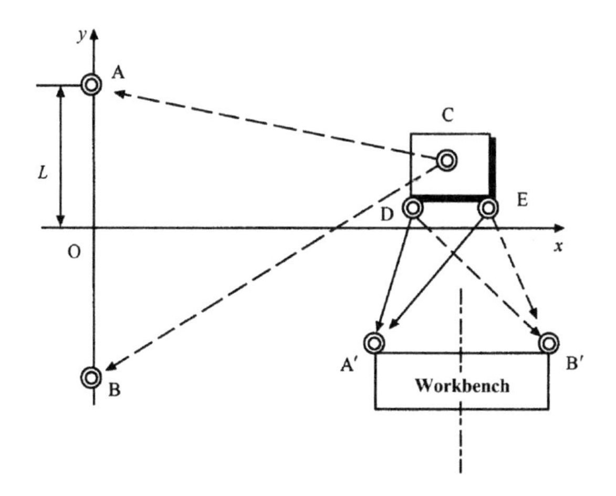

FIGURE 8 Principle measuring the positions of an AGV.

can be obtained and thus adapted to varied situations, including operations such as docking

d. False targets around the system cannot affect it, because they cannot respond to calling signals. Similarly, multipath interference will also disappear, because it will arrive at the receiver after direct signals.
e. The system has a function of frequency division multiple access (FDMA) to guide the AGV to operate at different navigating states as mentioned earlier. Of course, the frequencies in surveying obstacles may be suitably selected, which is unnecessary to change.

3. Principle of ultrasonic beacon navigation
The operating principle of the ultrasonic beacon navigation sensing system is shown in Fig. 8.

A triangulation scheme is adopted to conveniently estimate the position of an AGV. By measuring the ranges D_{CA} and D_{CB} between beacon C mounted on the AGV and beacons A and B placed at known positions, the location of the AGV can be determined.

$$y_C = \frac{D_{CB}^2 - D_{CA}^2}{4L}$$

$$x_C = \pm\sqrt{\frac{D_{CA}^2 + D_{CB}^2 - 2L^2 - 2y_C^2}{2}}$$

As shown in Fig. 8, accurate docking needs both high-precision localization and bearing measurements. Bearing measurement is performed based on the localization output. Because two points can define a line, by means of two beacons $E(x_E, y_E)$ and $D(x_D, y_D)$ mounted on the AGV and with beacons A′, B′ at respective workstations, we can measure the deviate angle α of the AGV relative to the workstation. The bearing calculation is as follows:

$$\alpha = \mathrm{arctg}\frac{y_D - y_E}{x_D - x_E}$$

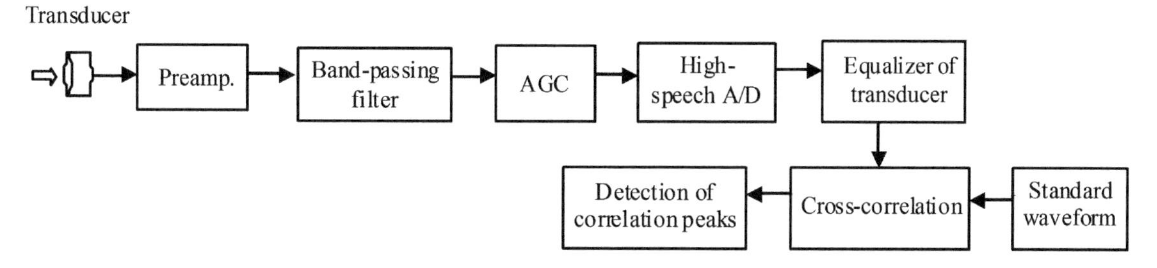

FIGURE 9 Principle block diagram of correlation receiver; *AGC*, automatic gain control.

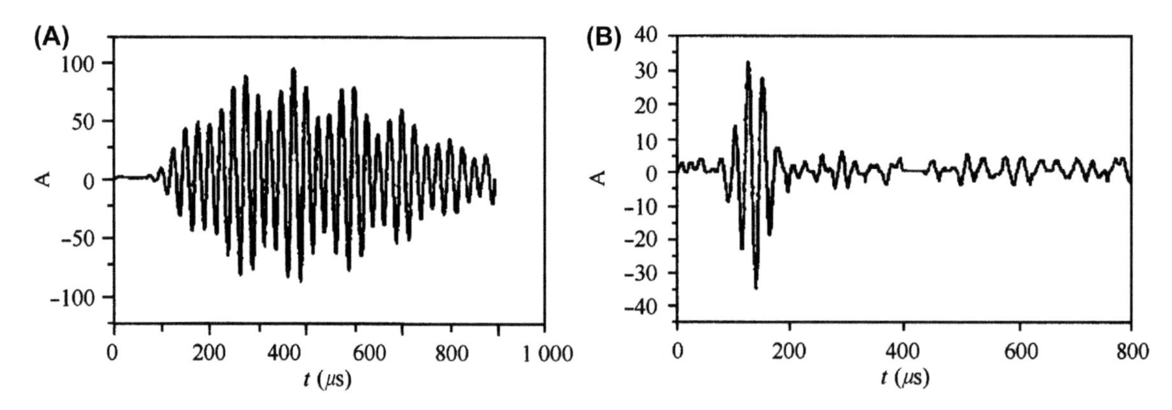

FIGURE 10 Received waveform (A) before and (B) after passing through the equalization transducer.

To ensure the accuracy of bearing measurement, the beacon *c* is first located on the centerline of A, B′ point.

4. Selecting suitable signal processing schemes to improve ranging precision

We have known that the navigating precision of an AGV directly depends on the ranging precision. To obtain high-ranging precision and the excellent performance combating with interference, a cross-correlator is employed in the sensing system. A block diagram of the principle behind the receiver is shown in Fig. 9.

Because the reference signal being introduced in the cross-correlator is a standard waveform acquired at a suitable distance, it belongs, in fact, to the pulse-to-pulse correlation receiver; detection performances are thus improved.

There are some factors, such as air turbulence, humidity, temperature, and the bandwidth of the transducer, that limit the accuracy of ultrasonic ranging to be improved. Among them, the narrowband effect is a main reason. So that a transducer equalization scheme based on an adaptive algorithm, which is a finite impulse response (FIR) equalizer trained according to a least mean squares (LMS) algorithm, can be employed in the sensing system. Excellent results are obtained as shown in Fig. 10A and B. Moreover, the sound velocity is adjusted according to field temperature.

5. Experimental results for the ultrasonic navigating sensing system

An experimental prototype of the AGV is employed in relative-navigating tests. The typical experimental results are as follows:

a. The tracking results for fixed and free paths are shown in Figs. 11 and 12.
b. The measuring results of the deviate angles α are shown in Table 9
c. The tests of avoiding obstacles

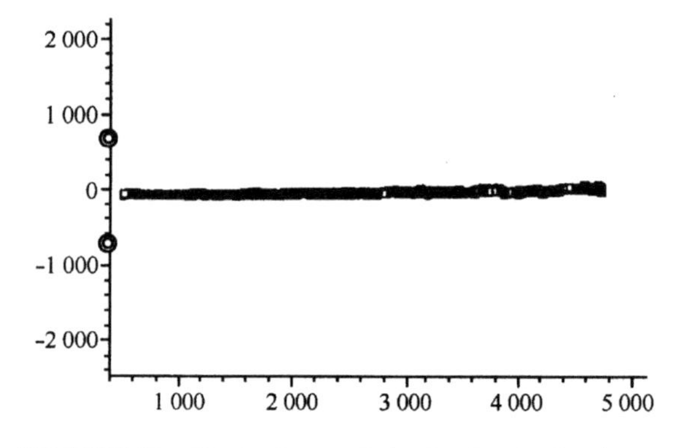

FIGURE 11 Navigation result for fixed paths (mm).

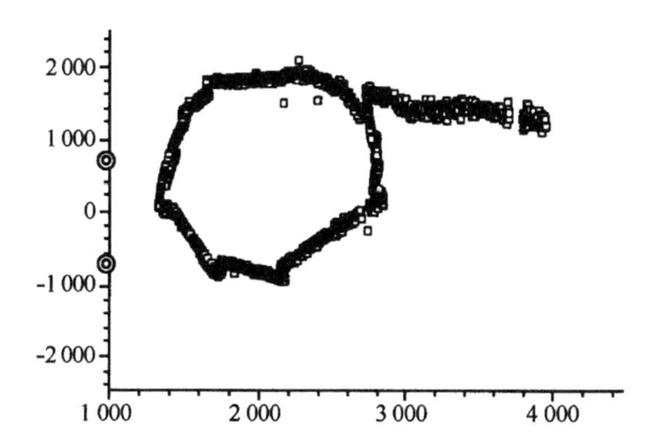

FIGURE 12 Navigation result for free paths (mm).

The output signal-to-noise ratio (SNR) of preamplifier for echo signals reflected from a plank having a area of 35 × 35 cm^2 and a human body are arranged in Tables 10 and 11, respectively.

To sum up, the ultrasonic navigation sensing system has the following main specifications:

1. The positioning precision at axis line is ±1 cm in the range of 5 m; ±3 cm in the range of 10 m.
2. The navigating precision of free paths in an operating area of 25 m^2: ±5 cm in transversal direction; ±3 cm in longitudinal direction.
3. Docking precision: ±3 mm in transversal direction; ±2 mm in longitudinal direction, and ±1 degree in deviate angle α.
4. Surveying obstacles: ±30 degrees at the range of 5 m in the front of an AGV.

The indexes would fit some practical requirements of AGV navigation.

TABLE 9 Measured Deviate Angles α

Actual α (degree)	−31.2	−17.2	−9.1	−3.1	0	15.6	23.0
Standard Deviation (degree)	0.53	0.37	0.31	0.48	0.31	0.65	0.63

TABLE 10 Output SNR of Ultrasonic Echo Signal Reflected From a Plank

Distances (m)	Output SNR (dB)
6	13.5
8	7.5
10	3

4. SOME ASPECTS OF POSSIBLE EXTENDING APPLICATIONS WITH RESPECT TO THE ULTRASONIC SENSING SYSTEMS DEVELOPED

1. Ultrasonic sensing systems may be employed in a paver to surface a road with bituminous concrete, in which a high-precision ultrasonic ranging and bearing sensor would be employed in field construction to realize fully automatic hydraulic-pressure servo controls.
2. Ultrasonic navigating senor may be employed in fire robots. Because dense smoke and raging fire do not remarkably affect the sensor, it is necessary to make some modifications to the ultrasonic terrain obstacle sensor.
3. AGV ultrasonic navigating sensors may be extended to various mobile robots. This sensor has some outstanding advantages, such as low cost, flexible operations and arrangements, and merging multifunction sensing.
4. The ranging sensor has longer surveying distances; the bearing precision is also sufficiently high. That would be employed in ranging and bearing for some faces in situ,

TABLE 11 Output SNR of Ultrasonic Echo Signal Reflected From Human Body

Distances (m)	Output SNR (dB)
5	4
6	3
7	1.5

including water face, oil face, and solid boundary.
5. It is simple and efficient by extending the functions of avoiding obstacles to touch proof vehicles, ships, and so on.
6. The function of surveying obstacles would be extended to a precautionary system, in particular one that would still be efficient in adverse circumstances having rain, snow, smoke, fog, and so on.
7. It would be expected to develop some ultrasonic controlling toys by means of the principle of the ultrasonic navigational sensors employed in mobile robots.

References

[1] X. Tianzeng, X. Keping, et al., Study on the ultrasonic transmission and ultrasonic sensors, Journal of Xiamen University (Natural Science) 40 (2) (2001) 303–309.
[2] T. Feng, X. Tanzeng, A wide-scope high-precision ultrasonic ranging processing method for mobile robot navigation, Journal of Central South University of Technology 31 (2000) 292–294 (special issue).
[3] T. Feng, X. Tanzeng, X. lufen, An ultrasonic navigation system used for automatic guided vehicle, High Technology Letters 8 (2002) 47–50. Beijing.
[4] W. Qingchi, W. Zaixun, Study on the ultrasonic transducer used in the system for detecting and sending the obstacles, Journal of Xiamen University (Natural Science) 36 (2) (1997) 227–231.

Index

'*Note*: Page numbers followed by "f" indicate figures and "t" indicate tables.'

E

F

T

U

Edwards Brothers Inc.
Ann Arbor MI. USA
December 1, 2017